Detection, Control, Risk Assessment, and Prevention of Foodborne Microorganisms

Detection, Control, Risk Assessment, and Prevention of Foodborne Microorganisms

Guest Editors

Fernando Pérez-Rodríguez
Arícia Possas

MDPI
Basel • Beijing • Wuhan • Barcelona • Belgrade • Novi Sad • Cluj • Manchester

Guest Editors
Fernando Pérez-Rodríguez
Department of Food Science and Technology
University of Cordoba
Córdoba
Spain

Arícia Possas
Department of Food Science and Technology
University of Cordoba
Córdoba
Spain

Editorial Office
MDPI AG
Grosspeteranlage 5
4052 Basel, Switzerland

This is a reprint of the Special Issue, published open access by the journal *Foods* (ISSN 2304-8158), freely accessible at: https://www.mdpi.com/journal/foods/special_issues/detection_control_assessment_foodborne.

For citation purposes, cite each article independently as indicated on the article page online and as indicated below:

Lastname, A.A.; Lastname, B.B. Article Title. *Journal Name* **Year**, *Volume Number*, Page Range.

ISBN 978-3-7258-3929-2 (Hbk)
ISBN 978-3-7258-3930-8 (PDF)
https://doi.org/10.3390/books978-3-7258-3930-8

Contents

About the Editors

Fernando Pérez-Rodríguez

Dr. Fernando Pérez Rodríguez is a Full Professor of Food Microbiology and Hygiene at the University of Córdoba (UCO), Spain. He holds degrees in Biological Sciences and Food Science and Technology from UCO, where he also completed his PhD in 2007, focusing on quantitative microbiological risk assessment. His research spans predictive microbiology, food modeling, and risk assessment, with over 100 peer-reviewed publications and several reference books in the field.

He has led and participated in numerous national and international projects, including BioFreshCloud, AgroVal4PackFood, and BioMedPack, and currently coordinates the OneHealth4Food network and contributes to the COST Action FoodWaStop. He is a co-creator of MicroHibro and List-RA, and collaborates with EFSA, FAO/WHO, and AECOSAN as an expert in microbial risk assessment.

Arícia Possas

Dr. Arícia Possas is a Research Scientist and Assistant Professor at the University of Córdoba, Spain. Her research specializes in Predictive Microbiology and Quantitative Microbial Risk Assessment (QMRA), with a focus on novel strategies for microbial inactivation in foods. She holds a PhD in Food Science and Technology from the University of Córdoba (2019), a Master's in Food Science jointly awarded by the University of Córdoba and the University of Campinas (Brazil), and a degree in Food Engineering from the Federal University of Viçosa (Brazil).

Her academic and professional trajectory is marked by an interdisciplinary approach and a strong commitment to international collaboration. Dr. Possas has authored over 40 peer-reviewed scientific articles in prestigious journals and has collaborated with more than 100 co-authors from 15 countries. She has participated in several national and international research projects and is one of the developers and curators of MicroHibro, a widely used open-access software for predictive microbiology.

Preface

Enteric foodborne diseases continue to pose a major threat to public health worldwide, despite decades of investment in prevention and control strategies. Global changes in consumer behavior, food supply chains, and population demographics, along with the increasing complexity of food systems, have contributed to the emergence and persistence of microbial risks. In this context, innovative tools and interdisciplinary approaches are essential to address the evolving landscape of foodborne pathogens.

This Special Issue, originally published in *Foods*, brought together a selection of cutting-edge research articles focusing on quantitative and molecular strategies to detect, control, and better understand enteric foodborne microorganisms. The collection reflects a broad range of scientific advances, including the application of molecular typing methods, predictive modeling, genomic surveillance, risk assessment frameworks, and data-driven tools to enhance microbial food safety.

The included contributions showcase novel insights into pathogens such as *Salmonella*, *Listeria monocytogenes*, *Campylobacter*, and Shiga toxin-producing *Escherichia coli*, among others. Several articles highlight the integration of molecular techniques with quantitative microbiology to study pathogen behavior in diverse food matrices and under varying environmental conditions. Others explore the use of advanced modeling techniques to support decision-making in food safety management.

Altogether, this reprint offers a comprehensive overview of current approaches and future perspectives in the fight against enteric foodborne pathogens. We hope this collection serves as a valuable resource for researchers, risk assessors, and professionals working toward safer and more resilient food systems.

Fernando Pérez-Rodríguez and Arícia Possas
Guest Editors

foods

MDPI

Editorial

Detection, Control, Risk Assessment, and Prevention of Foodborne Microorganisms

Arícia Possas * and Fernando Pérez-Rodríguez

Department of Food Science and Technology, UIC Zoonosis y Enfermedades Emergentes ENZOEM, CeiA3, Universidad de Córdoba, 14014 Córdoba, Spain; b42perof@uco.es

* Correspondence: g12mepoa@uco.es

Citation: Possas, A.; Pérez-Rodríguez, F. Detection, Control, Risk Assessment, and Prevention of Foodborne Microorganisms. *Foods* **2024**, *13*, 1551. https://doi.org/10.3390/foods13101551

Received: 29 April 2024
Accepted: 11 May 2024
Published: 16 May 2024

Despite significant efforts from government and industry, enteric foodborne diseases continue to pose a substantial public health challenge worldwide. In the European Union, the number of deaths resulting from foodborne outbreaks in 2022 reached the highest level recorded since 2012 [1]. *Listeria monocytogenes* was identified as the primary cause of these deaths, followed by *Salmonella* spp. [1]. Furthermore, recent shifts in consumer behavior, the globalization of commerce, advancements in food processing technologies, and climate change have contributed to the emergence and re-emergence of foodborne diseases [2,3].

In response to these challenges, this Special Issue sought to gather original research that addresses these critical issues. The fifteen articles featured in this Special Issue cover a wide range of topics, from the development of advanced detection methods to the implementation of risk assessment frameworks using computational tools. The featured articles highlight the importance of interdisciplinary collaboration and the integration of cutting-edge technologies to advance our understanding of foodborne microorganisms and improve food safety practices. Moreover, the articles included herein offer valuable knowledge on the complex dynamics of foodborne disease emergence and provide practical solutions for microbial detection, prevention, and control.

In line with the goal of improving the detection of foodborne pathogens, Luo et al. (contribution 1) introduced a microfluidic chip integrating loop-mediated isothermal amplification and CRISPR/Cas12a systems for detecting *Salmonella*, addressing issues of aerosol pollution in DNA amplification. This innovative chip facilitates amplification at 65 °C for 20 min, followed by fluorescent signal production at 43 °C for 30 min, achieving a detection sensitivity of 118 pg/μL with 100% accuracy. Application of the microfluidic chip in salmon and chicken samples spiked with *Salmonella* showed stable detection capabilities.

Furthermore, through their study, Niu et al. (contribution 2) made advancements in the detection of antimicrobial resistance mechanisms against quinolone and fluoroquinolone in foodborne pathogens by developing stable plasmid DNA reference materials. DNA fragments of 11 target genes were successfully synthesized, inserted into plasmid vectors, and transferred into recipient cells. Genetic stability, limit of detection, homogeneity, and storage stability were evaluated. All target DNA remained stable and detectable during subculturing, whereas plasmid DNA remained detectable after storage at various temperatures for different durations without mutations occurring. The materials developed in the study meet standard requirements and can be effectively used to detect resistance mechanisms in foodborne pathogens.

This Special Issue also presents studies on novel biopreservation strategies for controlling foodborne pathogens. Resendiz-Moctezuma et al. (contribution 3) screened the antimicrobial potential of organic acids and essential oils (EOs) as antimicrobials against *Salmonella* Typhimurium in pork loin. Their findings revealed significant reductions in the pathogen's prevalence caused by lactic acid, formic acid, cumin, peppermint, and spearmint, although no interactions between these antimicrobial candidates were found in pork loin. Similarly, Gál et al. (contribution 4) investigated the effectiveness of sage

EO and heat in inactivating *L. monocytogenes* in sous vide processed beef tenderloin. The study samples were cooked sous vide at different temperatures, and bacterial counts were assessed over 12 days. Both *L. monocytogenes* and coliform bacteria levels increased over time, with *Pseudomonas fragi* and *L. monocytogenes* being the most common isolated organisms. Notably, the addition of sage EO showed promise in ensuring the safety of sous vide beef tenderloin.

In addition to the use of strategies for reducing microbial loads in different types of food, other strategies for preventing microbial contamination and shedding have also been assessed. This is illustrated in the study by Jiménez et al. (contribution 5) on the effectiveness of physically removing lymph nodes from pork products prior to grinding in mitigating *Salmonella* and reducing indicator organisms in the final ground products. Three treatment groups were assigned in a commercial pork processing facility, with varying levels of lymph node removal. The results of their study showed a significant reduction in the presence of *Salmonella* and indicator organisms when topical and internal lymph nodes were removed before grinding. Their findings underscore the importance of implementing lymph node removal strategies to prevent contamination in pork products that undergo further processing.

The studies by Pasquali et al. (contribution 6) and Wiatrowski et al. (contribution 7) shed light on the crucial role of hygiene in ensuring food safety in different food processing environments. Pasquali et al. (contribution 6) highlight how variability in physicochemical parameters impacts the microbial quality and safety of Italian artisanal salami. The authors found that high enterobacteria levels in the meat mixture used for salami elaboration were related to bacterial pathogen occurrence. In addition, suboptimal salami ripening conditions favored the presence of *Staphylococcus aureus* and *L. monocytogenes* in different products and processing environments. Conversely, Wiatrowski et al. (contribution 7) assessed hygiene conditions in food trucks by using various methods including Petrifilm TM and bioluminescence. Swabs and prints from a total of 20 food trucks in Poland were analyzed. The study results highlight the need for detailed hygiene regulations and certified training for food truck personnel to mitigate the risk of bacterial contamination and foodborne infections.

Regarding microbial prevalence and its implications for human health, Wiktorczyk-Kapischke et al. (contribution 8) investigated the presence of *L. monocytogenes* in a salmon processing environment, identifying 38 genetically different strains among 62 isolates, including 6 persistent strains. The authors also identified serogroup 1/2a-3a as the dominant serogroup. Persistent strains showed higher tolerance to disinfectants and higher capacity for biofilm formation. The findings of this study provide information on the phenotypic characteristics of *L. monocytogenes* strains in salmon processing environments. On a related note, Harrison et al. (contribution 9) evaluated potential sources of extraintestinal pathogenic *Escherichia coli* infections using the genomic data of isolates from five U.S. government organizations. Virulence gene analysis of 38,032 isolates categorized into 40 virulence groups revealed associations between sequence types and human disease risk. Medium- and high-risk groups showed a higher prevalence of human-associated sequence types, including ST-131. The food source isolates mostly belonged to low-risk groups, while companion animal isolates predominantly belonged to medium- or high-risk groups.

This Special Issue also covers the application of predictive modeling to assess the efficacy of inactivation treatments to mitigate the presence of microorganisms in food. Cuggino et al. (contribution 10) collected information on steps, processing parameters, and controls applied in the ready-to-eat leafy vegetable processing industry in Argentina and applied predictive models to estimate *Salmonella* concentrations alongside the production process and distribution chains of fresh-cut lettuce, including the use of chlorine washing as a disinfection method. The findings of their study aid the development of informed risk-based sampling programs and the determination of optimal process parameters for mitigating *Salmonella* spp. in ready-to-eat leafy vegetables. Conversely, González-Tenedor et al. (contribution 11) applied predictive models for assessing the thermal inactivation of *Listeria innocua* in coconut water under isothermal and dynamic conditions, crucial for ensuring product safety in the growing coconut water market. The authors concluded that mild heat treatments offer a viable option for preserving the quality and safety of coconut water but that this form of treatment requires the careful selection of heating conditions to prevent microbial stress adaptation under dynamic conditions.

Pulsrikarn et al. (contribution 12) integrated predictive models into a risk assessment framework to evaluate the health impact of antimicrobial-resistant *Salmonella* spp. in retail pork sold in Thailand. More specifically, the authors assessed the health risks associated with susceptible and quinolone-resistant (QR) *Salmonella* contamination in pork. The probability of illness and mortality rates were estimated for both susceptible and resistant strains, with QR strains showing higher prevalence and lower mean concentrations. Monte Carlo simulations yielded annual mortality rates for QR salmonellosis, aligning with previous reports on the adverse health effects of antimicrobial resistance. Their findings underscore the relevance of addressing antimicrobial resistance in microbial risk assessments.

Concerning antimicrobial resistance, the authors of the studies included in this Special Issue also examined the prevalence and implications of resistance genes in different contexts. Regecová et al. (contribution 13) investigated antimicrobial resistance and the genes encoding staphylococcal enterotoxins in *Staphylococcus warneri* strains, a pathogen linked to inflammatory diseases in immunosuppressed patients. A total of 45 isolates were obtained from various meat samples and 22% of them displayed multidrug resistance, evidencing the urgent need for effective management strategies. Similarly, Pakbin et al. (contribution 14) explored antibiotic-resistant genes and foodborne pathogens in sweet samples from local markets in Iran. Their study identified *Staphylococcus aureus*, *Cronobacter sakazakii*, *Shigella* spp., *Campylobacter jejuni*, and *Campylobacter coli* at varying degrees of prevalence; *S. aureus* and *Shigella* spp. were noted as being the most prevalent pathogens. Preventive strategies such as the automation of food processing, monitoring the hygiene standards of food handlers, and regular testing for antibiotic resistance are recommended by the authors.

In a related context, Wang et al. (contribution 15) investigated *Bacillus cereus* prevalence, antimicrobial resistance, and virulence gene profiles in infant formula sourced from supermarkets in Beijing, China. Among 88 isolates recovered from 68 infant formula samples, the prevalence rates in domestic and imported samples were 70.6% and 52.9%, respectively. Most strains carried at least one virulence gene, with similar occurrences of certain genes being noted between domestic and imported brands. Antimicrobial susceptibility analysis showed varied resistance rates, with the rapid growth of *B. cereus* in infant formula prepared at room temperature. The authors highlighted the need for monitoring guidelines to establish accepted levels of *B. cereus* in infant formula.

In conclusion, this Special Issue evidences the collaborative efforts of researchers, policymakers, and industry stakeholders in advancing our understanding of foodborne diseases and implementing practical solutions to enhance food safety worldwide. By addressing novel detection methods, antimicrobial resistance, biopreservation strategies, predictive modeling, and risk assessment, this Special Issue presents advances to reduce the incidence of foodborne diseases. We thank all of the authors for sharing their findings in this Special Issue, as well as the reviewers and editorial team for their hard work in ensuring the high quality of the papers published herein.

Author Contributions: Conceptualization, A.P. and F.P.-R.; writing—review and editing, A.P. and F.P.-R. All authors have read and agreed to the published version of the manuscript.

Acknowledgments: The authors thank the research project BioFreshCloud, with reference PRIMA-S2-2019-PCI2020-112015, which is part of the PRIMA program supported by the European Union and funded by MCIN/AEI/10.13039/501100011033 and the European Union "NextGener- ationEU/PRTR".

Conflicts of Interest: The authors declare no conflicts of interest.

List of Contributions

1. Luo, Y.; Shan, S.; Wang, S.; Li, J.; Liu, D.; Lai, W. Accurate Detection of *Salmonella* Based on Microfluidic Chip to Avoid Aerosol Contamination. *Foods* **2022**, *11*, 3887. https://doi.org/10.3390/foods11233887.
2. Niu, Q.; Su, X.; Lian, L.; Huang, J.; Xue, S.; Zhou, W.; Zhao, H.; Lu, X.; Cui, S.; Chen, J.; et al. Developing Qualitative Plasmid DNA Reference Materials to Detect Mechanisms of Quinolone and Fluoroquinolone Resistance in Foodborne Pathogens. *Foods* **2022**, *11*, 154. https://doi.org/10.3390/foods11020154.
3. Resendiz-Moctezuma, C.; Fonville, A.P.L.; Harsh, B.N.; Stasiewicz, M.J.; Miller, M.J. Use of Doehlert Matrix as a Tool for High-Throughput Screening of Organic Acids and Essential Oils on Miniaturized Pork Loins, Followed by Lab-Scale Validation That Confirmed Tested Compounds Do Not Show Synergistic Effects against *Salmonella* Typhimurium. *Foods* **2023**, *12*, 4034. https://doi.org/10.3390/foods12214034.
4. Gál, R.; Čmiková, N.; Prokopová, A.; Kačániová, M. Antilisterial and Antimicrobial Effect of *Salvia officinalis* Essential Oil in Beef Sous-Vide Meat during Storage. *Foods* **2023**, *12*, 2201. https://doi.org/10.3390/foods12112201.
5. Jiménez, R.L.; Brashears, M.M.; Bueno López, R.; Vargas, D.A.; Sanchez-Plata, M.X. Mitigation of *Salmonella* in Ground Pork Products through Gland Removal in Pork Trimmings. *Foods* **2023**, *12*, 3802. https://doi.org/10.3390/foods12203802.
6. Pasquali, F.; Valero, A.; Possas, A.; Lucchi, A.; Crippa, C.; Gambi, L.; Manfreda, G.; De Cesare, A. Variability in Physicochemical Parameters and Its Impact on Microbiological Quality and Occurrence of Foodborne Pathogens in Artisanal Italian Organic Salami. *Foods* **2023**, *12*, 4086. https://doi.org/10.3390/foods12224086.
7. Wiatrowski, M.; Rosiak, E.; Czarniecka-Skubina, E. Surface Hygiene Evaluation Method in Food Trucks as an Important Factor in the Assessment of Microbiological Risks in Mobile Gastronomy. *Foods* **2023**, *12*, 772. https://doi.org/10.3390/foods12040772.
8. Wiktorczyk-Kapischke, N.; Wałecka-Zacharska, E.; Skowron, K.; Kijewska, A.; Bernaciak, Z.; Bauza-Kaszewska, J.; Kraszewska, Z.; Gospodarek-Komkowska, E. Comparison of Selected Phenotypic Features of Persistent and Sporadic Strains of *Listeria monocytogenes* Sampled from Fish Processing Plants. *Foods* **2022**, *11*, 1492. https://doi.org/10.3390/foods11101492.
9. Harrison, L.; Tyson, G.H.; Strain, E.; Lindsey, R.L.; Strockbine, N.; Ceric, O.; Fortenberry, G.Z.; Harris, B.; Shaw, S.; Tillman, G.; et al. Use of Large-Scale Genomics to Identify the Role of Animals and Foods as Potential Sources of Extraintestinal Pathogenic *Escherichia coli* That Cause Human Illness. *Foods* **2022**, *11*, 1975. https://doi.org/10.3390/foods11131975.
10. Cuggino, S.G.; Possas, A.; Posada-Izquierdo, G.D.; Theumer, M.G.; Pérez-Rodríguez, F. Unveiling Fresh-Cut Lettuce Processing in Argentine Industries: Evaluating *Salmonella* Levels Using Predictive Microbiology Models. *Foods* **2023**, *12*, 3999. https://doi.org/10.3390/foods12213999.
11. González-Tejedor, G.A.; Garre, A.; Iguaz, A.; Wong-Zhang, R.; Fernández, P.S.; Possas, A. Dynamic Thermal Treatments in Green Coconut Water Induce Dynamic Stress Adaptation of *Listeria innocua* That Increases Its Thermal Resistance. *Foods* **2023**, *12*, 4015. https://doi.org/10.3390/foods12214015.
12. Pulsrikarn, C.; Kedsin, A.; Boueroy, P.; Chopjitt, P.; Hatrongjit, R.; Chansiripornchai, P.; Suanpairintr, N.; Nuanualsuwan, S. Quantitative Risk Assessment of Susceptible and Ciprofloxacin-Resistant *Salmonella* from Retail Pork in Chiang Mai Province in Northern Thailand. *Foods* **2022**, *11*, 2942. https://doi.org/10.3390/foods11192942.
13. Regecová, I.; Výrostková, J.; Zigo, F.; Gregová, G.; Pipová, M.; Jevinová, P.; Becová, J. Detection of Resistant and Enterotoxigenic Strains of *Staphylococcus warneri* Isolated from Food of Animal Origin. *Foods* **2022**, *11*, 1496. https://doi.org/10.3390/foods11101496.
14. Pakbin, B.; Amani, Z.; Rahimi, Z.; Najafi, S.; Familsatarian, B.; Khakpoor, A.; Brück, W.M.; Brück, T.B. Prevalence of Foodborne Bacterial Pathogens and Antibiotic Resistance Genes in Sweets from Local Markets in Iran. *Foods* **2023**, *12*, 3645. https://doi.org/10.3390/foods12193645.
15. Wang, P.; Zhao, X.; Qu, T.; Liang, L.; Ji, Q.; Chen, Y. Insight into *Bacillus cereus* Associated with Infant Foods in Beijing. *Foods* **2022**, *11*, 719. https://doi.org/10.3390/foods11050719.

References

1. European Food Safety Authority. The European Union One Health 2022 Zoonoses Report. *EFSA J.* **2023**, *21*, e8442. [CrossRef]
2. Smith, J.L.; Fratamico, P.M. Emerging and Re-Emerging Foodborne Pathogens. *Foodborne Pathog. Dis.* **2018**, *15*, 737–757. [CrossRef]
3. Liguori, J.; Trübswasser, U.; Pradeilles, R.; Le Port, A.; Landais, E.; Talsma, E.F.; Lundy, M.; Béné, C.; Bricas, N.; Laar, A.; et al. How do food safety concerns affect consumer behaviors and diets in low- and middle-income countries? A systematic review. *Glob. Food Sec.* **2022**, *32*, 100606. [CrossRef]

foods

Article

Developing Qualitative Plasmid DNA Reference Materials to Detect Mechanisms of Quinolone and Fluoroquinolone Resistance in Foodborne Pathogens

Qinya Niu [1], Xiumin Su [1], Luxin Lian [1], Jinling Huang [1], Shutong Xue [1], Wei Zhou [2], Hongyang Zhao [3], Xing'an Lu [3], Shenghui Cui [4], Jia Chen [5] and Baowei Yang [1,*]

1 College of Food Science and Engineering, Northwest A&F University, Xianyang 712100, China; nqy736461094@163.com (Q.N.); sxm10182021@163.com (X.S.); llxin1116@163.com (L.L.); xncyxyxgb@163.com (J.H.); pcyxst20000714@163.com (S.X.)
2 Hebei Food Inspection and Research Institute, Hebei Food Safety Key Laboratory, Shijiazhuang 050091, China; zhouwei0311@163.com
3 Chinese Academy of Inspection and Quarantine, Beijing 100176, China; zhy8707@163.com (H.Z.); luxa@sina.com (X.L.)
4 National Institutes for Food and Drug Control, Beijing 100050, China; cuishenghui@aliyun.com
5 College of Chemical Technology, Shijiazhuang University, Shijiazhuang 050035, China; chenjia_0311@163.com
* Correspondence: ybwsheng@nwafu.edu.cn; Tel./Fax: +86-29-8709-2486

Abstract: The aim of this study was to develop homogeneous and stable plasmid DNA reference materials for detecting the mechanisms of resistance to quinolones and fluoroquinolones in foodborne pathogens. The DNA fragments of 11 target genes associated with quinolone and fluoroquinolone resistance were artificially synthesized, inserted into plasmid vectors, and transferred into recipient cells. PCR and sequencing of DNA were performed to assess the genetic stability of the target DNA in recombinant *Escherichia coli* DH5α cells during subculturing for 15 generations. The limit of detection (LOD) of the target DNA was determined using PCR and real-time qualitative PCR (qPCR). The homogeneity and storage stability of plasmid DNA reference materials were evaluated in terms of plasmid DNA quantity, PCR-measured gene expression, and qPCR threshold cycle. All 11 target DNAs were successfully synthesized and inserted into vectors to obtain recombinant plasmids. No nucleotide mutations were identified in the target DNA being stably inherited and detectable in the corresponding plasmids during subculturing of recombinant strains. When the target DNA was assessed using PCR and qPCR, the LOD was $\leq 1.77 \times 10^5$ and 3.26×10^4 copies/μL, respectively. Further, when the reference materials were stored at 37 °C for 13 days, 4 °C for 90 days, and −20 °C for 300 days, each target DNA was detectable by PCR, and no mutations were found. Although the threshold cycle values of qPCR varied with storage time, they were above the LOD, and no significant differences were found in the quantity of each plasmid DNA at different timepoints. Further, the homogeneity and stability of the materials were highly consistent with the requirements of standard reference materials. To summarize, considering that our plasmid DNA reference materials conformed to standard requirements, they can be used to detect the mechanisms of quinolone and fluoroquinolone resistance in foodborne pathogens.

Keywords: antibiotic resistance; plasmid DNA reference material; limit of detection; homogeneity; stability

Citation: Niu, Q.; Su, X.; Lian, L.; Huang, J.; Xue, S.; Zhou, W.; Zhao, H.; Lu, X.; Cui, S.; Chen, J.; et al. Developing Qualitative Plasmid DNA Reference Materials to Detect Mechanisms of Quinolone and Fluoroquinolone Resistance in Foodborne Pathogens. *Foods* **2022**, *11*, 154. https://doi.org/10.3390/foods11020154

Academic Editors: Fernando Pérez-Rodríguez and Arícia Possas

Received: 3 December 2021
Accepted: 5 January 2022
Published: 7 January 2022

1. Introduction

According to the World Health Organization, nearly one-tenth of the global population get sick from foodborne diseases each year [1]. This consequently leads to 420,000 deaths annually, with the main reason being consumption of food contaminated with foodborne pathogens [2,3]. Antibiotics are the standard and the most direct, effective method to treat foodborne diseases, but antibiotic resistance is becoming an increasingly serious issue.

Quinolones (i.e., nalidixic acid) and fluoroquinolones (i.e., ciprofloxacin) are important synthetic antibiotics that are commonly used to treat diseases caused by foodborne pathogens. However, mutations in the quinolone resistance-determining region of DNA gyrase subunits (GyrA and GyrB) and topoisomerase IV subunits (ParC and ParE) have been associated with quinolone and fluoroquinolone resistance. Moreover, the presence of *qnrA*, *qnrB*, *aac(6′)-Ib-cr*, *oqxAB*, and *qnrS* genes, which encode hydrolases and antibiotic-inactivating enzymes, in the genome of pathogens has been associated with fluoroquinolone resistance [4–7]. As a part of food safety inspection and for diagnosing clinical diseases caused by foodborne pathogens, homogeneous and stable plasmid DNA reference materials are required to ensure data accuracy when detecting the mechanisms of quinolone and fluoroquinolone resistance.

At present, >1000 biological reference materials exist in the list of the National Institute for Biological Standards and Control (https://www.nibsc.org/, accessed on 4 January 2021), the leading international standard and reference material producer and distributor in the world. However, serums, vaccines, and antibodies are the most prevalent reference materials [8]; the prevalence of viruses and genetically modified products as reference materials remains scarce. In China, despite there being some nucleic acid reference materials, they are mainly related to viruses and genetically modified products (https://www.ncrm.org.cn/Web/Home/Index, accessed on 4 January 2021). In recent years, although many studies in China have reported the development of reference materials to detect the DNA of bacterial pathogens, the adoption of these materials for technical purposes has been limited due to traceability and practicality. Consequently, the application of these materials for technical purposes has been limited [9–11]. To the best of our knowledge, to date, there exists no DNA reference material and/or certified reference material that can be used to identify genes responsible for antibiotic resistance and/or to assess pertinent mechanisms.

In this study, we developed 11 qualitative plasmid DNA reference materials to study mechanisms associated with quinolone and fluoroquinolone resistance in foodborne pathogens for food safety inspection and molecular diagnosis of clinical diseases caused by foodborne pathogens.

2. Materials and Methods

2.1. Target DNA Synthesis and Recombinant Plasmid and Strain Construction

We screened the following: five genes (*qnrA*, *qnrB*, *qnrS*, *aac(6′)-Ib*, and *oqxA*) that are commonly found in plasmids and six genes (*gyrA* with the single mutations Asp87Asn, Ser83Tyr, and Ser83Phe in GyrA; *gyrA* with the double mutations Ser83Phe/Asp87Gly and Ser83Phe/Asp87Ala in GyrA; and *parC* with the single mutation Ser80Arg in ParC) that are associated with quinolone and fluoroquinolone resistance from the GenBank nucleic acid sequence database of the National Center for Biotechnology Information (NCBI; http://www.ncbi.nlm.nih.gov, accessed on 4 December 2018). The sequences of each gene were downloaded from NCBI website and compared using BioEdit (BioEdit Inc., Manchester, NH, USA). The gene fragment with the same DNA sequence was defined as the target DNA. The DNA fragment of the target DNA was synthesized by Beijing AuGCT Co., Ltd. (Beijing, China), and subsequently inserted into pEASY and pUC57 to construct recombinant plasmids (Figure 1). The recombinant plasmids were then transferred into *Escherichia coli* DH5α cells to obtain recombinant strains.

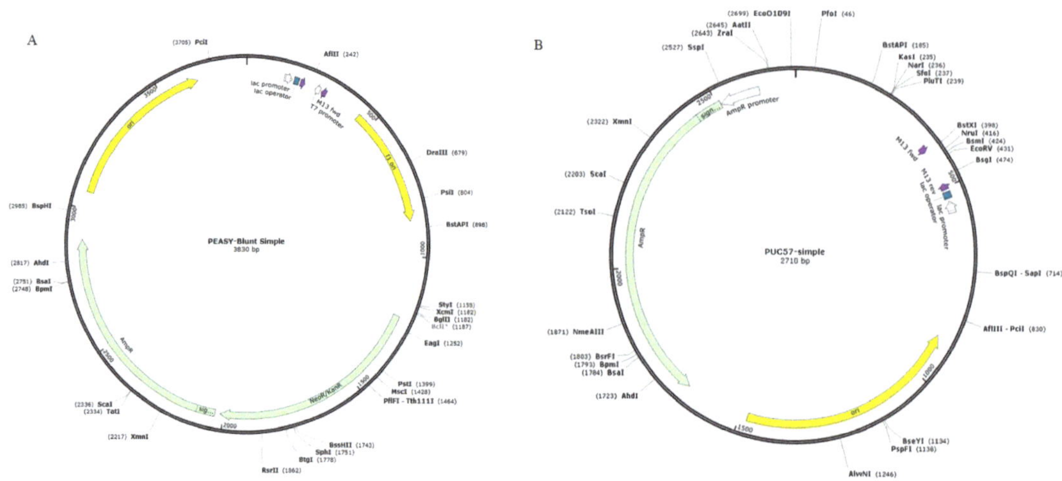

Figure 1. Schematic diagram of plasmid vectors. (**A**) pEASY. (**B**) pUC57.

2.2. Genetic Stability Test of the Target DNA in Recombinant Strains

All recombinant strains were plated on Luria-Bertani agar (CM1552, Beijing Land Bridge Technology Co., Ltd., Beijing, China). After incubation at 37 °C ± 0.5 °C for 12–18 h, a single colony was chosen and continuously subcultured for 15 generations on Luria-Bertani agar plates. Target DNA stability in the recombinant strains was determined by performing PCR every three generations. DNA sequencing and online BLAST alignment were performed to determine whether the target DNA was mutated.

Template DNA was prepared as previously described [12]. PCR was performed on a MyCircle PCR system (Bio-Rad, Hercules, CA, USA) in a 25-μL reaction mixture containing 13.15 μL of double-distilled (dd) H_2O, 0.3 μL of 50 ng/mL primer each, 2.5 μL of 10×PCR buffer (25 mM without Mg^{2+}; R001AM, TaKaRa, Dalian, China), 2 μL of 2.5 mM dNTP mixture (TaKaRa, Dalian, China, 0.25 μL of 5 U/μL Taq DNA polymerase (TaKaRa, Dalian, China), 1.5 μL of 25 mM $MgCl_2$ (TaKaRa, Dalian, China), and 5 μL of template DNA. The cycling conditions were as follows: initial denaturation at 94 °C for 5 min; followed by 35 cycles of 94 °C for 30 s, pertinent annealing temperature for 30 s, and 72 °C for 1 min; and final extension at 72 °C for 10 min [13]. Table 1 lists all primers and annealing temperatures for each target DNA.

Table 1. PCR primers and annealing temperatures for target DNA detection.

Gene	Primer	Sequence (5′–3′)	Annealing Temperature (°C)	Product Size (bp)	Reference
qnrA	*qnrA*-F *qnrA*-R	AGAGGATTTCTCACGCCAGG TGCCAGGCACAGATCTTGAC	56	579	[14]
qnrB	*qnrB*-F *qnrB*-R	GGCATTGAAATTCGCCACTG TTTGCTGCTCGCCAGTCGAA	56	263	[14]
qnrS	*qnrS*-F *qnrS*-R	GCAAGTTCATTGAACAGGGT TCTAAACCGTCGAGTTCGGCG	56	427	[14]
oqxA	*oqxA*-F *oqxA*-R	GACAGCGTCGCACAGAATG GGAGACGAGGTTGGTATGGA	56	339	[15]
aac(6′)-Ib	*aac(6′)-Ib*-F *aac(6′)-Ib*-R	TTGCGATGCTCTATGAGTGGCTA CTCGAATGCCTGGCGTGTTT	55	482	[16]
gyrA	*gyrA*-F *gyrA*-R	CCGTACCGTCATAGTTATCC CGTTGGTGACGTAATCGGTA	56	251	[17]
parC	*parC*-F *parC*-R	TAACAGCAGCTCGGCGTATT CTATGCGATGTCAGAGCTGG	54	262	[18]

The amplicons obtained were electrophoresed and visualized under UV light, and they were then transferred at low temperatures for sequencing to AuGCT Biotech (Beijing, China). The DNA sequence was aligned with the original sequence using the online BLAST program to determine whether the gene(s) had mutated in the subcultures. To detect mutations of *gyrA* and *parC*, the DNA sequence was translated into the corresponding amino acid sequence using Primer Premier 5 (Premier Biosoft, San Francisco, CA, USA) and aligned to ascertain that the preset mutation sites were stably inherited during the subculture.

2.3. Extraction of Plasmids Carrying Antibiotic Resistance-Encoding Genes

Plasmid extraction was performed using a Plasmid Mini Kit I (D6943-01*, OMEGA, Norcross, GA, USA), according to manufacturer's instructions. Plasmid DNA concentration was measured with Qubit$^{TM\,4}$ (Thermo Fisher Scientific, Waltham, MA, USA), and plasmid DNA was stored at −20 °C until needed.

2.4. Assessment of Limit of Detection (LOD) for the Target DNA

We serially diluted 10 μL of the plasmid suspension plus target DNA using 90 μL of sterile water by 10-fold each time (10^{-1}, 10^{-2}, and so on) until the suspension was diluted to 10^{-n} concentration. All these different dilutions served as template DNA for PCR and real-time quantitative PCR (qPCR). The LOD of PCR was calculated via the initial concentration of the target DNA solution divided by the maximum dilution times when the amplicon on the electrophoresis gel appeared markedly dark, unclear, or invisible. The PCR amplification system and conditions were the same as those described earlier (under *Genetic Stability of the Target DNA in Recombinant Strains*). Table 1 lists pertinent primers and annealing temperatures used for PCR.

qPCR was performed on a Bio-Rad iQ5 thermal cycler in a 25-μL reaction mixture comprising 8.5 μL of ddH_2O, 12.5 μL of 2 × SYBR Green *Pro Taq* HS Premix (AG11701, AG, Changsha, China), 1 μL of 50 ng/mL primer each, and 2 μL of template with different concentrations of the recombinant plasmid. The cycling conditions were as follows: 95 °C for 30 s; 40 cycles at 95 °C for 5 s and 60 °C for 30 s; 1 cycle at 95 °C for 15 s and 60 °C for 30 s; and 71 cycles with the temperature increasing from 60 °C to 95 °C. Primers used for qPCR are listed in Table 2. The LOD of qPCR was determined as the concentration of the recombinant plasmid solution (i.e., the template) that did not cause any further increase in the threshold cycle (Ct) value [19].

Table 2. qPCR primers for the target DNA.

Gene	Primer	Sequence (5′–3′)	Product Size (bp)
qnrA	*qnrA*-F *qnrA*-R	TGCTTTGGCATAGAGTTCAGG GGCATTGCTCCAGTTGTTTT	192
qnrB	*qnrB*-F *qnrB*-R	GGCATTGAAATTCGCCACTG TTTGCTGCTCGCCAGTCGAA	263
qnrS	*qnrS*-F *qnrS*-R	TCGTCGCTGCCACTTTGAT ATGCACCCGCTAGGTTCGTT	296
oqxA	*oqxA*-F *oqxA*-R	GACAGCGTCGCACAGAATG GGAGACGAGGTTGGTATGGA	339
aac(6′)-Ib	*aac(6′)-Ib*-F *aac(6′)-Ib*-R	CCGACACTTGCTGACGTACA GTTTCTTCTTCCCACCATCC	155
gyrA	*gyrA*-F *gyrA*-R	CCGTACCGTCATAGTTATCC CGTTGGTGACGTAATCGGTA	251
parC	*parC*-F *parC*-R	TAACAGCAGCTCGGCGTATT CTATGCGATGTCAGAGCTGG	262

The copy number of plasmid DNA that determined LOD was calculated using the following formula:

$$\text{Number of copy (copy/}\mu\text{L)} = (\text{Concentration} \times 10^{-9} \times 6.02 \times 10^{23})/(\text{Length} \times 660)$$

where Concentration represents plasmid DNA concentration measured using a NanoDrop One spectrophotometer (Thermo Fisher Scientific, Waltham, MA, USA; ng/μL) and Length represents template DNA length (bp).

2.5. Preparation of Plasmid DNA Reference Materials

The recombinant plasmid with the target DNA was extracted from the recombinant strain using a Plasmid Mini Kit I (OMEGA, Norcross, GA, USA). The concentration and OD260/280 and OD260/230 values of the plasmid DNA suspension were determined using the NanoDrop One spectrophotometer. The suspension was subsequently subpackaged in a 1.5-mL Eppendorf tube; the quantity of the recombinant plasmid in each tube was approximately 300 ng. The DNA was then vacuum dried to prepare reference materials.

2.6. Homogeneity Test of Plasmid DNA Reference Materials

According to the guidelines of China National Standard GB/T 15000.3-2008 "Directives for the work of reference materials (3)—Reference materials—General and statistical principles for certification" and ISO Guide 35:2006 "Reference materials—General and statistical principles for certification, IDT", when the total number of the units of reference materials is less than 500, the unit number selected for the homogeneity test should not be less than 10. We consequently selected 12 Eppendorf tubes of reference material samples at random for the homogeneity test, with each tube serving as a sample unit. After re-dissolution, the concentration of the plasmid DNA sample was determined using the NanoDrop One spectrophotometer. The quantity of DNA sample was calculated as plasmid DNA concentration × aqueous solution volume.

2.7. Storage Stability Test of Plasmid DNA Reference Materials

After vacuum drying, plasmid DNA was stored at 37 °C, 4 °C, and −20 °C to evaluate its storage stability. The stability of plasmid DNA reference materials stored at 37 °C was tested over 13 days (short term), and random samples were taken every week. Similarly, the stability of plasmid DNA reference materials stored at 4 °C was tested over 90 days (short term); samples were taken every week for the first 2 weeks (1, 7, and 13 days) and every month for the remaining period (30, 60, and 90 days). Finally, the stability of plasmid DNA reference materials stored at −20 °C was tested over 12 months (long term); from the first to the sixth month of storage, plasmid DNA was randomly sampled every month, and from the seventh to the twelfth month of storage, it was sampled every 2 months. All samples stored at different temperatures were sampled and tested in triplicate.

The indicators for the storage stability test were DNA quantity, PCR-measured gene expression, qPCR Ct value, DNA sequence, and amino acid mutation. PCR/qPCR primers, amplification system, and conditions were the same as those described earlier (see Sections 2.2 and 2.4, respectively).

2.8. Data Analysis

Microsoft Office Excel v2010 (Microsoft Corp., Redmond, WA, USA) was used for the basic processing of experimental data. IBM SPSS Statistics v22 (IBM, New York, NY, USA) was used for statistical analysis of variance (ANOVA; Duncan's method, $p \leq 0.05$ indicating the difference being statistically significant). RStudio v3.4.4 (RStudio Inc., Boston, MA, USA) was used for drawing graphs.

3. Results

3.1. Recombinant Plasmid and Strain Construction

The target DNA fragments of *qnrA*, *qnrB*, *qnrS*, *aac(6′)-Ib*, *oqxA*, *gyrA* with the single mutations Asp87Asn, Ser83Tyr, and Ser83Phe (i.e., *gyrA*1, *gyrA*2, and *gyrA*3, respectively) in GyrA, *gyr*A with the double mutations Ser83Phe/Asp87Gly and Ser83Phe/Asp87Ala (i.e., *gyrA*4 and *gyrA*5, respectively) in GyrA, and *parC* with the single mutation Ser80Arg in ParC were all successfully synthesized and ligated into pUC57 and pEASY-T. All recombinant plasmids were successfully transformed into *E. coli* DH5α cells and corresponding recombinant strains were obtained. DNA sequencing results indicated that the identities and coverage rates of the target DNA in the 11 recombinant plasmids were 100%, as anticipated (Table S1). The nucleotide sequences of all target DNA fragments were submitted to GenBank and issued an accession number to ensure traceability of the reference materials (Table S1).

3.2. Genetic Stability

The target DNA in all recombinant strains was stably inherited across all 15 generations, and no mutations were found (Figures S1–S7). Further, the single and double mutations in *gyrA* and *parC* were stably transferred from the first to the last (n = 15) generation (Figure 2).

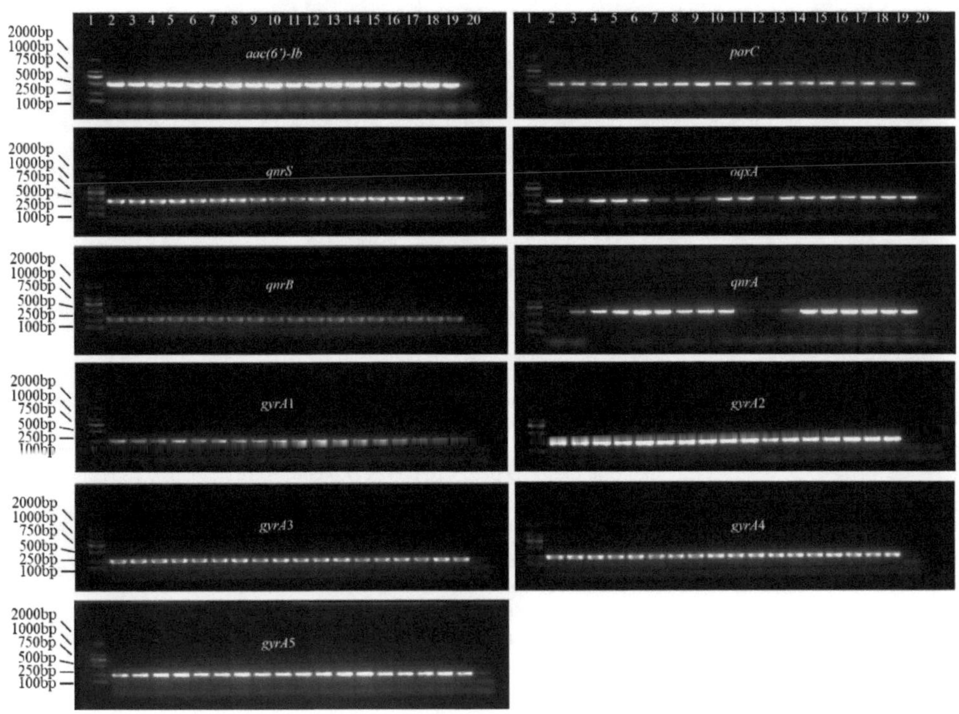

Figure 2. Genetic stability of the target DNA in recombinant strains of in *E. coli* DH5α. In all gels, lane 1, DL 2000 DNA marker; lanes 2–4, amplicons of the target DNA in the original recombinant strains; lanes 5–7, 8–10, 11–13, 14–16, and 17–19, amplicons of the target DNA in the third, sixth, ninth, twelfth, and fifteenth generations of recombinant strains, respectively; and lane 20, double-distilled H_2O, which served as the blank control.

3.3. LOD of PCR and qPCR

Spectrophotometric data indicated that all OD260/280 values for the plasmid DNA ranged from 1.8 to 2.0 and all OD260/230 values were between 2.0 and 2.2, implying that the purity of plasmid DNA extracted from the recombinant strains adequately met the requirements of standard reference materials. With regard to the LOD of PCR, it was 1.85×10^3 copies/μL for *aac(6′)-Ib*, 1.37×10^4 copies/μL for *gyrA*2, 1.44×10^4 copies/μL for *gyrA*3, 1.88×10^4 copies/μL for *parC*, 1.97×10^4 copies/μL for *gyrA*4, 2.02×10^4 copies/μL for *gyrA*1, 2.13×10^4 copies/μL for *qnrA*, 2.19×10^4 copies/μL for *qnrS*, 3.26×10^4 copies/μL for *oqxA*, 1.74×10^5 copies/μL for *qnrB*, and 1.77×10^5 copies/μL for *gyrA*5 (Figure 3). The LOD and copy number of all plasmid DNAs decreased with an increase in dilution times. These results indicated that when the target DNA was assessed using PCR, the LOD was $\leq 10^5$ copies/μL.

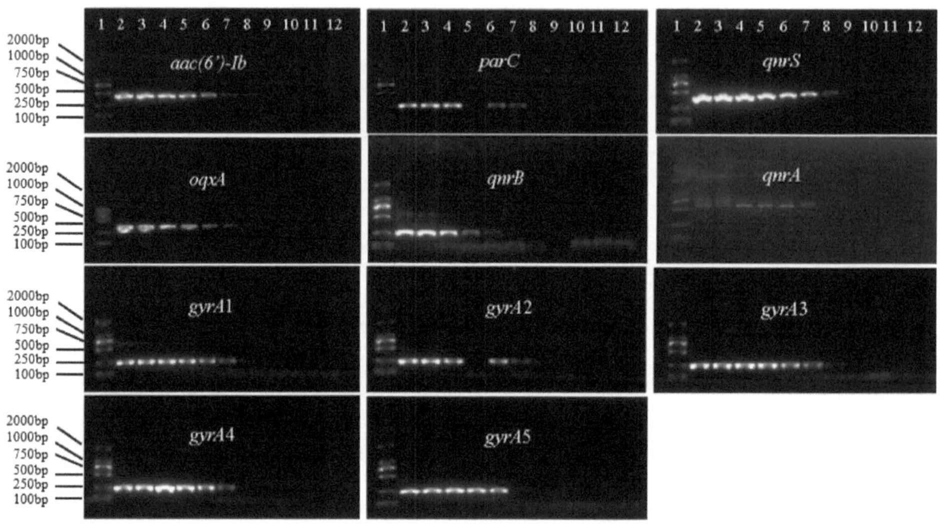

Figure 3. Limit of detection of PCR for the target DNA. In all gels: lane 1, DL 2000 DNA marker; lanes 2–11, amplicons obtained using template DNA (10^1–10^{10}-fold diluted); lane 12, double-distilled H_2O, which served as the blank control.

With regard to the LOD of qPCR, it was 17.4 copies/μL for *qnrB*, 21.3 copies/μL for *qnrA*, 1.44×10^2 copies/μL for *gyrA*3, 1.85×10^2 copies/μL for *aac(6′)-Ib*, 1.88×10^2 copies/μL for *parC*, 1.37×10^3 copies/μL for *gyrA*2, 1.77×10^3 copies/μL for *gyrA*5, 1.97×10^3 copies/μL for *gyrA*4, 2.02×10^3 copies/μL for *gyrA*1, 2.19×10^4 copies/μL for *qnrS*, and 3.26×10^4 copies/μL for *oqxA*. These results indicated that based on qPCR, the LOD of the target DNA was $\leq 10^4$ copies/μL. For qPCR, the standard curve was constructed using Ct values (Figure 4). We found that there was an excellent linear relationship between the template DNA concentration and Ct value, with all regression coefficients (R^2) being >0.99.

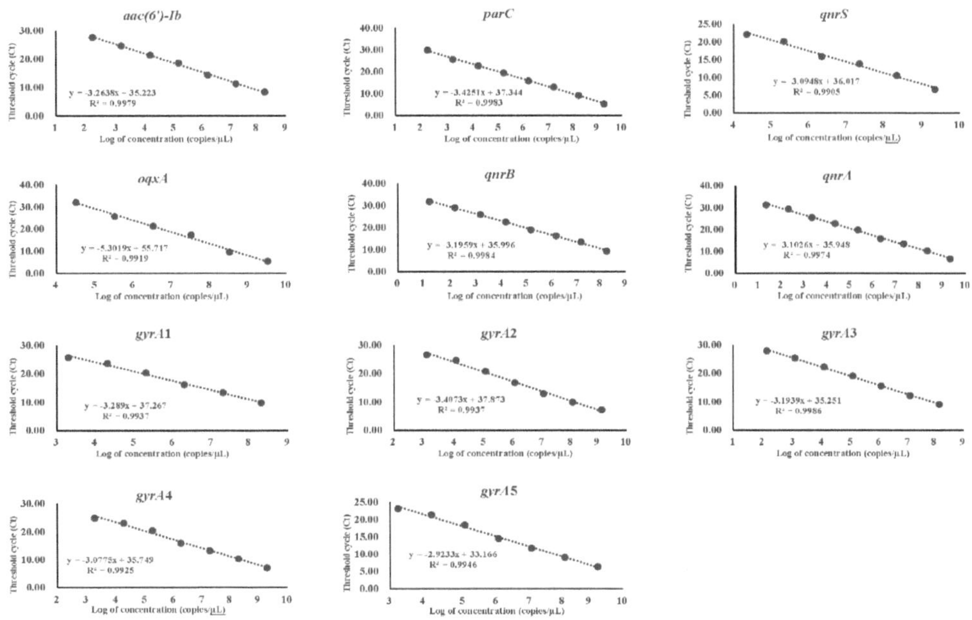

Figure 4. Standard curve and Ct values of qPCR for each target DNA. X axis represents logarithm of concentration, and Y axis represents Ct values for different concentrations of template DNA.

3.4. Homogeneity of Plasmid DNA Reference Materials

Gel electrophoresis results indicated that the target genes in the plasmid DNA could be successfully amplified, and the amplicons were of expected size (Figure S8). Sequencing data revealed that no mutations were present in the target DNA (Figures S9–S15). According to statistical analyses, the quantity of plasmid DNA harboring the target gene in each tube had an F value that was less than the F-critical value under 95% confidence interval; this result indicated that there were no significant differences in plasmid DNA quantity in each tube (Table 3). Collectively, these data showed that the homogeneity of plasmid DNA reference materials adequately met the requirements of standard reference materials.

Table 3. Homogeneity parameters of plasmid DNA reference materials.

Gene	Difference	SS	Df	MS	F-Value	*p*-Value	F-Critical Value
aac(6′)-Ib	interblock	8.60	11.00	0.78	2.52	0.06	2.72
	intraclass	3.73	12.00	0.31			
qnrA	interblock	4.63	11.00	0.42	2.05	0.12	2.72
	intraclass	2.47	12.00	0.21			
qnrB	interblock	11.60	11.00	1.05	2.65	0.05	2.72
	intraclass	4.77	12.00	0.40			
qnrS	interblock	11.87	11.00	1.08	2.64	0.05	2.72
	intraclass	4.91	12.00	0.41			
oqxA	interblock	8.48	11.00	0.77	1.68	0.19	2.72
	intraclass	5.50	12.00	0.46			
parC	interblock	1.66	11.00	0.15	0.86	0.60	2.72
	intraclass	2.12	12.00	0.18			
gyrA1	interblock	4.31	11.00	0.39	1.18	0.39	2.72
	intraclass	3.98	12.00	0.33			

Table 3. *Cont.*

Gene	Difference	SS	Df	MS	F-Value	*p*-Value	F-Critical Value
*gyrA*2	interblock	4.08	11.00	0.37	2.64	0.05	2.72
	intraclass	1.69	12.00	0.14			
*gyrA*3	interblock	6.05	11.00	0.55	1.41	0.28	2.72
	intraclass	4.69	12.00	0.39			
*gyrA*4	interblock	5.59	11.00	0.51	2.23	0.09	2.72
	intraclass	2.74	12.00	0.23			
*gyrA*5	interblock	3.29	11.00	0.30	2.47	0.07	2.72
	intraclass	1.45	12.00	0.12			

Note: "SS" represents stdev square, "Df" represents degree of freedom, and "MS" represents mean square.

3.5. Storage Stability of Plasmid DNA Reference Materials

With regard to the short-term storage stability of plasmid DNA reference materials, no significant differences in plasmid DNA quantity were found for *aac(6′)-Ib*, *qnrA*, *qnrS*, *gyrA*1, *gyrA*2, *gyrA*3, *gyrA*4, and *gyrA*5 upon storage at 37 °C for 7 days. Moreover, no significant differences in plasmid DNA quantity were observed for *oqxA*, *parC*, and *qnrB* upon storage for 13 days at 37 °C (Figure 5). All the 11 target DNAs were detectable by PCR within 13 days, and no mutations were found (Figures S16–S23). Although qPCR indicated that the Ct values for each target DNA varied in these 13 days, no significant differences were found in the Ct values for any target DNA (Table 4). To summarize, the storage stability of plasmid DNA reference materials was excellent when stored at 37 °C for at least 1 week, they can serve as positive standard samples to study genes and mechanisms associated with quinolone and fluoroquinolone resistance.

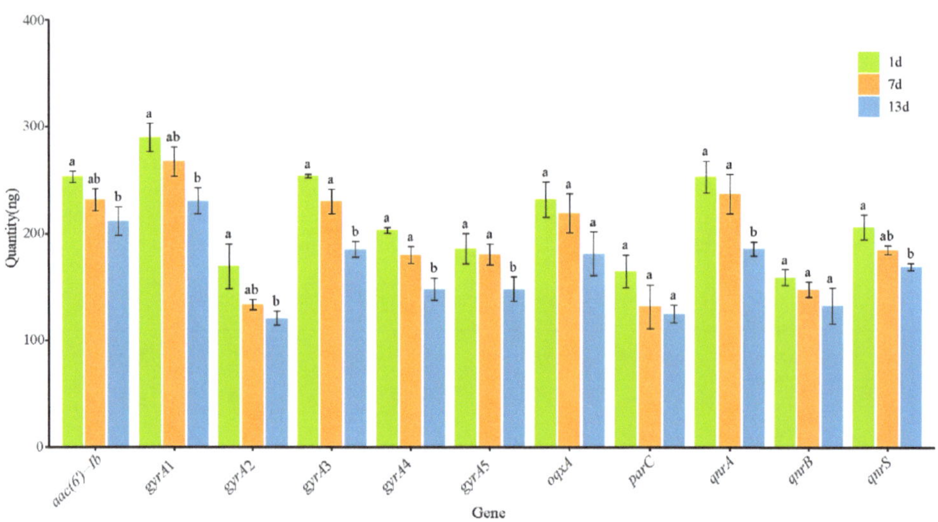

Figure 5. Short-term stability of the quantity of plasmid DNA reference materials stored at 37 °C. For each gene, columns labeled with the same letter indicates that no significant difference was found in plasmid DNA quantity upon storage for different durations.

Table 4. Ct values of qPCR for plasmid DNA reference materials stored at 37 °C (Mean ± SD).

Gene	1 Day	7 Days	13 Days
	Ct Value	Ct Value	Ct Value
aac(6')-Ib	7.33 ± 0.42	8.31 ± 0.57	7.79 ± 0.54
parC	7.66 ± 0.58	7.80 ± 0.82	9.24 ± 0.90
qnrS	7.26 ± 0.69	6.76 ± 0.25	7.36 ± 0.38
oqxA	7.89 ± 0.87	8.39 ± 1.22	7.98 ± 1.55
qnrB	8.62 ± 0.15	7.53 ± 0.28	7.75 ±0.61
qnrA	8.00 ± 0.63	7.94 ± 0.40	7.85 ± 0.86
gyrA1	9.12 ± 1.14	7.13 ± 0.32	8.33 ± 0.23
gyrA2	7.55 ± 0.44	8.78 ± 0.80	8.91 ± 1.33
gyrA3	8.44 ± 0.51	7.33 ± 0.38	6.39 ± 0.47
gyrA4	8.55 ± 0.78	8.16 ± 0.41	8.22 ± 0.85
gyrA5	8.42 ± 0.75	7.87 ± 0.77	7.77 ± 0.63

When plasmid DNA reference materials were stored at 4 °C for 90 days, no significant differences were detected in plasmid DNA quantity for all target genes, with the exception of *aac(6')-Ib* (Figure 6). Furthermore, all target genes were detectable by PCR within 90 days, and no mutations were identified (Figures S24–S31). There was no abnormal variation in the Ct values of qPCR after the samples were 10-fold serially diluted, and they remained within the LOD for each gene (Figure 7A). In summary, considering that most plasmid DNA reference materials exhibited excellent stability when stored for 90 days at 4 °C, they can serve as positive standard samples for studying genes and mechanisms associated with antibiotic resistance.

When plasmid DNA reference materials were stored at −20 °C for 1 year, the plasmid DNA quantity significantly ($p \leq 0.05$) declined for all target genes, except *par*C (Figure 8). However, all target genes were detectable by PCR within 360 days, and no mutations were found (Figures S32–S39). The Ct values of qPCR considerably varied after plasmid DNA reference materials were 10-fold serially diluted, but they did not exceed the LOD for each gene (Figure 7B). Therefore, despite the decline in quantity upon storage for 360 days at −20 °C, all plasmid DNA reference materials can still serve as qualitative standard reference samples.

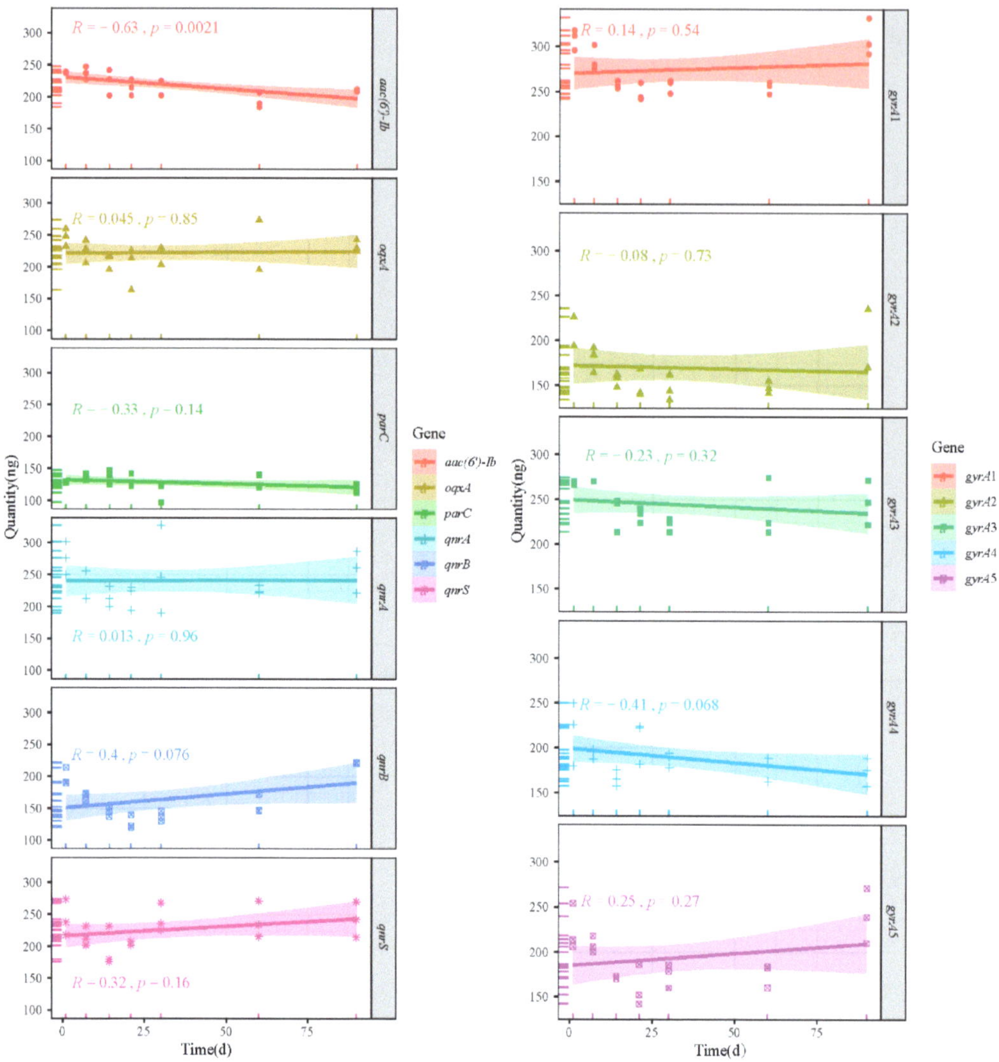

Figure 6. Short-term stability of the quantity of plasmid DNA reference materials stored at 4 °C. Note: "R" represents the correlation coefficient, "p" represents the significant difference, and the shadow represents the 95% confidence interval.

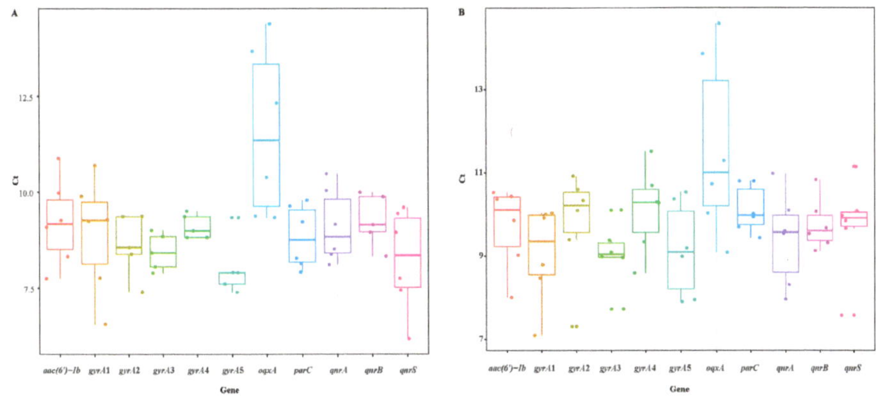

Figure 7. Ct values of qPCR with plasmid DNA reference materials stored at 4 °C and −20 °C. (**A**) Short-term storage stability at 4 °C. (**B**) Long-term storage stability at −20 °C.

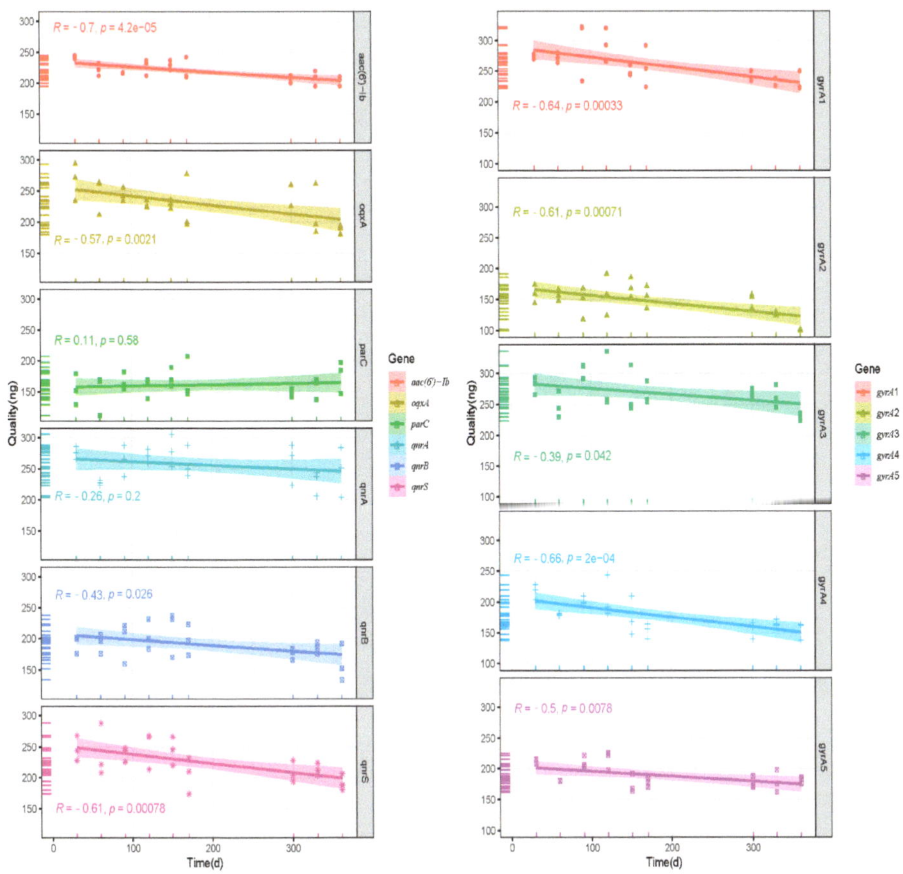

Figure 8. Long-term stability of the quantity of plasmid DNA reference materials stored at −20 °C. Note: "R" represents the correlation coefficient, "*p*" represents the significant difference, and the shadow represents the 95% confidence interval.

4. Discussion

Many molecular detection techniques, such as conventional PCR, qPCR, and loop-mediated isothermal amplification, are widely used to identify foodborne pathogens, determine antibiotic resistance dissemination, and study the potential mechanisms of action based on food safety perspective [20–24]. To obtain accurate results, using reference materials is advisable during pathogen resistance gene detection and testing. At present, available reference materials used for microbial detection are mainly for viruses, whereas those for DNA detection are mainly in the field of transgenic food and crops (https://www.nibsc.org/). Although the demand for standard and certified reference materials has been constantly increasing to achieve food safety and for consumer health protection, no studies have as yet reported the development of plasmid DNA reference materials to identify antibiotic resistance-encoding genes and to elucidate pertinent mechanisms in foodborne pathogens. Herein we developed 11 qualitative plasmid DNA reference materials to study genes and mechanisms associated with quinolone and fluoroquinolone resistance. We believe that our findings should facilitate the detection and prevention of antibiotic resistance in foodborne pathogens.

Previously, the development of few reference strains belonging to *Salmonella*, *Shigella*, and *Cronobacter* has been reported to detect antibiotic resistance [9,25–27]. However, such pathogenic reference materials are a potential threat to food safety in some food production environments. Therefore, developing DNA reference materials for detecting some antibiotic resistance-encoding genes and associated mechanisms becomes pivotal. In the present study, 11 recombinant plasmids and strains associated with quinolone and fluoroquinolone resistance were successfully constructed, and antibiotic resistance-encoding genes in the recombinant plasmids and strains were found to be stably inherited even after subculturing for 15 generations. Moreover, no nucleotide mutations were detected in the target DNA, indicating that these plasmids and strains could potentially serve as reference materials.

PCR is commonly used for the rapid detection of antibiotic resistance-encoding genes in pathogenic bacteria. During reference material development, it is thus essential to determine the LOD of PCR for detecting target DNA in recombinant plasmids and it is also the premise for ensuring successful detection. Xia et al. developed a plasmid DNA reference material to detect pathogenic *E. coli*, and they found the LOD of PCR for *escV*, *stx2*, and *hlyA* to be 3.93×10^6, 2.41×10^5, and 2.14×10^5 copies/μL, respectively [11]. Moreover, Ma et al. developed plasmid DNA reference materials to detect *Listeria monocytogenes*, and they found the LOD of PCR for *hlyA*, *prfA*1, and *prfA*2 to be 8.2×10^7, 1.1×10^6, and 1.24×10^5 copies/μL, respectively [28]. In the present study, when plasmid DNA reference materials were detected using PCR, the LOD was $\leq 10^5$ copies/μL, being lower than the LOD of PCR reported in previous studies. To explain, if target DNA fragments are obtained with different methods or DNA purification kits, the template DNA quantity is expected to vary, resulting in variances in results and LOD despite the detection method being the same [29]. This accordingly prompted us to herein use a kit with assured quality to ensure DNA purity, so that the reference materials could be reliably used for further research.

The qPCR technique is also commonly used to study antibiotic resistance-encoding genes and pertinent mechanisms as this technique has several advantages, such as high specificity and sensitivity, no involvement of a dye, and much shorter turnaround time as compared to PCR [30,31]. To evaluate the suitability of our plasmid DNA reference materials for qPCR, we determined the LOD of qPCR for each target DNA. We found that the LOD of qPCR for the 11 target DNAs ranged from 1.74×10^1 to 3.26×10^4 copies/μL. For all cases, the LODs were less than or equal to the DNA concentration in the target DNA solution diluted 10^6 times (or 10^9 times for some cases). Similarly, Dorlass et al. reported that when SARS-CoV-2 RNA was diluted 10^7 times, the positive detection rate of SYBR Green-based qPCR was 98.42% [30]. Furthermore, Fábio et al. developed a plasmid DNA reference material to quantify genetically modified common bean embrapa 5.1 and found that the lowest amount that could be reliably detected by qPCR was 10^3 copies per reaction [32]. Wu et al. also developed a general plasmid reference material for screening genetically modified organisms by qPCR, and they found that the sensitivity of screening and taxon-specific assays ranged from 5 to 10 copies of pBI121-Screening plasmid [33]. In addition, when we assessed the reference materials using qPCR, a good linear relationship was found between the template DNA concentration and Ct value, indicating that plasmid DNA reference materials developed in this study may even be used as universal calibrators.

Our data further indicated that the homogeneity and stability of all plasmid DNA reference materials met the standard requirements formulated by China National Standard and ISO Guide. In a previous study, Zhang et al. prepared some recombinant pseudovirus particles carrying specific St. Louis encephalitis virus genes; the pseudovirus particles showed excellent thermal stability upon storage at 37 °C for 20 days, room temperature for 30 days, 4 °C for 60 days, and –20 °C for 90 days [34]. Furthermore, Junichi et al. prepared a DNA reference material for quality control of PCR testing and found that the reference DNA molecule did not show rapid degradation when the material was stored at 37 °C for 1 week [35]. Herein even we found that our reference materials were relatively stable; they were in fact more stable than those developed in previous studies upon storage at 37 °C for 13 days, 4 °C for 90 days, and −20 °C for 360 days. However, long-term stability test results showed that the quantity of our plasmid DNA reference materials deteriorated after storage for 300–360 days at −20 °C, although no significant differences were found within the first 300 days. Although the DNA quantity decreased after 300 days of storage, the long-term stability of our plasmid DNA reference materials was still much better than that of the materials developed in previous studies. For example, Vallejo et al. developed a genomic DNA reference material for *Salmonella enteritidis* detection; on storage at 4 °C and −20 °C for 9 months, high concentration dispersion and DNA quantity deterioration were detected over time [36]. In addition, although our qPCR data showed that the Ct values for each plasmid DNA reference material varied between different timepoints of storage, the Ct values were still within the LOD for a particular gene in the material before storage. Similar to the results of our study, Zhou et al. prepared and characterized a pseudoviral positive control for the nucleic acid detection of MERS-CoV, and they found that when the samples were stored at 4 °C, −20 °C, and −70 °C for 1 week, the Ct values acquired via qPCR showed variation [37]. Based on these observations, the homogeneity and stability of our plasmid DNA reference materials were found to be highly consistent with the requirements of standard reference materials.

Considering the reliability of our plasmid DNA reference materials, we approached seven laboratories to jointly certify and validate our results. As anticipated, all of the laboratories reported the same or similar results. To establish our plasmid DNA reference materials as certified reference materials, we plan to obtain a Chinese standard reference material number (GSB series) for them; once the GSB number is issued, we will try to obtain equivalent international mutual recognition, which should standardize the use of these materials to explore the mechanisms underlying resistance to quinolones and fluoroquinolones.

As reference materials, they can meet all requirements for non-standard methods validation, new methods evaluation, laboratory testing personnel assessment, laboratory testing capacity evaluation, and inter-laboratory comparison of reference materials [26]. As reference materials of antibiotic resistance genes, they are suitable for the validation of PCR and qPCR detection methods for antibiotic resistance genes identification in foodborne pathogens. In addition, these reference materials can serve as positive standard samples for genes and mechanisms associated with quinolone and fluoroquinolone resistance.

To conclude, herein we developed 11 plasmid DNA reference materials that showed excellent genetic stability, homogeneity, and storage stability. The materials can thus be used to detect and explore the mechanisms underlying quinolone and fluoroquinolone resistance in foodborne pathogens. Moreover, they can serve as qualitative and positive controls in future studies.

Supplementary Materials: The following are available online at https://www.mdpi.com/article/10.3390/foods11020154/s1, Supplementary Table S1: Gene information for development of plasmid DNA reference materials; Supplementary Figures S1–S7: Genetic stability of *aac(6′)-Ib*, *parC*, *qnrS*, *oqxA*, *qnrB*, *qnrA*, and *gyrA*, respectively; Supplementary Figure S8: PCR results for homogeneity of plasmid DNA reference materials; Supplementary Figures S9–S15: Homogeneity of *aac(6′)-Ib*, *parC*, *qnrS*, *oqxA*, *qnrA*, *qnrB*, and *gyrA* in plasmid DNA reference materials, respectively; Supplementary Figure S16: PCR results for storage stability of plasmid DNA references materials stored at 37 °C; Supplementary Figures S17–S23: Sequencing results of *aac(6′)-Ib*, *parC*, *qnrS*, *oqxA*, *qnrB*, *qnrA*, and *gyrA* for stability of plasmid DNA references materials stored at 37 °C, respectively; Supplementary Figure S24: PCR results for storage stability of plasmid DNA references materials stored at 4 °C; Supplementary Figures S25–S31: Sequencing results of *aac(6′)-Ib*, *parC*, *qnrS*, *oqxA*, *qnrB*, *qnrA*, and *gyrA* for stability of plasmid DNA references materials stored at 4 °C, respectively; Supplementary Figure S32: PCR results for storage stability of plasmid DNA references materials stored at −20 °C; Supplementary Figures S33–S39: Sequencing results of *aac(6′)-Ib*, *parC*, *qnrS*, *oqxA*, *qnrB*, *qnrA*, and *gyrA* for stability of plasmid DNA references materials stored at −20 °C, respectively (DOCX).

Author Contributions: Conceptualization, Q.N. and B.Y.; methodology, Q.N., X.S. and J.H.; software, Q.N., L.L. and S.X.; validation, W.Z. and J.C.; resources, H.Z., X.L. and S.C.; data curation, Q.N.; writing—original draft preparation, Q.N.; writing—review and editing, B.Y.; visualization, Q.N.; supervision, B.Y.; project administration, B.Y.; funding acquisition, B.Y. and S.C. All authors have read and agreed to the published version of the manuscript.

Funding: The study was supported by a grant (2017YFC1601400) from the Ministry of Science and Technology of the People's Republic of China.

Acknowledgments: We are grateful to the laboratories in Dandong Customs of People's Republic of China, Tianjin Customs of People's Republic of China, Zhejiang Academy of Agricultural Sciences, Shandong Academy of Agricultural Sciences, Shanghai Customs of People's Republic of China, Hebei Food Inspection and Research Institute, and Shanxi Food and Drug Inspection and Research Institute for their support to verify some characteristics of plasmid DNA reference materials developed in this study.

Conflicts of Interest: The authors declare no conflict of interest.

References

1. World Health Organization. Estimating the Burden of Foodborne Diseases. 2016. Available online: https://www.who.int/activities/estimating-the-burden-of-foodborne-diseases (accessed on 6 November 2020).
2. Centers for Disease Control and Prevention [CDC]. Foodborne Diseases Active Surveillance Network FoodNet 2013 Surveillance Report. 2014. Available online: https://www.cdc.gov/foodnet/pdfs/508-compliant-2013-foodnet-annual-report.pdf (accessed on 6 November 2020).
3. European Food Safety Authority (EFSA). The European Union Summary Report on Trends and Sources of Zoonoses, Zoonotic Agents and Food-Borne Outbreaks in 2016. 2017. Available online: https://efsa.onlinelibrary.wiley.com/doi/epdf/10.2903/j.efsa.2017.5077 (accessed on 6 November 2020).
4. Crump, J.A.; Sjölund-Karlsson, M.; Gordon, M.; Parry, C.M. Epidemiology, clinical presentation, laboratory diagnosis, antimicrobial resistance, and antimicrobial management of invasive salmonella infections. *Clin. Microbiol. Rev.* **2015**, *28*, 901–937. [CrossRef] [PubMed]
5. Horiyama, T.; Yamaguchi, A.; Nishino, K. TolC dependency of multidrug efflux systems in Salmonella enterica serovar Typhimurium. *J. Antimicrob. Chemother.* **2010**, *65*, 1372–1376. [CrossRef] [PubMed]
6. Azargun, R.; Barhaghi, M.H.S.; Kafil, H.S.; Oskouee, M.A.; Sadeghi, V.; Memar, M.Y.; Ghotaslou, R. Frequency of DNA gyrase and topoisomerase IV mutations and plasmid-mediated quinolone resistance genes among *Escherichia coli* and *Klebsiella pneumoniae* isolated from urinary tract infections in Azerbaijan, Iran. *J. Glob. Antimicrob. Resist.* **2019**, *17*, 39–43. [CrossRef] [PubMed]
7. Yassine, I.; Rafei, R.; Osman, M.; Mallat, H.; Dabboussi, F.; Hamze, M. Plasmid-mediated quinolone resistance: Mechanisms, detection, and epidemiology in the Arab countries. *Infect. Genet. Evol.* **2019**, *76*, 104020. [CrossRef] [PubMed]
8. Jing, R.; Wang, H.; Ju, S.; Cui, M. Reference materials for molecular diagnostics: Current achievements and future strategies. *Clin. Biochem.* **2018**, *56*, 11–17. [CrossRef] [PubMed]
9. Ke, L.; Dai, X.L.; Lin, J.; Huang, C.J.; Zeng, W.Y.; Zhang, W.L. Preparation of reference materials for *Shigella*. *Food Sci.* **2015**, *36*, 253–259. [CrossRef]
10. Ke, L.; Huang, X.R.; Lin, J.; Dai, X.L.; Zeng, W.Y.; Zhang, W.L. Study on the reference materials of *Salmonella* Enteritidis nucleic acid. *J. Food Sci. Biotechnol.* **2015**, *34*, 891–895. [CrossRef]
11. Xia, D.D.; Zhao, Y.Y.; Ma, P.P.; Wang, S.L.; Ma, C.Y.; Gao, S.F.; Kang, W.Y. Preparation of plasmid DNA reference material and its application in rapid detection of goji berries (*Lycium barbarum*) contaminated with *Escherichia coli*. *Food Sci.* **2020**, *41*, 267–274. [CrossRef]
12. Tikhonova, E.B.; Yamada, Y.; Zgurskaya, H.I. Sequential mechanism of assembly of multidrug efflux pump AcrAB-TolC. *Chem. Biol.* **2011**, *18*, 454–463. [CrossRef] [PubMed]

13. Wang, J.; Sheng, H.; Xu, W.; Huang, J.; Meng, L.; Cao, C.; Zeng, J.; Meng, J.; Yang, B. Diversity of serotype, genotype, and antibiotic susceptibility of salmonella prevalent in pickled ready-to-eat meat. *Front. Microbiol.* **2019**, *10*, 2577. [CrossRef] [PubMed]
14. Cattoir, V.; Poirel, L.; Rotimi, V.; Soussy, C.-J.; Nordmann, P. Multiplex PCR for detection of plasmid-mediated quinolone resistance qnr genes in ESBL-producing enterobacterial isolates. *J. Antimicrob. Chemother.* **2007**, *60*, 394–397. [CrossRef] [PubMed]
15. Chen, X.; Zhang, W.; Pan, W.; Yin, J.; Pan, Z.; Gao, S.; Jiao, X. Prevalence of *qnr*, *aac(6′)-Ib-cr*, *qepA*, and *oqxAB* in *Escherichia coli* isolates from humans, animals, and the environment. *Antimicrob. Agents Chemother.* **2012**, *56*, 3423–3427. [CrossRef]
16. Park, C.H.; Robicsek, A.; Jacoby, G.A.; Sahm, D.; Hooper, D.C. Prevalence in the United States of *aac(6′)-Ib-cr* encoding a ciprofloxacin-modifying enzyme. *Antimicrob. Agents Chemother.* **2006**, *50*, 3953–3955. [CrossRef]
17. Zhang, Z.F.; Meng, X.F.; Yang, B.W.; Xia, X.D.; Wang, X.; Xi, M.L. Resistance of chickenborne *Salmonella* to quinolone and fluoroquinolones and related genes. *J. Chin. Food Sci. Technol.* **2015**, *15*, 158–165. [CrossRef]
18. Eaves, D.J.; Randall, L.; Gray, D.T.; Buckley, A.; Woodward, M.J.; White, A.P.; Piddock, L.J.V. Prevalence of mutations within the quinolone resistance-determining region of *gyrA*, *gyrB*, *parC*, and *parE* and association with antibiotic resistance in quinolone-resistant salmonella enterica. *Antimicrob. Agents Chemother.* **2004**, *48*, 4012–4015. [CrossRef]
19. Miao, X.; Jin, C.; Zhong, Y.; Feng, J.; Yan, C.; Xia, X.; Zhang, Y.; Peng, X. Data-independent acquisition-based quantitative proteomic analysis reveals the protective effect of apigenin on palmitate-induced lipotoxicity in human aortic endothelial cells. *J. Agric. Food Chem.* **2020**, *68*, 8836–8846. [CrossRef]
20. Verdier, V.; Mosquera, G.; Assigbetse, K. Detection of the cassava bacterial blight pathogen, *Xanthomonas axonopodis* pv. manihotis, by Polymerase Chain Reaction. *Plant Dis.* **1998**, *82*, 79–83. [CrossRef]
21. Verdier, V.; Mosquera, G. Specific detection of *Xanthomonas axonopodis* pv. manihotis with a DNA hybridization probe. *J. Phytopathol.* **1999**, *147*, 417–423. [CrossRef]
22. Ogunjobi, A.A.; Fagade, O.E.; Dixon, A.G.O. Comparative analysis of genetic variation among *Xanthomonas axonopodis* pv. manihotis isolated from the western states of Nigeria using RAPD and AFLP. *Indian J. Microbiol.* **2010**, *50*, 132–138. [CrossRef]
23. David, R.L.; Marta, H. Next-day *Salmonella* spp. detection method based on Real-Time PCR for foods. *Methods Mol. Biol.* **2021**, *2182*, 1–6. [CrossRef]
24. Domesle, K.J.; Yang, Q.R.; Hammack, T.S.; Ge, B.L. Validation of a *Salmonella* loop-mediated isothermal amplification assay in animal food. *Int. J. Food Microbiol.* **2018**, *264*, 63–76. [CrossRef]
25. Magnus, U.; Daniel, G.; Leonor, S.B.; Yonatan, G.; Susanne, J.; Makoto, O.; Monica, M.L.; Athena, L.; Aleksandra, E.S.; Teodora, W.; et al. The novel 2016 WHO *Neisseria gonorrhoeae* reference strains for global quality assurance of laboratory investigations: Phenotypic, genetic and reference genome characterization. *J. Antimicrob. Chemother.* **2016**, *71*, 3096–3108. [CrossRef]
26. Luo, H.P.; Qu, H.R.; Shen, J.Y.; Ren, X.; Chen, Y.W.; Xie, G.D.; Yu, W.; Liu, N.; Ding, H.; Cui, S.H. Development of microbial reference materials for *Cronobactersakazakii*. *J. Food Saf. Qual.* **2019**, *10*, 54–59.
27. Ke, L.; Lin, J.; Dai, X.L.; Huang, C.J.; Huang, X.R. The reference materials of preparation about *Salmonella* Enteritidis. *Food Res. Dev.* **2015**, *36*, 111–115. [CrossRef]
28. Ma, P.P. *Research and Development of Plasmid DNA Qualitative Reference Material of Listeria Monocytogenes and Its Application in the Detection of Tremella*; Henan University: Zhengzhou, China, 2020.
29. Debode, F.; Janssen, E.; Berben, G. Development of 10 new screening PCR assays for GMO detection targeting promoters (pFMV, pNOS, pSSuAra, pTA29, pUbi, pRice actin) and terminators (t35S, tE9, tOCS, tg7). *Eur. Food Res. Technol.* **2013**, *236*, 659–669. [CrossRef]
30. Dorlass, E.G.; Monteiro, C.O.; Viana, A.O.; Soares, C.P.; Machado, R.R.G.; Thomazelli, L.M.; Araujo, D.B.; Leal, F.B.; Candido, E.D.; Telezynski, B.L.; et al. Lower cost alternatives for molecular diagnosis of COVID-19: Conventional RT-PCR and SYBR Green-based RT-qPCR. *Braz. J. Microbiol.* **2020**, *51*, 1117–1123. [CrossRef] [PubMed]
31. Yoshii, Y.; Shimizu, K.; Morozumi, M.; Chiba, N.; Ubukata, K.; Uruga, H.; Hanada, S.; Wakui, H.; Ito, S.; Takasaka, N.; et al. Identification of pathogens by comprehensive real-time PCR versus conventional methods in community-acquired pneumonia in Japanese adults. *Infect. Dis.* **2016**, *48*, 782–788. [CrossRef] [PubMed]
32. Fábio, C.A.B.; Andréia, Z.D.; Deise, J.K.; Josias, C.F.; Ana, C.M.A. Development of plasmid DNA reference material for the quantification of genetically modified common bean Embrapa 5.1. *J. Agric. Food Chem.* **2013**, *61*, 4921–4926. [CrossRef]
33. Wu, Y.H.; Li, J.; Wang, Y.L.; Li, X.F.; Li, Y.J.; Zhu, L.; Li, J.; Wu, G. Development and application of a general plasmid reference material for GMO screening. *Plasmid* **2016**, *87–88*, 28–36. [CrossRef]
34. Zhang, N.N.; Deng, Y.Q.; Nian, Q.G.; Kang, X.P.; Yang, Y.H.; Qin, C.F. Construction and characterization of recombinant pseudovirus particles carrying specific genomic RNA of St. Louis encephalitis virus. *Mil. Med. Sci.* **2017**, *41*, 194–198. [CrossRef]
35. Junichi, M.; Shuko, H.; Satoshi, F.; Junji, Y.; Hiroki, N.; Shigehiro, N.; Reona, T.; Kazumi, K. Development of a reference material of a single DNA molecule for the quality control of PCR testing. *Anal. Chem.* **2014**, *86*, 8621–8627. [CrossRef]
36. Vallejo, C.V.; Tere, C.P.; Calderon, M.N.; Arias, M.M.; Leguizamon, J.E. Development of a genomic DNA reference material for *Salmonella* enteritidis detection using polymerase chain reaction. *Mol. Cell Probes* **2021**, *55*, 101690. [CrossRef] [PubMed]
37. Zhou, D.G.; Li, Y.P.; Luo, J.; Li, J.H.; Yu, J.P.; Wei, H.P. Preparation and characterization of pseudoviral positive control for nucleic acid detection of MERS-CoV. *Chin. J. Front. Health Quar.* **2019**, *42*, 233–238. [CrossRef]

foods

MDPI

Article

Insight into *Bacillus cereus* Associated with Infant Foods in Beijing

Ping Wang, Xiaomei Zhao, Tianming Qu, Lijiao Liang, Qinglong Ji and Ying Chen *

Chinese Academy of Inspection and Quarantine, Beijing 100176, China; wangp_129@163.com (P.W.); xiaomei200413@126.com (X.Z.); qtm110506@163.com (T.Q.); lianglijiao96@163.com (L.L.); jql81070305@aliyun.com (Q.J.)
* Correspondence: chenyingcaiq@163.com; Tel.: +86-10-53897910

Abstract: This study was undertaken to investigate the prevalence, antimicrobial resistance, and virulence gene profiles of *Bacillus cereus* in different brands of infant formula in Beijing supermarkets. Eighty-eight *Bacillus cereus* isolates were recovered in sixty-eight infant formulas of five domestic brands and fourteen imported brands. The prevalence rate in domestic and imported samples were 70.6% and 52.9%, respectively. Lower mean prevalence level was found in domestic samples (1.17 MPN/g) compared with the imported samples (3.52 MPN/g). Twenty-four virulence gene profiles were found, and most strains carried at least one virulence gene. The prevalence of *nheA*, *nheB*, *nheC*, *cytK*, *bceT*, and *entFM* in domestic and imported brand samples was similar. The occurrence of enterotoxin genes *hblA*, *hblC*, and *hblD* in domestic samples were 22.2%, 27.8%, and 22.2%, respectively, which was significantly higher than imported samples. Antimicrobial drugs-susceptibility analysis showed that all isolates were susceptible to gentamincin, amikacin, and ciprofloxacin; 38%, 7%, and 2.3% were resistant to rifampin, tetracycline, and chloramphenicol, respectively; and only one isolate was resistant to trimethoprim-sulfamethoxazole. Moreover, the cell numbers of *Bacillus cereus* in prepared infant formula increased rapidly at room temperature. Thus, monitoring guidelines are needed for accepted levels of *Bacillus cereus* in infant formula.

Keywords: infant formula; *Bacillus cereus*; virulence genes; antimicrobial drugs susceptibility

Citation: Wang, P.; Zhao, X.; Qu, T.; Liang, L.; Ji, Q.; Chen, Y. Insight into *Bacillus cereus* Associated with Infant Foods in Beijing. *Foods* **2022**, *11*, 719. https://doi.org/10.3390/foods11050719

Academic Editors: Fernando Pérez-Rodríguez and Arícia Possas

Received: 28 January 2022
Accepted: 25 February 2022
Published: 28 February 2022

1. Introduction

Infant formula is a major source of nutrition for children before they can digest other foods. The immune system of babies is weak, and any pathogen present in their food may cause illness. Therefore, the hygienic quality of infant formula is important to protect the health of infants and to diminish risks and associated with its consumption. As one of the high-protein foods, dairy products are conducive to foodborne pathogen proliferation. A growing research has suggested that *Bacillus cereus* is a common pathogen in milk or milk-related products [1,2].

Bacillus cereus, a spore-forming bacterium, is an opportunistic human food-borne pathogen, which is widely distributed in environment and frequently isolated as a contaminant of cereals, processed milk products, and other foods [3,4]. The spores of *B. cereus* are specifically troublesome in the food industry because they can be intractable to pasteurization, radiation, disinfectants, and desiccation, and their hydrophobic nature allows them to adhere to the surface of solid materials [5,6]. In addition to gastroenteritis, *B. cereus* can also cause systemic and local infection in immunologically compromised individuals [7].

Bacillus cereus is capable of producing emesis toxin (ETE) and three enterotoxins, including hemolytic enterotoxin (HBL), nonhemolytic enterotoxin (NHE), and enterotoxin K (EntK), of which HBL and NHE are important enterotoxins that cause diarrhea-based food poisoning. There are five virulence genes associated with the production of these enterotoxins: hemolysin BL gene (*hbl*), non-hemolytic enterotoxin gene (*nhe*), enterotoxin FM gene (*entFM*), enterotoxin T gene (*becT*), and cytotoxin K gene (*cytK*). HBL requires all three genes positive for *hblA*, *hblC*, and *hblD* to be toxic, while NHE is most toxic when all three genes are positive for *nheA*, *nheB*, and *nheC*. The virulence gene expressing vomitoxin is *ces* gene, which encodes a heat-resistant toxin that is not easily cleaved [4,8]. A total of 24 virulence gene carriage patterns of *B. cereus* were identified, among which the *nhe* gene had the highest carriage rate of 92.98%, followed by the *entFM* gene (71.93%), and 70.18% of strains carried both *nhe* and *entFM* genes. The subtyping results showed that the carriage rates of *nheA*, *nheB*, and *nheC* genes were 88.72%, 88.72%, and 49.12%, respectively. The hemolysin BL gene carriage rates were 24.56% for *hblA*, 22.81% for *hblC*, 17.54% for *hblD*, and 22.81% for *cytK* [8,9].

Besides the enterotoxic toxins, it has been found that some *B. cereus* isolates show resistance to antimicrobial drugs [10]. A variety of antimicrobial drugs are widely used, and the problem of bacterial resistance is getting worse, causing great threat to human health. With antimicrobial agents widely used in farmed-animal industries, the food chain constitutes an important source of antimicrobial resistance [11]. There is evidence that resistant microorganisms can spread to humans via the food chain or indirect contact from farm animal waste [12,13]. Therefore, it is necessary to evaluate the antibiotic resistance of *B. cereus* in infant formula.

Up to now, many countries have stipulated the limit standard of *B. cereus* in infant formula to control this bacterium. In Australia and New Zealand, it was ($n = 5$, $c = 0$, $m = 100$ cfu/g (mL)); in Canada, it was ($n = 10$, $c = 1$, $m = 100$ cfu/g (mL), $M = 10,000$ cfu/g (mL)); and in the European Union, it was ($n = 5$, $c = 1$, $m = 100$ cfu/g (mL), $M = 500$ cfu/g (mL)). While in China, the national standard of infant formula, GB 10765-2010, does not stipulate the limit standard of *B. cereus* in infant formula, the control and inspection of *B. cereus* in infant formula is mainly according to a prevalence level less than 100, 000 cfu/g (mL). Therefore, it is important to determine the prevalence and also enumerate *B. cereus* in babies' food in order to assess its safety.

Therefore, the present study was undertaken to evaluate the prevalence and potential risk of *B. cereus* in infant formulas in Beijing. The research was further extended to carry out the virulence genes profile, and antimicrobial-resistance profiles of the isolates were evaluated.

2. Materials and Methods

2.1. Sample Collection

A total of 68 infant formula samples were collected from supermarkets in Beijing during the period from June to July 2014. Among the 68 samples, 17 were from different domestic brands, and the other 51 samples involved 14 different imported brands. The samples of every brand were from different batches (1 to 5). The infant formulas selected in this study included most of the popular brands in the Chinese retail market.

2.2. Microbiological Analysis

The *B. cereus* in infant formula was detected by using a most probable number (MPN) procedure method described by Tallent et al. [14]. Briefly, 25 g of each sample was suspended in 225 mL of 0.85% saline and beaten for 1 min by a homogenizer (BAG-MIXER100, interscience, France). Then, 10 mL of homogenate was serially diluted (10-fold) in 0.85% physiological saline with selected 3 continuous dilution, and then, 3 × 10 mL of each dilution was inoculated into 3 tubes with 10 mL double trypticase soy polymyxin broth (TSPB, Difco, 7 Loveton Circle, Sparks, MD, USA), followed by incubation at 30 °C for 48 h. The culture was then taken from tubes streak and applied onto mannitol yolk polymyxin (MYP) agar (Difco, 7 Loveton Circle, Sparks, MD, USA) and then incubated at 30 °C for 24 h. The typical colonies from plates that contain suspicious colonies were cultured on blood plates (Oxoid, Hampshire, UK) incubated at 30 °C for 24 h. The colony on the blood plates were confirmed by BD PhoenixTM-100 (Becton, Dickinson and Company, 7 Loveton Circle Sparks, USA), and the *gyrB* gene was detected using a previously described PCR method for the identification of *Bacillus cereus* [15]. The strains were conserved in Inspection and Quarantine of China microbial culture collection management center (IQCC).

2.3. Detection of Virulence Genes

The strains isolated from infant formula samples were screened by PCR for the ten virulence genes, including *hblA/C/D*, *nheA/B/C*, *cytK*, *bceT*, *ces*, and *entFM* [9]. For extraction of DNA template, bacteria were plated on Tryptone soya agar (TSA, Oxoid, Hampshire, England) and incubated at 30 °C for 24 h; then, 5 colonies were taken in 300 μL of Tris-EDTA buffer. The lysis of bacteria was performed by incubation at 100 °C for 10 min, and debris was removed by centrifugation at 15,000× *g* for 3 min. The DNA-containing supernatant was transferred to a new tube and stored at −20 °C.

For amplification, Ex Taq kit (TaKaRa, Otsu City, Japan) was used, and 25 μL mixtures were as follows: 2.5 μL of 10×Ex Taq buffer, 0.25 μL Ex Taq DNA polymerase (5 U/μL), 2 μL dNTP mixture (2.5 mM), 2 μL template DNA, 0.5 μL of each primer (10 μM), and 17.25 μL ddH_2O. The reactions were performed on a Veriti 96-Well Thermal Cycler (Applied Biosystems, Carlsbad, CA, USA) under the following conditions: 95 °C for 10 min; 35 cycles at 95 °C for 30 s, annealing temperature for 30 s, and 72 °C for 30 s; and a final extension at 72 °C for 10 min. The *B. cereus* ATCC 11,778, ATCC 33,019 strains were used as positive control, and *Salmonella typhimurium* IQCC 10,503 without virulence genes as detected in this study was used as negative control. PCR products were analyzed by electrophoresis in 2% agarose and stained with ethidium bromide (EB). Gel images were captured on a Versadoc Imager (Bio-Rad, Hercules, CA, USA).

2.4. Antimicrobial Drugs Susceptibility Test

Antimicrobial drugs susceptibility of the *B. cereus* strains was determined by means of the agar disk diffusion method as recommended by the Clinical Laboratory Standards Institute [16]. Nine antimicrobial drugs often used in clinical and farmed-animal industries area were evaluated, namely gentamicin (GM) (10 μg/disk), tetracycline (TET) (30 μg/disk), erythromycin (ERY) (15 μg/disk), chloramphenicol (CHL) (30 μg/disk), amikacin (AMK) (30 μg/disk), and ciprofloxacin (CIP) (5 μg/disk), clindamycin (CC) (2 μg/disk), trimethoprim-sulfamethoxazole (SXT) (1.25/23.75 μg/disk), and rifampin (RIF) (5 μg/disk). Inhibition zones were measured in millimeters and interpreted according to the CLSI instruction.

2.5. Simulated Survive Test

Two strains, both isolated from infant formula samples, were randomly selected and used to investigate whether formula supports *B. cereus* survival, and simulated samples were prepared. Briefly, two *B. cereus* isolates were respectively inoculated onto infant formula powder (25 g) at finial concentration at 10 MPN/g, and 225 mL DDW were added and then incubated at room temperature (22 °C) and 37 °C for 24 h. During incubation, simulated samples were tested every hour by the MPN procedure method described above.

2.6. Statistical Analysis

The rates of recovery of *B. cereus* among the domestic brand and the imported brand formula were compared using Student's *t*-test. Student's *t*-test was also performed to compare the prevalence of virulence genes.

3. Results

3.1. Prevalence of B. cereus in Infant Formula Samples

A total of 68 infant formula samples, including 51 imported brands samples and 17 domestic brands samples, were analyzed. Among 51 imported brands samples, 27 samples were positive for *B. cereus*, with the prevalence rate of 52.9% and the mean prevalence level of 3.52 MPN/g (Table 1), while in 17 domestic brands samples, 12 samples were positive, with prevalence rate of 70.6% and the mean prevalence level 1.17 MPN/g. Among the samples positive for *B. cereus*, thirty-three had prevalence level < 3 MPN/g, four of them measured between 3~10 MPN/g, and only two of samples were measured to a level of 11~50 MPN/g. Despite the prevalence rate of *B. cereus* in domestic brands, the sample was higher than the imported brands sample; however, the difference was statistically not significant at a 5% level.

Table 1. Prevalence and contamination level of *B. cereus* in retail infant formulas in Beijing.

Groups	Detection Rate %	Prevailed Distribution of *B. cereus* (MPN/g)				Means of Prevalence (MPN/g)	Range of Prevalence (MPN/g)
		<3	3~10	11~50	>51		
Domestic brands (n = 17)	70.6 a	10	2	0	0	1.17	0.36–4.3
Imported brands (n = 51)	52.9 a	23	2	2	0	3.52	0.36–46

a, same letter in the same column indicates no significant difference ($p < 0.05$).

3.2. Virulence Genes Profiles of B. cereus

Eighty-eight isolates were identified as *B. cereus*, the multiplex PCR was carried to detect the virulence genes, and most isolates carried *nheC* gene, followed by *nheB* and *nheA*, which were 84.1%, 63.6%, and 54.5%, respectively. The occurrence of three genes of hemolytic enterotoxin complex (HBL) of *hblA*, *hblC*, and *hblD* were 5.68%, 9.09%, and 5.68%, respectively. The frequency of *cytK* gene was 21.6%. The *ces* gene encoding emetic toxin was only found in three (3.41%) isolates (Table 2).

Table 2. Virulence genes carried by *B. cereus* isolated from infant formulas in Beijing.

Virulence Genes	Detectable Strains	Carrying Rates (%)
hblA	5	5.68
hblC	8	9.09
hblD	5	5.68
nheA	48	54.5
nheB	56	63.6
nheC	74	84.1
entFM	19	21.6
bceT	19	21.6
cytK	19	21.6
ces	3	3.41
hblA + *hblC* + *hblD*	4	4.55
nheA + *nheB* + *nheC*	47	53.4
hblA/C/D + *nheA/B/C*	4	4.55

About 24 virulence gene profiles of *B. cereus* were detected (Table 3), and the predominant profile was XII (22.73%), in which NHE complex genes were positive and the other virulence genes were negative by PCR test. In profile XII, seventeen strains were isolated from imported formula samples. Additionally, there were 53.4% (47/88) *B. cereus* isolates with *nheA*, *nheB*, and *nheC* simultaneously, while 4.55% (4/88) strains carried three *hbl* genes. Three isolates include all three types of both *nhe* and *hbl*. However, thirteen strains were without any of the virulence genes. Statistical analysis suggested the prevalence of *nheA*, *nheB*, *nheC*, *cytK*, *bceT*, and *entFM* in domestic brand samples were similar with imported samples ($p > 0.05$), ranging from 18.6% to 64.3%, while *hblA*, *hblC*, and *hblD* in domestic samples were 22.2%, 27.8%, and 22.2% significantly higher than in imported samples ($p < 0.05$).

Table 3. Patterns of virulence genes of *B. cereus* isolated from infant formulas.

Virulence Gene Patterns	*hbla*	*hblc*	*hbld*	*nhea*	*nheb*	*nhec*	*entFM*	*bceT*	*cytK*	*ces*	Carrying Rate (%)	Number of Isolates
I	–	–	–	–	–	–	–	–	–	–	14.77	13
II	–	+	–	–	–	–	–	–	–	–	1.14	1
III	–	–	–	–	–	+	–	–	–	–	13.64	12
IV	–	–	–	–	–	+	+	–	–	–	2.27	1
V	–	–	–	–	–	+	–	+	–	–	3.41	3
VI	–	–	–	–	+	+	–	–	–	–	4.55	4
VII	–	–	–	–	+	+	+	–	–	–	1.14	1
VIII	–	–	–	–	+	+	–	+	–	–	1.14	1
IX	–	–	–	–	+	+	+	+	–	–	1.14	1
X	–	–	–	–	+	+	–	–	+	–	2.27	1
XI	–	–	–	+	–	+	–	–	–	–	1.14	1
XII	–	–	–	+	+	+	–	–	–	–	22.73	20
XIII	–	–	–	+	+	+	+	–	–	–	7.95	7
XIV	–	+	–	+	+	+	–	–	–	–	1.14	1
XV	–	–	–	+	+	+	+	+	–	–	1.14	1
XVI	–	–	–	+	+	+	–	+	–	–	1.14	1
XVII	–	–	–	+	+	+	–	–	+	–	5.68	4
XVIII	–	–	–	+	+	+	+	+	+	–	4.55	4
XIX	–	–	–	+	+	+	–	+	+	–	1.14	1
XX	–	+	–	+	+	+	–	+	+	–	1.14	1
XXI	–	+	+	+	+	+	–	+	+	–	1.14	1
XXII	+	–	–	+	+	+	–	+	+	–	1.14	1
XXIII	+	+	+	+	+	+	–	+	+	–	1.14	1
XXIV	+	+	+	+	+	+	+	+	+	+	3.41	3

Result of samples tested by virulence gene primer pairs: positive +; negative –.

3.3. Resistance to Antimicrobials

All of the 88 isolates were susceptible to GM, AMK, and CIP; seven strains were resistant to TET; two strains resistant against CHL; and only one isolate resistant against SXT. The strains had low sensitivity to RIF (Table 4). Furthermore, seven TET resistant isolates were also resistant to RIF. Seven strains with both TET resistance and RIF resistance were detected in this study, and no multiple antibiotic-resistant isolates were detected.

Table 4. Antimicrobial drugs susceptibility of *Bacillus cereus* strains isolated from infant formulas in Beijing.

Antimicrobial Drugs	Type	Conc. (μg/disk)	Resistant	Intermediate	Susceptible
GM	Aminoglycosides	10	0	0	88
TET	Tetracycline	30	7	1	80
ERY	Macrolides	15	0	19	69
CHL	Chloramphenicol	30	2	1	85
AMK	Aminoglycosides	30	0	0	88
CIP	Quinolones	5	0	0	88
CC	Lincomycin	2	0	15	73
SXT	Sulfa	1.25/23.75	1	0	87
RIF	Rifampin	5	38	38	12

3.4. Simulated Survive Test

To investigate whether prepared infant formula supports *B. cereus* growth, two strains were randomly selected for the simulated survive test. *B. cereus* MPN levels were tested every hour for 24 h. At room temperature, the MPN reached 139/g and 127/g in two prepared simulated infant formula samples within ten hours. After 24 h incubation, *B. cereus* levels reached 1000 MPN/g, while increasing the temperature to 37 °C resulted in increased of *B. cereus* levels to 1000 MPN/g following only 6 h of incubation and 10,000 MPN/g after 24 h incubation.

4. Discussion

Foodborne diseases caused by foodborne pathogens have become one of the major safety issues that threaten human health. In particular, *Salmonella, E. coli, Staphylococcus aureus,* and *Listeria monocytogenes* are considered to be the most important and serious foodborne pathogenic bacteria [17]. *Bacillus cereus* is a neglected foodborne pathogen because of its strong environmental tolerance, spore-forming ability, and ability to produce toxins. It is often found in raw milk and dairy products.

In China, it is generally considered that imported brands of infant food are better than the domestic. However, the results of the present study proved that infant formulas were frequently found to include *B. cereus* although there was no significant difference between the imported and domestic infant formula samples ($p > 0.05$) for the prevalence of *B. cereus*. The result of the overall prevalence of *B. cereus* in infant formula was higher than a previously study [18], which reported that 14.08% of samples were contaminated with *B. cereus* from infant formula samples. The reason may be that *B. cereus* was examined by the plate count method in their study, while the MPN method was adopted in our study. The MPN method has lower detection limit than the plate count method and frequently was used when the level of *Bacillus cereus* ≤ 1000 CFU/g.

Infant formulas are prepared with warm water before consumption, and this organism may be activated by a normal process. If the temperature is adequate, one can assume that spores of *B. cereus* will be activated. The temperature of the water used to prepare the infant formula is advised to be 35 °C~60 °C, and this temperature may facilitate the activation of spores. Therefore, it is not surprising that some cases of *B. cereus* poisoning were linked to the infant formula [19,20]. Investigation was also conducted to study the possibility that large number of *B. cereus* could be ingested through the consumption of contaminated infant formula. Although most infant formula products in China have directions for use, it is possible that they may be consumed outside the instructions. Results of room temperature (22 °C) were examined to reflect situations such as those in some villages and towns or small childcare centers, where remaining infant formula food may not remain under refrigeration temperature. At room temperature, in simulated prepared infant formula samples, with an increase in cell number to an MPN of 100/g for 10 h after 24 h incubation, *B. cereus* numbers reached an MPN of 1000/g. However, increasing the temperature to 37 °C resulted in increases of *B. cereus* to an MPN of 1000/g after only 6 h of incubation and an MPN of 10,000/g after 24 h incubation. These data raise the concern that contaminated infant formula products available in China pose a potential risk to infants and raise the possibility that these products have already been a cause of illness in the past. Considering the prevalence level of *B. cereus* in foods that caused illness in the past, mostly ≥100,000 cfu/g, the level in determined in this study was quite low, indicating that the samples were relatively safe in terms of prevalence level at the time of purchase. However, the infant formula food can be safe only if it is consumed as per the guidelines given by the brands. In prepared infant formula that supports growth, the cell number is expected to increase at the time of consumption, especially when unfinished, prepared products are not properly maintained under refrigeration.

Virulence genes are thought to be linked with food spoilage, diarrhea, emesis, and other complications caused by *B. cereus* [21]. The diarrheal form of the syndrome has been associated mainly with hemolysin BL (*Hbl*), non-hemolytic enterotoxin (*Nhe*), and the cytotoxin K (*CytK*) [22,23]. For the detection virulence genes, we observed 75 isolates carried at least one virulence gene, and of them, *nhe* genes were the greatest (98.67%). This was consistent with what was reported in previous studies [9], which showed that *B. cereus* isolated from foods showed more frequent detection of NHE complex genes. HBL enterotoxin complex consists of B, L1, and L2, and its enterotoxic activity appears when all these components of the HBL complex are present [4,24]. Regarding the occurrence of the hemolytic BL genes, *hblCDA* among food strains isolated in Korea was 81.8% [24], while *hblC, hblD*, and *hblA* genes among *B.cereus* isolated in Chinese pasteurized, full-fat milk occurred with frequencies between 37.0% and 71.7% [25]. In this study, we found these three genes detection rates was 4.55%, lower than those mentioned above. Three isolates harbored all detected virulence genes; according to the labels, we found these strains isolated from two domestic-brand infant formulas. It is suspected that these isolates could be more virulent to humans, and the possibility of severe food poisoning case caused by these virulence genes might exist in China.

Foodborne pathogens that are resistant to a variety of antibiotics have become a major health concern [10]. Aminoglycosides, macrolides, and chloramphenicol antibiotics are usually recommended as the drugs of choice against *B. cereus* infections. In this study, some isolates were found to be resistant to macrolide and chloramphenicol antibiotics, which may be due to their identical action, acting on the 50S subunit of the bacterial ribonucleoprotein bodies and blocking protein synthesis. Seven multiple antibiotic isolates were also found, a result similar to previous studies, and *B. cereus* isolated from food was also found to be resistant to multiple antibiotics [9,11]. The isolation of these resistant *B.cereus* strains from infant formula is worrying. It is suggested that the inappropriate use of antibiotics in veterinary medicine contribute to a potential prevalence of raw materials of infant formula.

It is also found that strains isolated from one sample had different virulence gene profiles and antibiotic resistance. This might indicate the natural environment is reservoir for *B. cereus,* and food products are easily contaminated, and there is more than one contaminant source of infant formula. More isolated from infant formula, other foods and even some samples from the production environment are needed for subtyping and for infectious resource tracing and control.

5. Conclusions

The current findings suggest that the prevalence of *Bacillus cereus* in infant formula remains high and that antibiotic-resistance genes and virulence genes are present. Moreover, our data suggest that *B. cereus* may be an important pathogen of infant formula food poisoning in Beijing and needs to be controlled in some way. While in China, the regulation of *B. cereus* is currently insufficient in dairy products, these data could be useful for establishing microbiological safety rules for food, including infant formulas. Due to sampling volume and regional limitations, our study could not cover a larger area, but this study can be used as a basis In the future, the research on *Bacillus cereus* prevalence can be carried out on a large scale, such as increasing the number of samples and collection sites and conducting more in-depth research on its pathogenicity.

Author Contributions: Conceptualization, P.W. and Y.C.; methodology, X.Z.; formal analysis, P.W., T.Q., L.L. and Q.J.; data curation, P.W.; writing—original draft preparation, P.W.; writing—review and editing, P.W. and Y.C.; project administration, P.W. and Y.C. All authors have read and agreed to the published version of the manuscript.

Funding: This work was supported by national Key Research and Development Program of China (Grant Number 2018YFC1603800).

Informed Consent Statement: Informed consent was obtained from all subjects involved in the study.

Data Availability Statement: The datasets generated for this study are available on request to the corresponding author.

Conflicts of Interest: The authors declare no conflict of interest.

References

1. Bartoszewicz, M.; Hansen, B.; Swiecicka, I. The members of the *Bacillus cereus* group are commonly present contaminants of fresh and heat-treated milk. *Food Microbiol.* **2008**, *25*, 588–596. [CrossRef]
2. da Silva Fernandes, M.; Fujimoto, G.; Schneid, I.; Kabuki, D.Y.; Kuaye, A.Y. Enterotoxigenic profile, antimicrobial susceptibility, and biofilm formation of *Bacillus cereus* isolated from ricotta processing. *Int. Dairy J.* **2014**, *38*, 16–23. [CrossRef]

3. Fang, T.; Chen, C.-Y.; Kuo, W.-Y. Microbiological quality and incidence of *Staphylococcus aureus* and *Bacillus cereus* in vegetarian food products. *Food Microbiol.* **1999**, *16*, 385–391. [CrossRef]
4. Schoeni, J.L.; LEE WONG, A.C. *Bacillus cereus* food poisoning and its toxins. *J. Food Prot.* **2005**, *68*, 636–648. [CrossRef]
5. Andersson, A.; Granum, P.E.; Rönner, U. The adhesion of *Bacillus cereus* spores to epithelial cells might be an additional virulence mechanism. *Int. J. Food Microbiol.* **1998**, *39*, 93–99. [CrossRef]
6. Kotiranta, A.; Haapasalo, M.; Kari, K.; Kerosuo, E.; Olsen, I.; Sorsa, T.; Meurman, J.H.; Lounatmaa, K. Surface structure, hydrophobicity, phagocytosis, and adherence to matrix proteins of *Bacillus cereus* cells with and without the crystalline surface protein layer. *Infect. Immun.* **1998**, *66*, 4895–4902. [CrossRef]
7. Bottone, E.J. *Bacillus cereus*, a volatile human pathogen. *Clin. Microbiol. Rev.* **2010**, *23*, 382–398. [CrossRef] [PubMed]
8. Ehling-Schulz, M.; Frenzel, E.; Gohar, M. Food–bacteria interplay: Pathometabolism of emetic *Bacillus cereus*. *Front. Microbiol.* **2015**, *6*, 704. [CrossRef] [PubMed]
9. Hansen, B.M.; Hendriksen, N.B. Detection of enterotoxic *Bacillus cereus* and *Bacillus thuringiensis* strains by PCR analysis. *Appl. Environ. Microbiol.* **2001**, *67*, 185–189. [CrossRef]
10. Chaves, J.Q.; Pires, E.S.; Vivoni, A.M. Genetic diversity, antimicrobial resistance and toxigenic profiles of *Bacillus cereus* isolated from food in Brazil over three decades. *Int. J. Food Microbiol.* **2011**, *147*, 12–16. [CrossRef]
11. Marshall, B.M.; Levy, S.B. Food animals and antimicrobials: Impacts on human health. *Clin. Microbiol. Rev.* **2011**, *24*, 718–733. [CrossRef] [PubMed]
12. Banerjee, M.; Sarkar, P.K. Antibiotic resistance and susceptibility to some food preservative measures of spoilage and pathogenic micro-organisms from spices. *Food Microbiol.* **2004**, *21*, 335–342. [CrossRef]
13. Liebana, E.; Carattoli, A.; Coque, T.M.; Hasman, H.; Magiorakos, A.-P.; Mevius, D.; Peixe, L.; Poirel, L.; Schuepbach-Regula, G.; Torneke, K. Public health risks of enterobacterial isolates producing extended-spectrum β-lactamases or AmpC β-lactamases in food and food-producing animals: An EU perspective of epidemiology, analytical methods, risk factors, and control options. *Clin. Infect. Dis.* **2013**, *56*, 1030–1037. [CrossRef] [PubMed]
14. Tallent, S.; Rhodehamel, E.; Harmon, S.; Bennett, R. *Bacillus cereus Bacteriological Analytical Manual*; US Food and Drug Administration: Washington, DC, USA, 2012; Volume 2012.
15. Tewari, A.; Singh, S.; Singh, R. Incidence and enterotoxigenic profile of *Bacillus cereus* in meat and meat products of Uttarakhand, India. *J. Food Sci. Technol.* **2015**, *52*, 1796–1801. [CrossRef] [PubMed]
16. CLSI. *Performance Standards for Antimicrobial Susceptibility Testing*; Twenty-second informational supplement; CLSI document M100-S23; Clinical and Laboratory Standards Institute: Wayne, PA, USA, 2013.
17. Ghaly, M.F.; Nasr, Z.M.; Abousaty, A.I.; Seadawy, H.G.; Shaheen, M.A.A.; Albogami, S.; Al-Sanea, M.M.; Bendary, M.M. Alternative and Complementary Therapies against Foodborne *Salmonella* Infections. *Antibiotics* **2021**, *10*, 1453. [CrossRef] [PubMed]
18. Li, Y.; Pei, X.; Yang, D.; Li, N. Occurrence of Bacillus cereus in infants and young children foods in 8 provinces in China. *J. Hyg. Res.* **2014**, *43*, 435–438.
19. Delbrassinne, L.; Botteldoorn, N.; Andjelkovic, M.; Dierick, K.; Denayer, S. An emetic *Bacillus cereus* outbreak in a kindergarten: Detection and quantification of critical levels of cereulide toxin. *Foodborne Pathog. Dis.* **2015**, *12*, 84–87. [CrossRef]
20. Duc, L.H.; Dong, T.C.; Logan, N.A.; Sutherland, A.D.; Taylor, J.; Cutting, S.M. Cases of emesis associated with bacterial contamination of an infant breakfast cereal product. *Int. J. Food Microbiol.* **2005**, *102*, 245–251. [CrossRef]
21. Granum, P.E.; Lund, T. *Bacillus cereus* and its food poisoning toxins. *FEMS Microbiol. Lett.* **1997**, *157*, 223–228. [CrossRef]
22. Stenfors Arnesen, L.P.; Fagerlund, A.; Granum, P.E. From soil to gut: *Bacillus cereus* and its food poisoning toxins. *FEMS Microbiol Rev.* **2008**, *32*, 579–606. [CrossRef] [PubMed]
23. Doll, V.M.; Ehling-Schulz, M.; Vogelmann, R. Concerted action of sphingomyelinase and non-hemolytic enterotoxin in pathogenic *Bacillus cereus*. *PLoS ONE* **2013**, *8*, e61404. [CrossRef] [PubMed]
24. Kim, J.B.; Kim, J.M.; Cho, S.H.; Oh, H.S.; Choi, N.J.; Oh, D.H. Toxin genes profiles and toxin production ability of *Bacillus cereus* isolated from clinical and food samples. *J. Food Sci.* **2011**, *76*, T25–T29. [CrossRef] [PubMed]
25. Zhou, G.; Liu, H.; He, J.; Yuan, Y.; Yuan, Z. The occurrence of *Bacillus cereus*, *B. thuringiensis* and *B. mycoides* in Chinese pasteurized full fat milk. *Int. J. Food Microbiol.* **2008**, *121*, 195–200. [CrossRef] [PubMed]

foods
MDPI

Article

Comparison of Selected Phenotypic Features of Persistent and Sporadic Strains of *Listeria monocytogenes* Sampled from Fish Processing Plants

Natalia Wiktorczyk-Kapischke [1], Ewa Wałecka-Zacharska [2,*], Krzysztof Skowron [1], Agnieszka Kijewska [3], Zuzanna Bernaciak [1], Justyna Bauza-Kaszewska [4], Zuzanna Kraszewska [1] and Eugenia Gospodarek-Komkowska [1]

1 Department of Microbiology, L. Rydygier Collegium Medicum in Bydgoszcz, Nicolaus Copernicus University in Toruń, 85-094 Bydgoszcz, Poland; natalia12127@gmail.com (N.W.-K.); skowron238@wp.pl (K.S.); zuza.benciak@gmail.com (Z.B.); zuzia95@op.pl (Z.K.); gospodareke@cm.umk.pl (E.G.-K.)
2 Department of Food Hygiene and Consumer Health, Wrocław University of Environmental and Life Sciences, 50-375 Wrocław, Poland
3 Department of Immunobiology and Environmental Biology, Institute of Maritime and Tropical Medicine, Medical University of Gdansk, 80-210 Gdańsk, Poland; agnieszka.p.kijewska@gumed.edu.pl
4 Department of Microbiology and Food Technology, Jan and Jędrzej Śniadecki University of Technology in Bydgoszcz, 85-029 Bydgoszcz, Poland; justynabauza@gazeta.pl
* Correspondence: ewa.walecka@upwr.edu.pl

Citation: Wiktorczyk-Kapischke, N.; Wałecka-Zacharska, E.; Skowron, K.; Kijewska, A.; Bernaciak, Z.; Bauza-Kaszewska, J.; Kraszewska, Z.; Gospodarek-Komkowska, E. Comparison of Selected Phenotypic Features of Persistent and Sporadic Strains of *Listeria monocytogenes* Sampled from Fish Processing Plants. *Foods* **2022**, *11*, 1492. https://doi.org/10.3390/foods11101492

Academic Editors: Fernando Pérez-Rodríguez and Arícia Possas

Received: 21 March 2022
Accepted: 17 May 2022
Published: 20 May 2022

Abstract: (1) Background: The main source of transmission of *Listeria monocytogenes* is contaminated food, e.g., fish and meat products and raw fruit and vegetables. The bacteria can remain for 13 years on machines in food processing plants, including fish plants. (2) Methods: A total of 720 swabs were collected from a salmon filleting line. The research material consisted of 62 (8.6%) *L. monocytogenes* isolates. Pulsed Field Gel Electrophoresis (PFGE) allowed detecting a pool of persistent strains. All persistent strains ($n = 6$) and a parallel group of strains collected sporadically ($n = 6$) were characterized by their ability to invade HT-29 cells, biofilm formation ability, and minimum bactericidal concentrations (MBC) of selected disinfectants. (3) Results: Among the obtained isolates, 38 genetically different strains were found, including 6 (15.8%) persistent strains. The serogroup 1/2a-3a represented 28 strains (73.7%), including the persistent ones. There were no significant differences in invasiveness between the persistent and sporadic strains. The persistent strains tolerated higher concentrations of the tested disinfectants, except for iodine-based compounds. The persistent strains initiated the biofilm formation process faster and formed it more intensively. (4) Conclusions: The presence of persistent strains in the food processing environment is a great challenge for producers to ensure consumer safety. This study attempts to elucidate the phenotypic characteristics of persistent *L. monocytogenes* strains.

Keywords: *Listeria monocytogenes*; persistent strains; sporadic strains; fish processing; resistance to disinfectants; biofilm; foodborne microorganism

1. Introduction

Listeria monocytogenes is a rod-shaped, Gram-positive, and widespread in the environment bacterium. *L. monocytogenes* adapts to unfavorable conditions in food processing plants. This pathogen is vulnerable to nutrient deficiency, heat shock, high osmolarity, and low pH [1–4]. The most common sources of *L. monocytogenes* are ready-to-eat (RTE) food and fish products, meat, poultry, raw milk, soft raw milk cheese, fresh and frozen vegetables, and packed salads. Fish is a common source of *L. monocytogenes*. The contamination of finished products may occur during the production process, during such activities as filleting, rinsing, and salting [2,5–9]. *L. monocytogenes* is the etiologic factor of a severe food-borne disease—listeriosis. Pregnant women, immunocompromised individuals, and the elderly are very sensitive groups to *Listeria* infections [1,10]. Despite its low incidence, the high fatality rate (15–20%) makes listeriosis a serious food-borne disease [11]. A recent listeriosis outbreak, with 200 deaths, was noted in Africa (2017–2018) and was due to RTE product consumption [12]. The presence of *L. monocytogenes* in food may be the effect of the contamination of raw materials or processed products at different stages of the production chain [13,14]. Point 22 of EU Regulation 2073/2005 aims at food safety assurance by testing samples from the food environment [15]. One serious problem is the emergence of persistent strains, despite the undertaken hygiene interventions. There is no unified, clear definition of the persistent strains and they are determined subjectively for individual research. So, if the subset of the bacterial population can survive exposure to a higher than usually used bactericidal drug concentration, it may be referred to as persistence, according to the definition given by Balaban et al. [16]. Presumably, the same strain recurrently detected over a specific period confirms the emergence of persistent strains [1]. The extreme period was noted by Fagerlund et al. [17] and lasted for 13 years. Stable strains of *L. monocytogenes* are indistinguishable pulsotypes (bacteria separated by pulsed field gel electrophoresis, PFGE), but their characteristics, both in terms of PFGE metric and serogroup, remain the same over time and maintain their properties, including a resistance to certain biocides [18]. In different studies, the persistence time varied from months to 12 years [3,19,20] and bacteria remained to be dangerous for potential consumers. For instance, a *L. monocytogenes* strain considered persistent was sampled from an Estonian company's premises, which produced cold smoked salmon and trout sold in countries of the European Union. The outbreak caused by the presence of this persistent strain of *L. monocytogenes* (sequence type (ST) 1247) in food included 22 cases of listeriosis in five EU countries [21].

Several concepts have been suggested to describe the strains' persistence. Resistance to stress factors is the most important, followed by biofilm formation and increased tolerance to disinfectants [4]. The most important factor influencing the persistence of bacteria is the high resistance to stress factors, such as pH, temperature, specificity and limitation of nutrient sources, and competition with other microorganisms. The pH of the fish is usually alkaline, and salt is also used as a preservative in the food industry [22–24]. The capability to grow at a low temperature and high salt concentrations promotes *L. monocytogenes* survival in the production environment.

The relevant source of the pathogen in the food processing environment is the reintroduction of persistent strains from external habitats [3]. *L. monocytogenes* has many adaptive mechanisms enabling its survival in adverse environmental conditions, including food processing [25]. Some studies suggest a higher adherence of persistent bacteria to food contact surfaces than the nonpersistent strains [26].

Researchers have documented higher biofilm formation ability among persistent, compared to nonpersistent, strains [27]. Biofilms are considered a source of persistent pathogenic microorganisms [10]. Moreover, the increased persistence of pathogenic bacteria can be the effect of co-existence with other, non-pathogenic microorganisms in multispecies biofilms [3]. *L. monocytogenes* in the biofilm revealed increased resistance to disinfecting agents compared to planktonic form. Adaptation and resistance to disinfectants, developed after *L. monocytogenes* exposition to their sublethal concentrations, can also affect the prolonged survival of the bacteria in the food processing environment [8,24,28].

The research aims to evaluate the differences in selected phenotypic properties between the persistent and sporadic strains of *L. monocytogenes* collected along the entirety of the fish processing line.

2. Materials and Methods

2.1. Sampling Procedure

From April to September 2019, 720 swabs (120 per month) were collected from machines and surfaces used for fish fillet production at one plant located in East-Central Europe, the north Poland. Samples were collected between work shifts, after cleaning procedures, when machines were not working.

To collect bacteria, the wet swab method was used. A sterile, flexible template limiting the tested area to 100 cm^2 was used. Samples were taken from the box pallets in the raw material warehouse (60 swabs), fish head remover machine (60 swabs), filleting machine (conveyor—60 swabs; knives—60 swabs), trim conveyor (conveyor rollers—60 swabs; worktops—60 swabs), pin bone remover (conveyor—60 swabs; trommel—60 swabs), fish skin remover machine (60 swabs), fillet washer (60 swabs) and portion machine (conveyor—60 swabs; knives—60 swabs). A sterile 50 cm^2 cellulose sponge (Enviroscreen, Technical Service Consultants Ltd., Lancashire, UK) soaked in 10 mL of a sterile 0.9% NaCl packed in a reinforced zip-bag was used for sampling.

2.2. Sampling and Identification of *L. monocytogenes* Isolates

The analysis of the samples was based on ISO 11290-1 procedures [29]. The swabs taken from the surface of machines were immersed in 100 mL of half-Fraser broth (Merck, Darmstadt, Germany) and incubated at 30 °C for 24 h. Secondary selective enrichment was performed for 48 h at 37 °C after transferring 0.1 mL of the culture into 9.9 mL of Fraser broth (Merck). Next (both after incubation in half-Fraser broth and Fraser broth), bacteria were plated on the selective agar medium according to Ottaviani and Agosti (ChromoCult® Listeria Selective Agar, ALOA®, Merck) and incubated for 24 h at 37 °C. Selected colonies, initially identified according to the manufacturer's recommendations as *Listeria* spp., were transferred to Columbia Agar with 5% sheep blood (bioMérieux, Marcy-l'Étoile, France).

Finally, the MALDI-TOF MS (Matrix-Assisted Laser Desorption and Ionization—Time of Flight Mass Spectrometry) technique was used to confirm if presumptive colonies belonged to the *L. monocytogenes* species. The acquisition and analysis of mass spectra were performed by a Microflex LT/SH mass spectrometer (Bruker, Billerica, MA, USA) using the MALDI Biotyper software package (version 4.1) with the Bruker Taxonomy reference database (Bruker). The ethanol–formic acid extraction procedure was applied for samples preparation. The bacterial test standard (BTS; Bruker) was used for validation according to the manufacturer's instructions.

The identified *L. monocytogenes* isolates were frozen in a brain–heart infusion broth (BHI, Merck) with 15% glycerol (Avantor, Gliwice, Poland) and stored at −80 °C.

2.3. Assessment of the Genetic Similarity of the Collected Isolates

The genetic similarity analysis of the confirmed *L. monocytogenes* strains was performed with the pulsed-field gel electrophoresis (PFGE), which is the golden standard to identify putative routes of contamination and persistent strains according to Dalmasso and Jordan [30]. The procedure for genotyping was performed following the Standard Operating Procedure for PulseNet PFGE of *Listeria monocytogenes* (PNL04, last updated April 2014) [31]. The ApaI enzyme was used in the study. The electrophoretic separation was performed with the following parameters: initial and final pulse duration: 4–40 s; voltage: 6 V/cm; pulse angle: 120°; temperature 14 °C; program duration: 17 h. The degree of genetic similarity between the analyzed *L. monocytogenes* isolates was evaluated using a phylogenetic dendrogram drawn in the CLIQS 1D Pro program (TotalLab, Newcastle upon Tyne, UK). Clustering analysis was performed using hierarchical clustering with the UPGMA technique and Dice's coefficient. The cut-off value to define the PFGE patterns was set at 80% similarity. The isolates were considered as genetically identical when identical pulsotypes were demonstrated for them by the PFGE method.

2.4. Isolation of Genomic DNA

Isolation of genomic DNA was performed using the Genomic Mini AX Bacteria Spin Kit (A&A Biotechnology, Gdańsk, Poland), according to the manufacturer's procedure, and the DNA was stored at −20 °C for further analyses.

2.5. Determination of Serological Groups

Multiplex PCR for the identification of the main *L. monocytogenes* serogroups (1/2a-3a, 1/2b-3b, 1/2c-3c, 4b-4d-4e) was performed as described by Doumith et al. [32]. The PCR was performed on a cycler Mastercycler® pro (Eppendorf, Hamburg, Germany) using: 1.5 × PCR buffer (Promega, Madison, WI, USA), 2 mM $MgCl_2$ (Promega), 1.25 mM dNTPs (Promega), 0.5 μM of each primer (Oligo.pl, Warszawa, Poland), 1 U GoTaq DNA polymerase (Promega), ultrapure water (Sigma Aldrich, Saint Louis, MO, USA), and the previously isolated genomic DNA. The amplicons were electrophoretically separated in 1.5% agarose gel (Sigma Aldrich) stained with Midori Green (NIPPON Genetics EUROPE GmbH, Düren, Germany) in 1 × TBE buffer (BioRad, Hercules, CA, USA), in the presence of a DNA size standard (GeneRulerTM1000 bp DNA Ladder) (Fermentas, Waltham, MA, USA) (90 V, 1 h).

For each PCR reaction, the four selected *L. monocytogenes* strains examined by Wałecka-Zacharska et al. [33] were used as positive control strains for serogroup identification. The negative control in each reaction was a sample without DNA.

2.6. Assessment of Drug Susceptibility of the Tested L. monocytogenes Strains

The selected antibiotics were among those frequently used in the first-line treatment of *L. monocytogenes* infections in humans and those also used in the veterinary treatment of farm animals. This is important as Poland is one of the largest producers and exporters of meat and dairy products in the European Union [34]. The evaluation of drug susceptibility was performed for genetically different isolates (62) using the disk diffusion method on the Mueller–Hinton agar with 5% defibrinated Horse Blood and 20 mg/L β-NAD (MH-F, bioMérieux). Disks with penicillin (1 IU), ampicillin (2 μg), meropenem (10 μg), erythromycin (15 μg), and cotrimoxazole (1.25–23.75 μg) were used. Antibiograms were incubated in an atmosphere enriched in 5% CO_2 at 35 °C for 18 h. The results were interpreted, according to the recommendations of EUCAST (European Committee on Antimicrobial Susceptibility Testing) v. 12.0. [35].

2.7. Comparison of the Selected Properties of Sporadic and Persistent L. monocytogenes Strains

At this stage of the study, 6 persistent strains (LMO-P1, LMO-P2, LMO-P3, LMO-P4, LMO-P5, and LMO-P6) and 6 sporadic strains (LMO 4, LMO 23, LMO 46, LMO 52, LMO 53, and LMO 61) were selected. The following tests were repeated in triplicate for each isolate. Each repetition consisted in the independent preparation of a new bacterial suspension for a given strain and the performance of all tests described in the methodology.

2.7.1. Assessment of Invasiveness against HT-29 Eukaryotic Cells

Tested strains were plated on Columbia Agar with 5% sheep blood (bioMérieux) and incubated for 24 h at 37 °C. Single colonies were transferred into 5 mL brain–heart infusion broth (BHI, Merck) and incubated in a thermoblock (TDB-100, Biosan, Józefów, Poland) at 37 °C (230 rpm, 6 h). Next, 5 μL of the bacterial suspension was transferred into 5 mL of BHI broth and incubated another 18 h until an OD600 of 2.4–2.6 was obtained (measured with the DU 8800D spectrophotometer). The bacteria of 5–6 log CFU (Colony Forming Units) were used to infect the human colon carcinoma HT-29 cell line (CLS, Germany).

HT-29 cells were seeded in 6-well polystyrene culture plates (Genoplast) and incubated to approx. 90% confluence in Dulbecco's Modified Eagle Medium (DMEM, Sigma-Aldrich), containing 10% fetal bovine serum (FBS, Gibco, Park Ridge Ln S Billings, MT, USA), 2 mM glutamine, and 100 IU/mL penicillin and 100 μg/mL streptomycin (Sigma-Aldrich). Before the cells' infection (24 h), the medium was changed to DMEM without antibiotics. The HT-29 cells were incubated with bacteria for 2 h (37 °C, 5% CO_2). The wells were then washed twice with a sterile PBS solution (Sigma-Aldrich) and incubated in DMEM containing 100 μg/mL gentamycin (Sigma-Aldrich) for 1.5 h (37 °C, 5% CO_2). Next, the wells were washed twice with PBS and overlaid with a medium containing 10 μg/mL gentamicin and 1.0% low melting point agarose (Prona, Gdańsk, Poland). After 48 h of incubation, the number of plaques was determined. Bacterial invasiveness was calculated as the quotient of the number of plaques (expressing the number of bacteria that entered HT-29 cells) and the number of bacteria introduced into the wells. Invasiveness was expressed as a percentage.

2.7.2. Determination of the Minimum Bactericidal Concentration (MBC) of Selected Disinfectants against Persistent and Sporadic Strains of *L. monocytogenes*

The evaluation of the minimum bactericidal concentrations (MBC) for the selected disinfectants was previously described by Skowron et al. [36]. Table 1 presents the disinfectants included in the study.

Table 1. Characteristics of the disinfectants used for the evaluation of the minimum bactericidal concentrations.

Group of Disinfectants	Trade Name	Active Substances	Manufacturer	Working Solution Concentration
Quaternary ammonium compounds	Sansept 0200	Didecyldimethylammonium chloride, benzyl-C12-16-alkyldimethyl chlorides	Sanechem	3 mL/L
Oxidizing agents	Peroxat	Peracetic acid, hydrogen peroxide	Agro-trade	5 mL/L
Chlorine compounds	Calcium hypochlorite	Hypochlorous acid calcium salt	Chem Point	2 g/L
Iodine compounds	Rapicid	Iodine	Pfizer	10 mL/L

The bacterial suspensions (100 μL) and 100 μL of the appropriate concentration of disinfectant were added to the 96-well polystyrene plate. The final concentrations of the disinfectants were: 100%, 90%, 80%, 70%, 60%, 50%, 40%, 30%, 20%, 10%, 5%, 1%, and 0.5% of the working solution. The negative control consisted of 200 μL of a sterile MHB (Mueller–Hinton Broth, Becton Dickinson) medium, and the positive control 200 μL of the bacterial suspension. After 5 min of the agent's action, 100 μL of each suspension was transferred into 900 μL of neutralizer (10 g Tween 80 (Sigma Aldrich), 1 g lecithin (Sigma Aldrich), 0.5 g histidine L (Sigma Aldrich), 2.5 g $Na_2S_2O_3$ (Avantor), 3.5 g $C_3H_3NaO_3$ (Avantor), and 1000 mL sterile water)). After 5 min of neutralization, samples were inoculated onto Columbia Agar with 5% sheep blood (bioMérieux). After 24 h incubation at 37 °C, the bacterial growth and the MBC value were assessed.

After determining the MBC range, the procedure was repeated with solutions at concentrations varying by 1% from the designated MBC value. This procedure allowed for the exact determination of the MBC value.

2.7.3. Assessment of the Rate of Initiation of Biofilm Formation

Biofilm formation ability was assessed on stainless steel coupons (1 cm × 2 cm, AISI 304 type). Coupons were washed in a commercial detergent, soaked for 5 min in 70% ethanol (Avantor), and autoclaved. For each strain, 5 coupons for one experiment were prepared. The research was carried out in triplicate.

Sterile steel coupons were placed in tubes containing 3 mL of bacterial suspension (0.5 McF) in BHI (Merck) and incubated in aerobic atmosphere at 37 °C for 1, 2, 3, 4 and 5 h, respectively. After incubation, the samples were rinsed with PBS solution and placed in a tube containing 3 cm^3 of this solution. Next, sonication (10 min, 30 kHz, 150 W) was performed using the Ultrasonic DU-4 (Nickel-Electro Ltd., Oldmixon Cres, Weston-super-Mare BS24 9BL, UK) sonicator.

After sonication, serial 10-fold dilutions of the obtained suspension in sterile PBS were prepared, plated on the Columbia Agar medium with 5% Sheep Blood (Becton Dickinson, Franklin Lakes, NJ, USA), and incubated for 24 h at 37 °C. For the group of persistent and sporadic strains, the mean number of *L. monocytogenes* recovered from the coupon surface after a given incubation time was calculated. The results were presented as the log CFU × cm^{-2}.

2.7.4. Assessment of the Intensity of Biofilm Formation

Sterile steel coupons were placed in tubes containing 3 mL of suspension of each strain (0.5 McF) in BHI of selected parameters (pH, salinity, and availability of nutrients), and were incubated for 72 h (Table 2). During incubation for each strain, except for experimental condition 1 (low temperature growth), the medium was replaced every 24 h with a fresh one and the coupons were rinsed with sterile PBS. For the variant 1 (4 °C) strain, the medium was replaced every 4 days and the incubation was extended to 12 days. As a negative control, steel coupons in a sterile BHI, in variables set up appropriately, were used. The first CFU's counting was performed after 1 day of incubation (in the case of low temperature, after the 4th day of incubation).

Table 2. Experimental conditions regarding the set of temperature, pH, salinity, and nutrient availability in individual variants of biofilm formation.

Environment Parameter	Experimental Conditions Set for Biofilm Formation	Temperature (°C)	pH	Salinity (% NaCl)	Nutrient Availability (BHI)
Temperature (°C)	1	4	7	0	1.0
	2	20	7	0	1.0
	3	**37**	**7**	**0**	**1.0**
pH	4	37	4	0	1.0
	5	**37**	**7**	**0**	**1.0**
	6	37	9	0	1.0
Salinity (% NaCl)	**7**	**37**	**7**	**0**	**1.0**
	8	37	7	5	1.0
	9	37	7	10	1.0
Nutrient availability (BHI)	10	37	7	0	0.5 *
	11	**37**	**7**	**0**	**1.0 ***
	12	37	7	0	1.5 *

BHI—brain heart infusion broth; * BHI 1.0—medium containing the amount recommended by the manufacturer; BHI 0.5—medium containing 50% of the amount recommended by the manufacturer; BHI 1.5—medium containing 150% of the amount recommended by the manufacturer. The control variant was marked with bold and the variable parameters with grey color.

After incubation, the samples were rinsed with a PBS solution and placed in a tube containing 3 mL of this solution. Next, sonication (10 min, 30 kHz, 150 W) was performed using the Ultrasonic DU-4 (Nickel-Electro Ltd.) sonicator.

After sonication, serial 10-fold dilutions of the obtained suspension in sterile PBS were prepared, plated on Columbia Agar medium with 5% Sheep Blood (Becton Dickinson), and incubated for 24 h at 37 °C. For the group of persistent and sporadic strains, the mean number of *L. monocytogenes* recovered from the coupon surface under given environmental conditions was calculated. The results are presented as the log CFU × cm^{-2}.

2.8. Statistical Analysis

The statistical analysis was carried out in the STATISTICA 13.0 PL (TIBCO Software, Palo Alto, CA, USA) software. With the use of general linear models (GLM) and ANOVA analysis, the statistical significance of differences was checked at the level of $\alpha = 0.05$. Based on one-way ANOVA, the differences in the percentages of all strains tested representing a given serogroup, resistance to a given antibiotic, and a given drug's susceptibility profile were checked. The significance of differences in invasiveness between the persistent and nonpersistent strains, MBC values for each of the tested disinfectants, and the intensity of biofilm formation over time were also checked. In turn, based on the multivariate ANOVA, the differences in the number of bacteria recovered from the biofilm between the persistent and nonpersistent strains, depending on the conditions of its formation, were checked.

3. Results

Out of 720 swabs taken from the surface of the fish processing machines, 62 isolates were identified as *L monocytogenes* (8.6%) (Table 3). We collected the highest number of isolates (14, 22.6%) from the filleting machine knives and the least (3, 4.8%) from box pallets in the raw material warehouse, the filleting machine conveyor, the worktops of the trim conveyor, the pin bone remover conveyor, the fillet washer, and the portion machine knives (Table 3). We obtained the highest number of isolates (14, 22.6%) in September and the lowest number (7, 11.3%) in May (Figure 1).

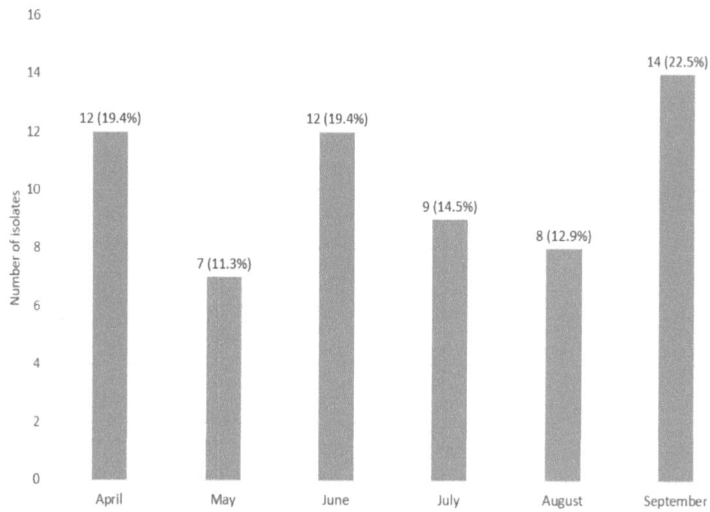

Figure 1. Number and percentage (%) of the total number of obtained isolates (62–100%) of *L. monocytogenes* according to months of sampling.

Table 3. Sources of *L. monocytogenes* isolates collected during the research.

Element of the Processing Line		Number (%) of All Isolates	Isolates *
Box pallets in the raw material warehouse		3 (4.8)	LMO 36, LMO 39, LMO 62
Head remover machine (cutting element)		7 (11.3)	**LMO-P2** (LMO 5, LMO 14, LMO 20, LMO 45) LMO 48, LMO 55, LMO 61
Filleting machine	Conveyor	3 (4.8)	LMO 40, LMO 49, LMO 52
	Knives	14 (22.6)	**LMO-P3** (LMO 7, LMO 16, LMO 33, LMO 41, LMO 54), **LMO-P4** (LMO 9, LMO 18, LMO 26, LMO 43, LMO 60) LMO 2, LMO 13, LMO 51, LMO 53
Trim conveyor	Conveyor rollers	7 (11.3)	**LMO-P1** (LMO 3, LMO 24, LMO 35, LMO 56) LMO 17, LMO 25, LMO 59
	Worktops	3 (4.8)	LMO 23, LMO 31, LMO 46
Pin bone remover	Conveyor	3 (4.8)	LMO 6, LMO 21, LMO 42
	Trommel	4 (6.5)	**LMO-P6** (LMO 22, LMO 38, LMO 44) LMO 58

Table 3. *Cont.*

Element of the Processing Line		Number (%) of All Isolates	Isolates *
Box pallets in the raw material warehouse		3 (4.8)	LMO 36, LMO 39, LMO 62
Skin remover machine		8 (12.9)	**LMO-P5** (LMO 15, LMO 27, LMO 47, LMO 50) LMO 8, LMO 32, LMO 34, LMO 57
Fillet washer		3 (4.8)	LMO 11, LMO 29, LMO 30
Portion machine	Conveyor	4 (6.5)	LMO 1, LMO 10, LMO 19, LMO 28
	Knives	3 (4.8)	LMO4, LMO 12, LMO 37

* In brackets are isolates clustered as belonging to each persistent strain written in bold letters.

3.1. *Assessment of the Genetic Similarity of Isolate Strains Classified as Persistent*

The assessment of the genetic similarity of the isolates allowed the selection of the persistent strains. Persistent strains were defined as strains represented by genetically identical isolates obtained from a particular part of the processing line at least three times over six months.

Among the 62 obtained isolates, 38 genetically different strains of *L. monocytogenes* were found (Figure 2). The cut-off level equal to 80% allowed to identify 14 strains with a different number of isolates. Twelve isolates corresponded to single-member clusters. The existence of 5 pairs of genetically identical isolates collected at the same time of sampling was demonstrated (Figure 2).

Six persistent strains meet the criterion adopted in the research methodology (Figure 2). The genetically identical isolates obtained at different times, but belonging to a given persistent strain, are listed in brackets:

- LMO-P1 (LMO 3 (April 2019), LMO 24 (June 2019), LMO 35 (July 2019), and LMO 56 (September 2019))—isolated from the trim conveyor rollers.
- LMO-P2 (LMO 5 (April 2019), LMO 14 (May 2019), LMO 20 (June 2019), and LMO 45 (August 2019))—isolated from the cutting element of the head remover machine.
- LMO-P3 (LMO 7 (April 2019), LMO 16 (May 2019), LMO 33 (July2019), LMO 41 (August 2019), and LMO 54 (September 2019))—isolated from the knives of the filleting machine.
- LMO-P4 (LMO 9 (April 2019), LMO 18 (May 2019), LMO 26 (June 2019), LMO 43 (August 2019), and LMO 60 (September 2019))—isolated from the knives of the filleting machine.
- LMO-P5 (LMO 15 (May 2019), LMO 27 (June 2019), LMO 47 (August 2019), and LMO 50 (September 2019))—isolated from the skin remover machine.
- LMO-P6 (LMO 22 (June 2019), LMO 38 (July 2019), and LMO 44 (August 2019))—isolated from the trommel of the pin bone remover.

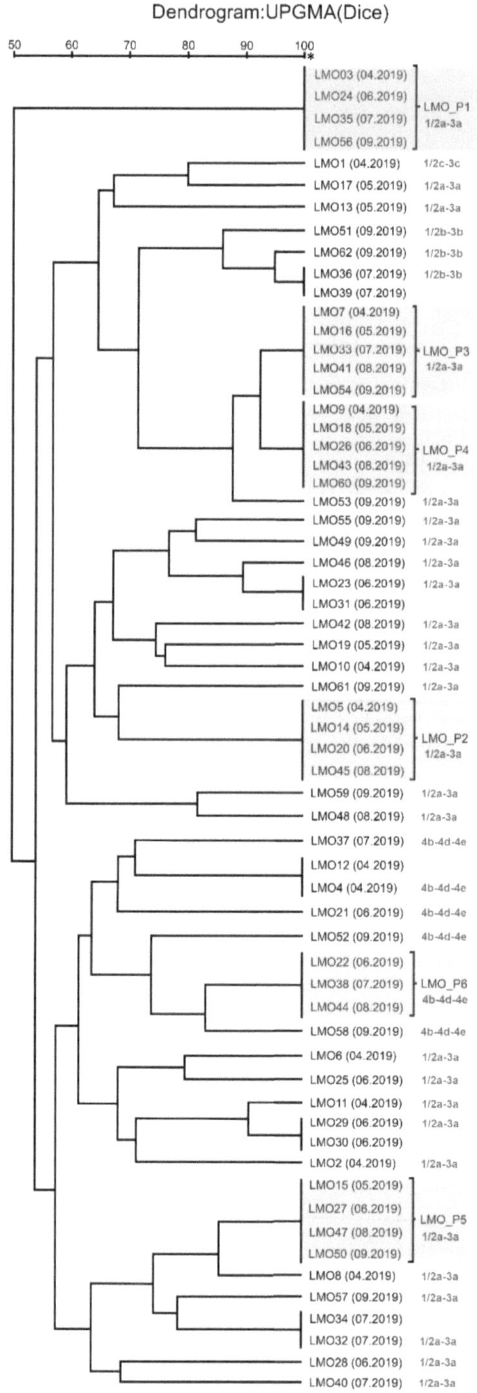

Figure 2. Genetic similarity dendrogram of each tested isolate with clusters of isolates belonging to each persistent strain (marked by gray blocks). * isolates genetically indistinguishable.

Each isolate constituting a given persistent strain was obtained from the same place in the processing line, but at different sampling times.

In this pool of strains, the LMO-P3 and LMO-P4 strains were genetically similar at 93%, while the LMO-P1 strain was the most genetically distant (Figure 2).

3.2. Molecular Serotyping of *L. monocytogenes* Strains

The analysis of DNA patterns on agarose gel showed 5 persistent strains belonging to the 1/2a-3a serogroup represented by 28 (73.7%) of all detected strains, while the 6th persistent strain (LMO-P6) belonged to 4b-4d-4e serogroup. In turn, only one strain, LMO-1 representing 2.6% of all strains, belonged to the 1/2c-3c serogroup (Figure 2).

3.3. Assessment of Drug Susceptibility of the Tested *L. monocytogenes* Strains

Among the *L. monocytogenes* isolates resistant to at least one antibiotic, the greatest number of isolates (13, 34.2%) were resistant to meropenem. In turn, resistance to penicillin was the least common (8, 21.1%) (Figure 3).

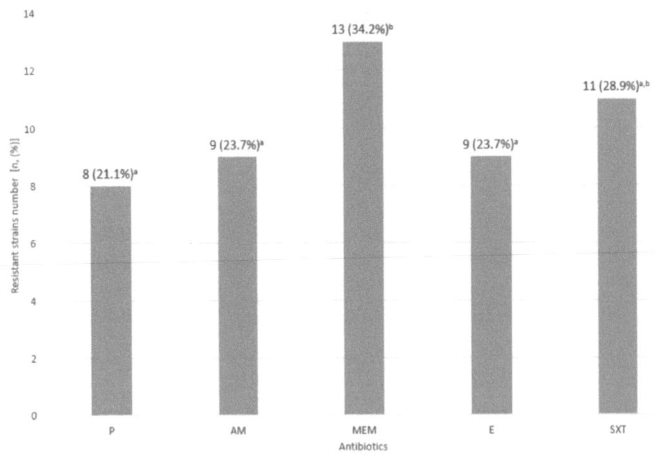

Figure 3. Resistance of 38 isolates of *L. monocytogenes* to antibiotics (P—penicillin; AM—ampicillin; MEM—meropenem; E—erythromycin; SXT—cotrimoxazole; ^{a},^{b}—values marked with different letters differ in a statistically significant way ($p \leq 0.05$). In brackets, the percentage of resistant isolates is given. The limited number of isolates is a result of the exclusion of genetically identical isolates.

The conducted experiment allowed for the identification of six antibiotic resistance profiles. The isolates representing profile no. 1 (18 isolates, 47.4%) (Table 4) were susceptible to all tested antibiotics. The remaining 20 (52.6%) individual isolates and those representing persistent strains showed resistance to at least one tested antibiotic. Counting isolates without persistent strains, 39 of all isolates were resistant to at least one antibiotic. Persistent strains belonged to profiles no. 3 (LMO-P1 and LMO-P2—resistant to meropenem, erythromycin, and cotrimoxazole), no. 4 (LMO-P6—resistant to penicillin and meropenem), and no. 5 (LMO-P3, LMO-P4, and LMO-P5—resistant to all tested antibiotics) (Table 4). Part of single isolates represented an identical resistance to antibiotics as persistent strains.

Table 4. Antibiotic resistance profiles.

Profile Number	Antibiotic Resistance Profile	Number (%) of Strains	Isolates and Strains Representing the Profile
1	R: — S: P, AM, MEM, E, SXT	18 (47.4) [a]	LMO 2, LMO 6, LMO 8, LMO 10, LMO 11, LMO 19, LMO 21, LMO 25, LMO 28, LMO 29, LMO 32, LMO 36, LMO 37, LMO 40, LMO 42, LMO 51, LMO 57, LMO 62
2	R: AM S: P, MEM, E, SXT	5 (13.2) [b]	LMO 1, LMO 13, LMO 17, LMO 49, LMO 55
3	R: MEM, E, SXT S: P, AM	5 (13.2) [b]	**LMO-P1 *, LMO-P2** LMO 23, LMO 46, LMO 61
4	R: P, MEM S: AM, E, SXT	4 (10.5) [b]	**LMO-P6** LMO 4, LMO 52, LMO 58
5	R: P, AM, MEM, E, SXT S: —	4 (10.5) [b]	**LMO-P3, LMO-P4, LMO-P5** LMO 53
6	R: SXT S: P, AM, MEM, E	2 (5.3) [b]	LMO 48, LMO 59

* bold—persistent strains; P—penicillin; AM—ampicillin; MEM—meropenem; E—erythromycin; SXT—cotrimoxazole; R—resistant; S—sensitive; [a],[b]—statistical significance with $p \leq 0.05$.

3.4. Assessment of Invasiveness against HT-29 Eukaryotic Cells

The invasiveness of the persistent *L. monocytogenes* strains ranged from 1.07% for LMO-P6 to 11.21% for LMO-P4. In turn, the invasiveness of sporadic strains ranged from 1.42% for LMO 4 to 7.99% for LMO 61. The mean invasiveness calculated for the persistent strains was slightly higher than for the sporadic strains (5.50% vs. 4.60%); however, the difference was not statistically significant (Table 5).

Table 5. Invasiveness of the persistent and sporadic strains of *L. monocytogenes* against HT-29 cells.

Persistent Strains		Sporadic Strains	
Strain	%	Strain	%
LMO-P1	$3.42 \pm 0.70^{a,b}$	LMO 4	1.42 ± 0.17^{a}
LMO-P2	2.05 ± 0.35^{a}	LMO 23	5.59 ± 1.14^{b}
LMO-P3	9.55 ± 4.37^{c}	LMO 46	4.25 ± 0.93^{b}
LMO-P4	11.21 ± 3.19^{d}	LMO 52	2.10 ± 0.69^{a}
LMO-P5	5.71 ± 1.66^{b}	LMO 53	$6.22 \pm 1.84^{b,e}$
LMO-P6	1.07 ± 0.39^{a}	LMO 61	$7.99 \pm 1.11^{c,e}$
Mean	5.50 ± 3.71 *	Mean	4.60 ± 2.29 *

[a],[b],[c],[d],[e], values marked with different letters differ statistically ($p \leq 0.05$). * values marked with this symbol do not differ statistically ($p \leq 0.05$).

3.5. Minimum Bactericidal Concentrations (MBC) of the Selected Disinfectants against the Persistent and Sporadic Strains of *L. monocytogenes*

For all strains, both persistent and sporadic, the effective concentrations of disinfectants were lower than the concentrations of the working solution recommended by the manufacturer (Table 6). MBC values depended on the properties of a particular strain and the type of disinfectant.

Table 6. Minimum bactericidal concentration value of the tested disinfectants against the persistent and sporadic strains of *L. monocytogenes*.

Disinfectant	PERSISTENT STRAINS							SPORADIC STRAINS						
	LMO-P1	LMO-P2	LMO-P3	LMO-P4	LOM-P5	LMO-P6	MEAN	LMO 4	LMO 23	LMO 46	LMO 52	LMO 53	LMO 61	MEAN
Sansept 0200 [mL/L]	0.81 ± 0.11 (27%)	0.81 ± 0.18 (27%)	1.17 ± 0.27 (39%)	1.47 ± 0.3[illegible] (49%)	0.93 ± 0.09 (31%)	0.72 ± 0.15 (24%)	0.99 ± 0.28 a (33%)	0.15 ± 0.04 (5%)	0.21 ± 0.10 (7%)	0.21 ± 0.06 (7%)	0.18 ± 0.07 (6%)	0.27 ± 0.02 (9%)	0.30 ± 0.12 (10%)	0.22 ± 0.06 b (7%)
Peroxat [mL/L]	3.50 ± 1.02 (70%)	3.10 ± 0.68 (62%)	4.10 ± 0.91 (82%)	4.60 ± 1.1[illegible] (92%)	3.85 ± 0.74 (77%)	3.10 ± 1.00 (62%)	3.71 ± 0.59 a (74%)	1.75 ± 0.51 (35%)	2.10 ± 0.43 (42%)	2.05 ± 1.40 (41%)	1.85 ± 0.82 (37%)	2.50 ± 0.99 (50%)	2.90 ± 0.38 (58%)	2.19 ± 0.43 b (44%)
Calcium hypochlorite [g/L]	0.68 ± 0.23 (34%)	0.54 ± 0.10 (27%)	0.78 ± 0.31 (39%)	0.90 ± 0.0[illegible] (45%)	0.80 ± 0.66 (40%)	0.60 ± 0.17 (30%)	0.72 ± 0.14 a (36%)	0.30 ± 0.08 (15%)	0.34 ± 0.13 (17%)	0.38 ± 0.20 (19%)	0.34 ± 0.14 (17%)	0.48 ± 0.07 (24%)	0.40 ± 0.23 (20%)	0.37 ± 0.06 b (19%)
Rapicid [mL/L]	3.10 ± 0.71 (31%)	3.20 ± 0.52 (32%)	3.80 ± 0.84 (38%)	4.00 ± 1.11 (40%)	3.80 ± 0.95 (38%)	2.70 ± 0.36 (27%)	3.43 ± 0.51 a (34%)	2.80 ± 0.29 (28%)	3.10 ± 0.41 (31%)	3.10 ± 0.37 (31%)	3.00 ± 0.19 (30%)	3.10 ± 0.28 (31%)	3.60 ± 0.37 (36%)	3.12 ± 0.26 a (31%)

A percentage of working solution (given in brackets) concentration, which is the MBC value; ^{a,b}, values marked with different letters differ statistically with $p \leq 0.05$.

For each strain belonging to the persistent strain group, the MBC values were higher than for strains from the sporadic strain group. For the persistent strains, the values ranged from 33% (0.99 mL/L) of the working solution concentration for Sansept 0200 to 74% (3.71 mL/L) of the working solution concentration for Peroxat (Table 6). In turn, for sporadic strains, the MBC values ranged from 7% (0.22 mL/L) of the working solution concentration for Sansept 0200 to 44% (2.19 mL/L) of the working solution concentration for Peroxat (Table 6). For each tested disinfectant, except Rapicid, statistically significant differences were found between the MBC values established for persistent and sporadic strains (Table 6). In the pool of persistent strains, the LMO-P4 strain turned out to be the most resistant, and the most susceptible one was LMO-P6 (Table 6). In turn, among sporadic strains, the most resistant was LMO 61 and the most sensitive was LMO 4 (Table 6).

Sansept 0200 turned out to be the most effective disinfectant for strains from both groups, and Peroxat the least effective (Table 6).

3.6. Assessment of the Rate of Initiation of Biofilm Formation

Persistent *L. monocytogenes* strains formed biofilm faster than sporadic strains (Figure 4). The number of *L. monocytogenes* recovered from the biofilm after 1, 2, 3, and 4 h was statistically significantly higher for persistent than for sporadic strains. No statistically significant differences were found at the 5th h of biofilm formation, although the number of rods recovered remained higher for persistent strains (Figure 4). We noticed the largest difference in the number of recovered *L. monocytogenes* between persistent and sporadic strains, amounting to 1.66 log CFU at the 2nd h of biofilm formation (Figure 4).

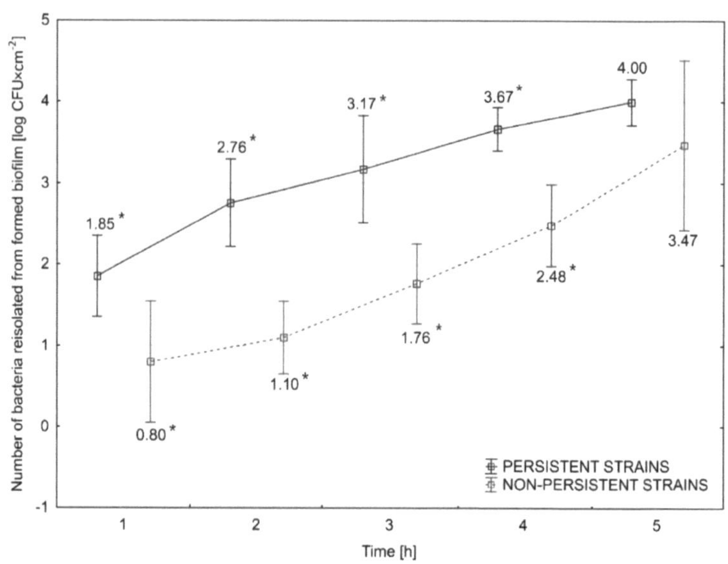

Figure 4. Assessment of the rate of initiation of biofilm formation for the persistent and nonpersistent *L. monocytogenes* strains. The horizontal shift of points relative to the timeline is only used to maintain the chart's readability. The starting density at the 0 h timepoint was 0.5 McF for each strain. *—values assigned to the same time point, marked with an asterisk, differ statistically significantly ($p \leq 0.05$).

3.7. Assessment of the Intensity of Biofilm Formation

The effect of environmental conditions on biofilm formation by both persistent and sporadic *L. monocytogenes* strains was shown (Figure 5). We observed statistically significant differences in the number of *L. monocytogenes* recovered from biofilm between the persistent and sporadic strains (6.85 vs. 6.09 log CFU $\times$ cm^{-2}) under the control conditions (37°C, pH 7, 0% NaCl, 1 BHI) (Figure 5).

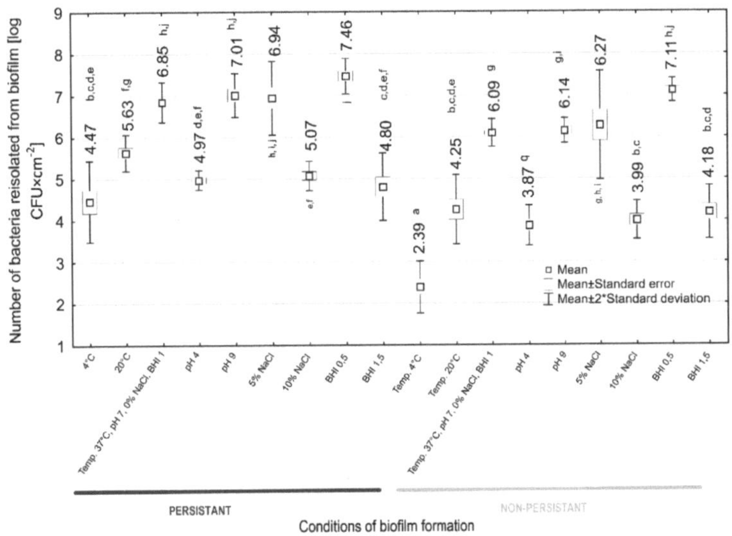

Figure 5. Assessment of the intensity of biofilm formation for the persistent and nonpersistent *L. monocytogenes* strains according to single variable changes being given along X-ordinate. Control parameters were 37 °C, pH 7, 0% NaCl, 1 BHI. a–j: values marked with different letters differ statistically ($p \leq 0.05$).

Analyzing the effect of temperature, we found that the strains from both groups showed a weaker biofilm-forming capacity at 4°C and performed best at 37 °C. The number of bacteria recovered from the biofilm under all temperature conditions was statistically significantly higher for the persistent strains (Figure 5).

Both persistent and sporadic strains formed a weak biofilm at pH 4 and a one stronger at pH 9 compared to the biofilm formation in the control conditions (Figure 5). At the tested pH values, the number of *L. monocytogenes* obtained from biofilms was statistically significantly higher in the case of persistent strains than sporadic ones (Figure 5).

All strains formed biofilms less intensely in increased salinity (5% and 10% NaCl) compared to what was formed in the control conditions (Figure 5). The bacteria performed better at creating biofilm at 5% NaCl than 10% NaCl (Figure 5). Regardless of salinity, the number of *L. monocytogenes* recovered from biofilm was higher for persistent strains, with a statistically significant difference only shown for 10% NaCl salinity (Figure 5).

The reduced availability of nutrients (0.5 BHI) increased the intensity of biofilm formation in both strain groups (Figure 5). In turn, the increased availability of nutrients (1.5 BHI) lowered the biofilm formation intensity (Figure 5). The persistent strains formed a biofilm slightly better at both 0.5 BHI and 1.5 BHI, but the observed differences were not statistically significant (Figure 5).

Collectively, strains from both groups formed the weakest biofilm at 4 °C, pH 7, 0% NaCl, and 1 BHI, and the strongest at 37 °C, pH 7, 0% NaCl, and 0.5 BHI (Figure 5).

4. Discussion

There is a growing demand among consumers for fresh and low-processed foods. The contamination of the fish processing environment with *L. monocytogenes* increases the epidemiological risk linked to fish product consumption. The ability of *L. monocytogenes* to survive in extreme conditions and to form biofilms on various surfaces is a significant challenge for food safety [9]. One of the factors affecting the distribution of pathogens in the facility is the type and quality of the equipment used for material processing. In our study, among the 62 *L. monocytogenes* isolates (8.6%) obtained from processing devices used for fish fillet production, the highest number (14, 22.6%) originated from the filleting machine knives, which directionally contacts with the fish flesh and thus being a probable source of bacteria. A study by Kurpas et al. [8] confirmed food of animal origin as a source of pathogenic *L. monocytogenes* for humans. The presence of persistent strains within the food processing environment entails the contamination of finished products, increasing the risk for future consumers, especially when the product is raw and ready-to-eat fish. Part of the devices included in the processing line have a continuous contact with processed fish. The pin bone remover or knives in the filleting machines can directly contaminate the food, frequently consumed raw or cold-smoked. Salmonidae fish are a popular species consumed without any heat treatment. That is why *L. monocytogenes* infections are enormously dangerous for consumers and should be monitored to prevent outbreaks of listeriosis. Similarly, the studies of Di Ciccio et al. [37] confirmed an overall *L. monocytogenes* prevalence rate of 16% in a cold-smoked salmon processing environment. The samples from working tables (43%) and slicing machines (37%) were the most contaminated [37]. Leong et al. [38] reported a lower *L. monocytogenes* prevalence in the seafood industry compared to the dairy, meat, and vegetable industries. The installation of new equipment increased *L. monocytogenes* occurrence in food production facilities (from 5 to 23% during a year) [38].

The fish processing environment, due to the amount of equipment used during the processing of the raw materials, increases the risk of the finished products being contaminated with *L. monocytogenes*. Despite the procedures used, cleaning and disinfection are becoming insufficient and not effective if we consider the fact of the frequent presence of *L. monocytogenes* persistent strains along the food processing lines. This is one of the most problematic properties of this bacteria, which easily adapts to a wide spectrum of unfavorable factors. They easily adapt to stress factors, form biofilms, and are resistant to disinfectants. In our study, we found six persistent strains of *L. monocytogenes*. Miettinen and Wirtanen [39], analyzing 81 isolates from 15 fish farms and fish processing plants, found 30 *L. monocytogenes* pulsotypes. Scientists observed the repetitive isolation of at least one pulsotype from the same facility, suggesting the presence of persistent *L. monocytogenes* strains in the processing environment. In Ramires et al. [40] research, two of the total four *L. monocytogenes* pulsotypes from salmon sushi were persistent. Aalto-Araneda et al. [41] sampled the same *L. monocytogenes* pulsotypes on separate sampling occasions in three of seven fish processing plants. Three of eight RAPD types of *L. monocytogenes*, found in raw fish and their products from Polish fish processing plants, were collected continually over 8–10 months [42]. Cruz and Fletcher [43] identified persistent *L. monocytogenes* strains in the mussel processing environment (in raw mussels and finished products). One of the persistent pulsotypes was linked to non-perinatal listeriosis cases.

The mechanism of bacterial persistence is poorly understood. Some authors suggest that biofilm-forming ability is an essential factor for its prolonged survival in the food production environment [40]. Our research showed a higher rate of biofilm formation initiation for persistent *L. monocytogenes* strains. Their number, recovered from the biofilm, was statistically significantly higher than for sporadic strains after 1, 2, 3, and 4 h, but not after 5 h of the process. Lundén et al. [26] reported that the persistent *L. monocytogenes* strains revealed higher adherence than the nonpersistent strains after 1 and 2 h of contact time. However, after 72 h, the adherence ability was comparable for persistent and nonpersistent strains [26]. Contrary to Lundén et al., Costa et al. [19] observed the significantly higher attachment abilities of nonpersistent *L. monocytogenes* isolates. In our study, the number of *L. monocytogenes* recovered from biofilms, formed in the control conditions (37 °C, pH 7, 0% NaCl, 1 BHI), for persistent strains was statistically significantly higher (6.85 log CFU $\times$ cm^{-2}) compared with sporadic strains (6.09 log CFU $\times$ cm^{-2}). The biofilm formation was the most intense at 37 °C, pH 9, 5% NaCl salinity, and a reduced nutrient availability (0.5 BHI), which confirms our previous studies [44]. Regardless of the thermal conditions and environmental pH, the number of bacteria recovered from the biofilm was statistically significantly higher for the persistent strains. We did not observe statistically significant differences between biofilms formed in environments with various levels of nutrient availability and at 5% NaCl salinity. The tolerance of *L. monocytogenes* to environmental stress factors contributes to a better ability to survive in food products and processing environments. Researchers have suggested that the adaptation of *L. monocytogenes* to stress factors is one of the main theories of the formation of persistent strains. Jensen et al. [45] have reported the enhanced adhesion of *L. monocytogenes* strains to a plastic surface at 37 °C after NaCl addition to the growth medium. Taylor and Stasiewicz [46] documented the growth of persistent and sporadic strains of planktonic *L. monocytogenes* cells at 5% and 10% salt concentration, as well as acidic (pH 5.2) and alkaline (pH 9.2) conditions. Nowak et al. [47] reported increased biofilm formation in such conditions. In turn, Assisi et al. [48] found that the persistence of *L. monocytogenes* in the environment is probably a matter of the poor sanitation of the facility and not the ability of isolates to form a biofilm and tolerance to disinfectants.

Microbial tolerance towards sanitizing agents may lead to the higher persistence of pathogens in the production environment [20]. The results of our study show that the applied disinfectants were effective against persistent and sporadic *L. monocytogenes* strains in concentrations lower than those recommended by the manufacturer. Their high efficacy could be because pathogen cells were in planktonic form. Moreover, no potentially protecting organic substances were present in the environment. Magalhães et al. [23] noted the reduction in persistent and nonpersistent *L. monocytogenes* isolates by commonly used disinfectants applied in concentrations lower than those recommended by the manufacturers. They found no relation between pathogen persistence and increased resistance to sanitizers. Costa et al. [19] also noticed no significant differences between persistent and nonpersistent *L. monocytogenes* isolates in their sensitivity to disinfectant treatments, suggesting no link between persistence and disinfectant susceptibility. In our study, we observed significant differences in susceptibility to Sansept 0200 (Didecyldimethylammonium chloride, benzyl-C12-16-alkyldimethyl chlorides), Peroxat (Peracetic acid, hydrogen peroxide), and Calcium hypochlorite between persistent and sporadic *L. monocytogenes* strains. According to the calculated MBC values, Sansept was the most effective disinfectant and Peroxat the least effective. However, the obtained results do not support that resistance to disinfectants is one of the hypotheses for the formation of persistent strains. Wang et al. [49] found no significant difference in disinfectant tolerance between the persistent and transient strains.

The study also assessed the belonging to serogroups and antimicrobial resistance. The highest number of collected *L. monocytogenes* isolates (28, 73.7%) was classified as 1/2a-3a serogroup. Additionally, Gambarin et al. [50] have shown a high percentage of strains related to serotype 1/2a (73.33%) in RTE seafood. In turn, Ramires et al. [40] have noted that all *L. monocytogenes* isolates from sushi establishments belonged to serotype 4b. Serotype 4b-4d-4e was the second numerous group (15.8% of *L. monocytogenes* isolates) in the present study and included the persistent LMO-P6 strain. The other persistent strains belonged to the 1/2a-3a serogroup. In our study, 18 (47.4%) isolates of *L. monocytogenes* strains were susceptible to all antibiotics tested. The highest number of strains (34.2%) was resistant to meropenem. Three of the six persistent strains were resistant to all tested antibiotics. Skowron et al. [51] have found the highest resistance to erythromycin (47.1%) and cotrimoxazole (47.1%) among *L. monocytogenes* strains isolated from the fish and fish processing plant.

The ability of *L. monocytogenes* to adhere, invade, and grow in intestinal cells is directly associated with the pathogen's virulence [33,52,53]. The invasiveness of *L. monocytogenes* isolates against HT-29 cells amounted to 7.99% for sporadic strain LMO 61 and 11.21% for persistent strain LMO-P4. We did not observe statistically significant differences between the mean invasiveness values for the persistent and sporadic strains. Moroni et al. [54] have reported an invasion ability of *L. monocytogenes* LSD348 against HT-29 cells of 45.49%. Jensen et al. [45] have observed low invasiveness of the four RAPD type 9 persistent strains (N53-1, H13-1, La111, and M103-1) against Caco-2 cells compared to the remaining strains. Wałecka-Zacharska et al. [33] have noted that *L. monocytogenes* strains of 1/2a serotype revealed a lower ability to invade epithelial cells than those of the 4b and 1/2b serotypes.

There are many controversies about the persistence mechanism of strains frequently and repetitively isolated from the production environment. Research results concerning the relationship between resistance to disinfectants, adherence ability, biofilm formation, and the long-term survival of *Listeria* spp. in food industry plants are contradictory [3,23,55]. According to the hypotheses of Carpentier and Cerf [1], the key factor of a strain's persistence is the specificity of harborage sites inhabited by the bacteria. The bacteria living in these areas, protected from environmental stresses, can survive for a longer time [19,46].

5. Conclusions

The presence of persistent strains of *L. monocytogenes* increases the risk of food cross-contamination. Our study aimed to characterize strains collected from fish processing plants and the different phenotypic responses of persistent and sporadic strains. The results indicate that the persistent strains of *L. monocytogenes* can form a stronger biofilm (also in unfavorable environmental conditions) and have a lower disinfectant susceptibility than sporadic strains. We think that future research should explore the genetic variation between persistent and sporadic strains to explain the molecular basis of persistence.

Author Contributions: Conceptualization, N.W.-K., K.S., A.K. and E.W.-Z.; methodology, N.W.-K., K.S., A.K. and E.W.-Z.; validation, N.W.-K.; formal analysis, K.S. and E.W.-Z.; investigation, N.W.-K., Z.B. and Z.K.; resources, K.S. and J.B.-K.; data curation, N.W.-K.; writing—original draft preparation, N.W.-K. and Z.K.; writing—review and editing, N.W.-K., K.S., A.K. and E.W.-Z.; visualization, Z.B. and J.B.-K.; supervision, N.W.-K. and E.G.-K.; project administration, N.W.-K. and E.G.-K.; funding acquisition, K.S. All authors have read and agreed to the published version of the manuscript.

Funding: This article was financed with a contest fund in the "Excellence Initiative Research University" program at Nicolaus Copernicus University in Toruń.

Institutional Review Board Statement: Not applicable.

Informed Consent Statement: Not applicable.

Data Availability Statement: The datasets generated for this study are available on request to the corresponding author.

Conflicts of Interest: The authors declare no conflict of interest.

References

1. Carpentier, B.; Cerf, O. Review—Persistence of *Listeria monocytogenes* in food industry equipment and premises. *Int. J. Food Microbiol.* **2011**, *145*, 1–8. [CrossRef] [PubMed]
2. Lundén, J. Persistent *Listeria monocytogenes* Contamination in Food Processing Plants. Ph.D. Thesis, University of Helsinki, Helsinki, Finland, January 2004. Available online: http://urn.fi/URN:ISBN:952-10-1507-1 (accessed on 12 November 2021).
3. Larsen, M.H.; Dalmasso, M.; Ingmer, H.; Langsrud, S.; Malakauskas, M.; Mader, A.; Møretrø, T.; Možina, S.S.; Rychli, K.; Wagner, M.; et al. Persistence of foodborne pathogens and their control in primary and secondary food production chains. *Food Control* **2014**, *44*, 92–109. [CrossRef]
4. Unrath, N.; McCabe, E.; Macori, G.; Fanning, S. Application of Whole Genome Sequencing to Aid in Deciphering the Persistence Potential of *Listeria monocytogenes* in Food Production Environments. *Microorganisms* **2021**, *9*, 1856. [CrossRef] [PubMed]
5. Axelsson, L.; Bjerke, G.A.; McLeod, A.; Berget, I.; Holck, A.L. Growth Behavior of *Listeria monocytogenes* in a Traditional Norwegian Fermented Fish Product (Rakfisk), and Its Inhibition through Bacteriophage Addition. *Foods* **2020**, *9*, 119. [CrossRef]
6. Shamloo, E.; Hosseini, H.; Moghadam, Z.A.; Larsen, M.H.; Haslberger, A.; Alebouyeh, M. Importance of *Listeria monocytogenes* in food safety: A review of its prevalence, detection, and antibiotic resistance. *Iran. J. Vet. Res.* **2019**, *20*, 241–254.
7. Szymczak, B.; Szymczak, M.; Trafiałek, J. Prevalence of Listeria species and L. monocytogenes in ready-to-eat foods in the West Pomeranian region of Poland: Correlations between the contamination level, serogroups, ingredients, and producers. *Food Microbiol.* **2020**, *91*, 103532. [CrossRef]
8. Kurpas, M.; Osek, J.; Moura, A.; Leclercq, A.; Lecuit, M.; Wieczorek, K. Genomic Characterization of *Listeria monocytogenes* Isolated from Ready-to-Eat Meat and Meat Processing Environments in Poland. *Front. Microbiol.* **2020**, *11*, 1412. [CrossRef]
9. Kurpas, M.; Wieczorek, K.; Osek, J. Ready-to-eat Meat Products As a Source of *Listeria monocytogenes*. *J. Veter. Res.* **2018**, *62*, 49–55. [CrossRef]
10. Cherifi, T.; Arsenault, J.; Pagotto, F.; Quessy, S.; Côté, J.-C.; Neira, K.; Fournaise, S.; Bekal, S.; Fravalo, P. Distribution, diversity and persistence of *Listeria monocytogenes* in swine slaughterhouses and their association with food and human listeriosis strains. *PLoS ONE* **2020**, *15*, e0236807. [CrossRef]

11. World Health Organization (WHO). Listeriosis. Available online: https://www.who.int/news-room/fact-sheets/detail/listeriosis (accessed on 21 April 2022).
12. Smith, A.M.; Tau, N.P.; Smouse, S.L.; Allam, M.; Ismail, A.; Ramalwa, N.R.; Disenyeng, B.; Ngomane, M.; Thomas, J. Outbreak of *Listeria monocytogenes* in South Africa, 2017–2018: Laboratory Activities and Experiences Associated with Whole-Genome Sequencing Analysis of Isolates. *Foodborne Pathog. Dis.* **2019**, *16*, 524–530. [CrossRef]
13. Abdallah, M.; Benoliel, C.; Drider, D.; Dhulster, P.; Chihib, N.-E. Biofilm formation and persistence on abiotic surfaces in the context of food and medical environments. *Arch. Microbiol.* **2014**, *196*, 453–472. [CrossRef]
14. Fagerlund, A.; Langsrud, S.; Møretrø, T. In-Depth Longitudinal Study of *Listeria monocytogenes* ST9 Isolates from the Meat Processing Industry: Resolving Diversity and Transmission Patterns Using Whole-Genome Sequencing. *Appl. Environ. Microbiol.* **2020**, *86*, e00579-20. [CrossRef]
15. Commission Regulation (EC) No 2073/2005. *Commission Regulation (EC) No 2073/2005 of 15 November 2005 on Microbiological Criteria for Foodstuffs (Text with EEA Relevance)*; European Commission: Brussels, Belgium, 2005.
16. Balaban, N.Q.; Helaine, S.; Lewis, K.; Ackermann, M.; Aldridge, B.; Andersson, D.I.; Brynildsen, M.P.; Bumann, D.; Camilli, A.; Collins, J.J.; et al. Definitions and guidelines for research on antibiotic persistence. *Nat. Rev. Genet.* **2019**, *17*, 441–448. [CrossRef]
17. Fagerlund, A.; Langsrud, S.; Schirmer, B.C.; Møretrø, T.; Heir, E. Genome Analysis of *Listeria monocytogenes* Sequence Type 8 Strains Persisting in Salmon and Poultry Processing Environments and Comparison with Related Strains. *PLoS ONE* **2016**, *11*, e0151117. [CrossRef]
18. Leong, D.; Alvarez-Ordonez, A.; Jordan, K. Monitoring occurrence and persistence of *Listeria monocytogenes* in foods and food processing environments in the Republic of Ireland. *Front. Microbiol.* **2014**, *5*, 436. [CrossRef]
19. Costa, A.; Bertolotti, L.; Brito, L.; Civera, T. Biofilm Formation and Disinfectant Susceptibility of Persistent and Nonpersistent *Listeria monocytogenes* Isolates from Gorgonzola Cheese Processing Plants. *Foodborne Pathog. Dis.* **2016**, *13*, 602–609. [CrossRef]
20. Fox, E.M.; Leonard, N.; Jordan, K. Physiological and Transcriptional Characterization of Persistent and Nonpersistent *Listeria monocytogenes* Isolates. *Appl. Environ. Microbiol.* **2011**, *77*, 6559–6569. [CrossRef]
21. European Centre for Disease Prevention and Control. Multi-country outbreak of *Listeria monocytogenes* clonal complex 8 infections linked to consumption of cold-smoked fish products. *EFSA Support. Publ.* **2019**, *16*, 1665E. [CrossRef]
22. Lundén, J.; Tolvanen, R.; Korkeala, H. Acid and heat tolerance of persistent and nonpersistent *Listeria monocytogenes* food plant strains. *Lett. Appl. Microbiol.* **2008**, *46*, 276–280. [CrossRef]
23. Magalhães, R.; Ferreira, V.; Brandão, T.; Palencia, R.C.; Almeida, G.; Teixeira, P. Persistent and non-persistent strains of *Listeria monocytogenes*: A focus on growth kinetics under different temperature, salt, and pH conditions and their sensitivity to sanitizers. *Food Microbiol.* **2016**, *53*, 103–108. [CrossRef]
24. Martínez-Suárez, J.V.; Ortiz, S.; López-Alonso, V. Potential Impact of the Resistance to Quaternary Ammonium Disinfectants on the Persistence of *Listeria monocytogenes* in Food Processing Environments. *Front. Microbiol.* **2016**, *7*, 638. [CrossRef]
25. Bucur, F.I.; Grigore-Gurgu, L.; Crauwels, P.; Riedel, C.U.; Nicolau, A.I. Resistance of *Listeria monocytogenes* to Stress Conditions Encountered in Food and Food Processing Environments. *Front. Microbiol.* **2018**, *9*, 2700. [CrossRef]
26. Lundén, J.M.; Miettinen, M.K.; Autio, T.J.; Korkeala, H. Persistent *Listeria monocytogenes* Strains Show Enhanced Adherence to Food Contact Surface after Short Contact Times. *J. Food Prot.* **2000**, *63*, 1204–1207. [CrossRef]
27. Nowak, J.; Visnovsky, S.B.; Pitman, A.R.; Cruz, C.D.; Palmer, J.; Fletcher, G.C.; Flint, S. Biofilm Formation by *Listeria monocytogenes* 15G01, a Persistent Isolate from a Seafood-Processing Plant, Is Influenced by Inactivation of Multiple Genes Belonging to Different Functional Groups. *Appl. Environ. Microbiol.* **2021**, *87*, e02349-20. [CrossRef]
28. Lundén, J.M.; Autio, T.J.; Sjöberg, A.-M.; Korkeala, H.J. Persistent and Nonpersistent *Listeria monocytogenes* Contamination in Meat and Poultry Processing Plants. *J. Food Prot.* **2003**, *66*, 2062–2069. [CrossRef]
29. *ISO 11290-1*; Microbiology of the food chain—Horizontal method for the detection and enumeration of Listeria mono-cytogenes and of *Listeria* spp.—Part 1: Detection method. ISO: Geneva, Switzerland.
30. Dalmasso, M.; Jordan, K. PFGE as a Tool to Track *Listeria monocytogenes* in Food Processing Facilities: Case Studies. Pulse Field Gel Electrophoresis. In *Methods in Molecular Biology*; Jordan, K., Dalmasso, M., Eds.; Humana Press: New York, NY, USA, 2015; Volume 1301, Available online: https://doi.org/10.1007/978-1-4939-2599-5_3 (accessed on 20 November 2021).
31. PNL04. Standard Operating Procedure for PulseNet PFGE of *Listeria monocytogenes*. Available online: https://pulsenetinternational.org/assets/PulseNet/uploads/pfge/PNL04_ListeriaPFGEProtocol.pdf (accessed on 25 November 2019).
32. Doumith, M.; Buchrieser, C.; Glaser, P.; Jacquet, C.; Martin, P. Differentiation of the Major *Listeria monocytogenes* Serovars by Multiplex PCR. *J. Clin. Microbiol.* **2004**, *42*, 3819–3822. [CrossRef]
33. Wałecka-Zacharska, E.; Kosek-Paszkowska, K.; Bania, J.; Karpíšková, R.; Stefaniak, T. Salt stress-induced invasiveness of major *Listeria monocytogenes* serotypes. *Lett. Appl. Microbiol.* **2013**, *56*, 216–221. [CrossRef]
34. Kuch, A.; Goc, A.; Belkiewicz, K.; Filipello, V.; Ronkiewicz, P.; Gołębiewska, A.; Wróbel, I.; Kiedrowska, M.; Waśko, I.; Hryniewicz, W.; et al. Molecular diversity and antimicrobial susceptibility of *Listeria monocytogenes* isolates from invasive infections in Poland (1997–2013). *Sci. Rep.* **2018**, *8*, 14562. [CrossRef]
35. European Committee on Antimicrobial Susceptibility Testing (EUCAST). Breakpoints Tables for Interpretation of MICs and Zones Diameters. Version 12.0. 2022. Available online: http://www.eucast.org (accessed on 15 April 2022).

36. Skowron, K.; Wałecka-Zacharska, E.; Grudlewska, K.; Białucha, A.; Wiktorczyk, N.; Bartkowska, A.; Kowalska, M.; Kruszewski, S.; Gospodarek-Komkowska, E. Biocidal Effectiveness of Selected Disinfectants Solutions Based on Water and Ozonated Water against *Listeria monocytogenes* Strains. *Microorganisms* **2019**, *7*, 127. [CrossRef]
37. Di Ciccio, P.; Meloni, D.; Festino, A.R.; Conter, M.; Zanardi, E.; Ghidini, S.; Vergara, A.; Mazzette, R.; Ianieri, A. Longitudinal study on the sources of *Listeria monocytogenes* contamination in cold-smoked salmon and its processing environment in Italy. *Int. J. Food Microbiol.* **2012**, *158*, 79–84. [CrossRef]
38. Leong, D.; NicAogáin, K.; Luque-Sastre, L.; McManamon, O.; Hunt, K.; Alvarez-Ordóñez, A.; Scollard, J.; Schmalenberger, A.; Fanning, S.; O'Byrne, C.; et al. A 3-year multi-food study of the presence and persistence of *Listeria monocytogenes* in 54 small food businesses in Ireland. *Int. J. Food Microbiol.* **2017**, *249*, 18–26. [CrossRef] [PubMed]
39. Miettinen, H.; Wirtanen, G. Ecology of Listeria spp. in a fish farm and molecular typing of *Listeria monocytogenes* from fish farming and processing companies. *Int. J. Food Microbiol.* **2006**, *112*, 138–146. [CrossRef] [PubMed]
40. Ramires, T.; Kleinubing, N.R.; Iglesias, M.A.; Vitola, H.R.S.; Núncio, A.S.P.; Kroning, I.S.; Moreira, G.M.S.G.; Fiorentini, M.; da Silva, W.P. Genetic diversity, biofilm and virulence characteristics of *Listeria monocytogenes* in salmon sushi. *Food Res. Int.* **2020**, *140*, 109871. [CrossRef] [PubMed]
41. Aalto-Araneda, M.; Lundén, J.; Markkula, A.; Hakola, S.; Korkeala, H. Processing plant and machinery sanitation and hygiene practices associate with *Listeria monocytogenes* occurrence in ready-to-eat fish products. *Food Microbiol.* **2019**, *82*, 455–464. [CrossRef]
42. Mędrala, D.; Dąbrowski, W.; Czekajło-Kołodziej, U.; Daczkowska-Kozon, E.; Koronkiewicz, A.; Augustynowicz, E.; Manzano, M. Persistence of *Listeria monocytogenes* strains isolated from products in a Polish fish-processing plant over a 1-year period. *Food Microbiol.* **2003**, *20*, 715–724. [CrossRef]
43. Cruz, C.D.; Fletcher, G. Prevalence and biofilm-forming ability of *Listeria monocytogenes* in New Zealand mussel (Perna canaliculus) processing plants. *Food Microbiol.* **2011**, *28*, 1387–1393. [CrossRef]
44. Skowron, K.; Wałecka-Zacharska, E.; Grudlewska, K.; Gajewski, P.; Wiktorczyk, N.; Wietlicka-Piszcz, M.; Dudek, A.; Skowron, K.J.; Gospodarek-Komkowska, E. Disinfectant Susceptibility of Biofilm Formed by *Listeria monocytogenes* under Selected Environmental Conditions. *Microorganisms* **2019**, *7*, 280. [CrossRef]
45. Jensen, A.; Larsen, M.H.; Ingmer, H.; Vogel, B.F.; Gram, L. Sodium Chloride Enhances Adherence and Aggregation and Strain Variation Influences Invasiveness of *Listeria monocytogenes* Strains. *J. Food Prot.* **2007**, *70*, 592–599. [CrossRef]
46. Taylor, A.; Stasiewicz, M.J. Persistent and sporadic *Listeria monocytogenes* strains do not differ when growing at 37 °C, in planktonic state, under different food associated stresses or energy sources. *BMC Microbiol.* **2019**, *19*, 257. [CrossRef]
47. Nowak, J.; Cruz, C.D.; Tempelaars, M.; Abee, T.; van Vliet, A.; Fletcher, G.C.; Hedderley, D.; Palmer, J.; Flint, S. Persistent *Listeria monocytogenes* strains isolated from mussel production facilities form more biofilm but are not linked to specific genetic markers. *Int. J. Food Microbiol.* **2017**, *256*, 45–53. [CrossRef]
48. Assisi, C.; Forauer, E.; Oliver, H.F.; Etter, A.J. Genomic and Transcriptomic Analysis of Biofilm Formation in Persistent and Transient *Listeria monocytogenes* Isolates from the Retail Deli Environment Does Not Yield Insight into Persistence Mechanisms. *Foodborne Pathog. Dis.* **2020**, *18*, 179–188. [CrossRef]
49. Wang, J.; Ray, A.J.; Hammons, S.R.; Oliver, H.F. Persistent and Transient *Listeria monocytogenes* Strains from Retail Deli Environments Vary in Their Ability to Adhere and Form Biofilms and Rarely HaveinlAPremature Stop Codons. *Foodborne Pathog. Dis.* **2015**, *12*, 151–158. [CrossRef]
50. Gambarin, P.; Magnabosco, C.; Losio, M.N.; Pavoni, E.; Gattuso, A.; Arcangeli, G.; Favretti, M. *Listeria monocytogenes* in Ready-to-Eat Seafood and Potential Hazards for the Consumers. *Int. J. Microbiol.* **2012**, *2012*, 497635. [CrossRef]
51. Skowron, K.; Kwiecińska-Piróg, J.; Grudlewska, K.; Świeca, A.; Paluszak, Z.; Bauza-Kaszewska, J.; Walecka-Zacharska, E.; Gospodarek-Komkowska, E. The occurrence, transmission, virulence and antibiotic resistance of *Listeria monocytogenes* in fish processing plant. *Int. J. Food Microbiol.* **2018**, *282*, 71–83. [CrossRef]
52. Pizarro-Cerdá, J.; Kühbacher, A.; Cossart, P. Entry of *Listeria monocytogenes* in Mammalian Epithelial Cells: An Updated View. *Cold Spring Harb. Perspect. Med.* **2012**, *2*, a010009. [CrossRef]
53. Povolyaeva, O.; Chalenko, Y.; Kalinin, E.; Kolbasova, O.; Pivova, E.; Kolbasov, D.; Yurkov, S.; Ermolaeva, S. *Listeria monocytogenes* Infection of Bat *Pipistrellus nathusii* Epithelial cells Depends on the Invasion Factors InlA and InlB. *Pathogens* **2020**, *9*, 867. [CrossRef]
54. Moroni, O.; Kheadr, E.; Boutin, Y.; Lacroix, C.; Fliss, I. Inactivation of Adhesion and Invasion of Food-Borne *Listeria monocytogenes* by Bacteriocin-Producing *Bifidobacterium* Strains of Human Origin. *Appl. Environ. Microbiol.* **2006**, *72*, 6894–6901. [CrossRef]
55. Porsby, C.H.; Vogel, B.F.; Mohr, M.; Gram, L. Influence of processing steps in cold-smoked salmon production on survival and growth of persistent and presumed non-persistent *Listeria monocytogenes*. *Int. J. Food Microbiol.* **2008**, *122*, 287–295. [CrossRef]

foods

MDPI

Article

Detection of Resistant and Enterotoxigenic Strains of *Staphylococcus warneri* Isolated from Food of Animal Origin

Ivana Regecová [1], Jana Výrostková [1,*], František Zigo [2], Gabika Gregová [3], Monika Pipová [1], Pavlina Jevinová [1] and Jana Becová [1]

[1] Department of Food Hygiene Technology and Safety, University of Veterinary Medicine and Pharmacy in Košice, Komenského 73, 041 81 Košice, Slovakia; ivana.regecova@uvlf.sk (I.R.); monika.pipova@uvlf.sk (M.P.); pavlina.jevinova@uvlf.sk (P.J.); Jana.Becova@uvlf.student.sk (J.B.)

[2] Department of Nutrition and Animal Husbandry, University of Veterinary Medicine and Pharmacy in Košice, Komenského 73, 041 81 Košice, Slovakia; frantisek.zigo@uvlf.sk

[3] Department of Public Veterinary Medicine and Animal Welfare, University of Veterinary Medicine and Pharmacy in Košice, Komenského 73, 041 81 Košice, Slovakia; gabriela.gregova@uvlf.sk

* Correspondence: jana.vyrostkova@uvlf.sk; Tel.: +421-907-185-658

Abstract: The topic of this work is the detection of antimicrobial resistance to *Staphylococcus warneri* strains and the genes encoding staphylococcal enterotoxins. It is considered a potential pathogen that can cause various—mostly inflammatory—diseases in immunosuppressed patients. The experimental part of the paper deals with the isolation of individual isolates from meat samples of *Oryctolagus cuniculus*, *Oncorhynchus mykiss*, *Scomber scombrus*, chicken thigh, beef thigh muscle, pork thigh muscle, and bryndza cheese. In total, 45 isolates were obtained and subjected to phenotypic (plasma coagulase activity, nuclease, pigment, hemolysis, lecithinase, and lipase production) and genotypic analyses to confirm the presence of the *S. warneri* species. The presence of genes encoding staphylococcal enterotoxins A (three isolates) and D (six isolates) was determined by PCR. Using the Miditech system, the minimum inhibitory concentration for various antibiotics or antibiotics combinations was determined, namely for ampicillin; ampicillin + sulbactam; oxacillin; cefoxitin; piperacillin + tazobactam; erythromycin; clindamycin; linezolid; rifampicin; gentamicin; teicoplanin; vancomycin; trimethoprim; chloramphenicol; tigecycline; moxifloxacin; ciprofloxacin; tetracycline; trimethoprim + sulfonamide; and nitrofurantoin. Resistance to ciprofloxacin and tetracycline was most common (73%). At the same time, out of a total of 45 isolates, 22% of the isolates were confirmed as multi-resistant. Isolates that showed phenotypic resistance to β-lactam antibiotics were subjected to *mecA* gene detection by PCR.

Keywords: antimicrobial resistance; enterotoxins; food; *mecA*; *Staphylococcus warneri*

Citation: Regecová, I.; Výrostková, J.; Zigo, F.; Gregová, G.; Pipová, M.; Jevinová, P.; Becová, J. Detection of Resistant and Enterotoxigenic Strains of *Staphylococcus warneri* Isolated from Food of Animal Origin. *Foods* **2022**, *11*, 1496. https://doi.org/10.3390/foods11101496

Academic Editors: Fernando Pérez-Rodríguez and Arícia Possas

Received: 22 March 2022
Accepted: 17 May 2022
Published: 20 May 2022

Publisher's Note: MDPI stays neutral with regard to jurisdictional claims in published maps and institutional affiliations.

1. Introduction

Staphylococcus warneri is a coagulase-negative staphylococcal (CNS) commonly present in the microbiota of human epithelium and mucous membranes. In the last two decades, *S. warneri*, like other species of the CNS, has been reported as a new emerging pathogen that can cause serious infections, usually in connection with the presence of implanted materials but sometimes even in the absence of a foreign body and in patients considered immunocompetent. At present, there is still a lack of scientific data on the pathogenesis and epidemiology of this species [1]. According to Bhardwaj et al. [2], *Staphylococcus warneri* is a common saprophyte on human skin, present in approximately 50% of the healthy adult population.

S. warneri creates a biofilm on the surface of various materials (e.g., air, walls, floors, and medical equipment). In addition, it possesses various virulence factors, such as adhesion to polymeric surfaces and metabolic changes in various situations. The ability to form biofilms in foreign bodies and medical devices allows this bacterium to be resistant to antimicrobials used to treat infections [3].

The CNS, including *S. warneri*, were previously considered food-related saprophytes [4,5]. *S. warneri*, as well as other CNSs, may possess or acquire mobile pathogenic factors (e.g., exotoxins, enterotoxins, adherence factors, leukocidins, and antibiotic resistance) in the form of transposons, pathogenicity islands (PIs), plasmids, and phages similar to those of *Staphylococcus aureus*, which may lead to a potential health risk for the consumer [5,6]. The most common contamination of food with this type of staphylococcus is secondary and comes from processing at food companies as well as from the poor handling of food during the production process [5–7]. Most recently, *S. warneri* was detected in chicken meat, ready-for-consumption fish dishes, and raw dairy products [8–10]. This species is also reported as a potential source of food enterotoxicosis [11,12].

The antibiotic resistance of microorganisms is a major health, social, and economic problem for all of us today. As consumers of animal products, humans should also be aware of the risks associated with consuming products that contain bacteria resistant to some antibiotics (chloramphenicol, streptomycin, enrofloxacin). When consuming such products, there is the possibility of transferring resistance genes to the microbial population in the human intestinal tract and possible future complications in the treatment of bacterial diseases, not to mention the spread of resistance genes further in the human population and environment [13].

Antimicrobial resistance in bacteria is not a new phenomenon. Estimates for the emergence of biosynthetic pathways have been published by Wright, showing that such resistance began hundreds of millions of years ago [14]. It has also been documented that bacteria share a pool of antimicrobial-resistant genes (AMRG) conferring antibiotic resistance before human discovery and the widespread use of antibiotics. In the permafrost of 30,000 years ago, genes encoding β-lactam resistance and resistance to other antibiotics were documented by D'Costa [15]. There are self-transferring plasmids carrying antibiotic resistance genes (ARGs) that can infect many phylogenetically distinct bacteria, forming a group of genes that can be shared [16]. Methicillin-resistant coagulase-negative staphylococci (MR-CoNS) are a major cause of infectious diseases [3].

The CNS, including *S. warneri*, are usually resistant to multiple drugs, have limited treatment options, cause incurable diseases, and become reservoirs of resistant strains in hospitals. The cause of antimicrobial resistance is multi-factorial, ranging from a lack of infection prevention to the irrational use of antimicrobials by health-care professionals and patients. Recent studies suggest that CNSs were highly resistant to penicillin, methicillin, vancomycin, oxacillin, and erythromycin. CNSs were less resistant to cephalosporins, aminoglycosides, and quinolones [17].

A significant proportion of *S. warneri* isolates belonging to the CNS group have a high degree of genetic diversity, indicating a high predisposition to antimicrobial resistance [18]. Biofilm isolates can carry multiple genes encoding resistance to beta-lactams, aminoglycosides, and macrolide–lincosamide antibiotics. Antibiotic resistance genes were found to be more common in biofilm-positive than biofilm-negative isolates of *S. warneri*. Biofilm structure, due to cell aggregation, may be ideal for horizontal gene transfer and thus facilitate the spread of antibiotic resistance [18].

A study by Chaves et al. [19] shows that CNS are capable of producing enterotoxins, confirming data previously published by other authors. Although current legislation only recommends counting coagulase-positive staphylococci, the study points to the possibility of reformulating existing microbiological standards. In addition, a new approach that correlates the levels and intervals of enterotoxin production could be used for a new and safer food policy.

The aim of our study was to point out the presence of resistant and enterotoxigenic isolates of *S. warneri* from various food commodities of animal origin. In particular, the study focused on the presence of genes encoding for the production of staphylococcal enterotoxins A–E. At the same time, the aim was to determine the MIC for different groups of antibiotics used in human and veterinary medicine and subsequently to detect methicillin-resistant staphylococci carrying the *mecA* gene.

2. Materials and Methods

Food samples, 4 from each species, investigated in this study derived from the following areas: Atlantic mackerel muscle (*Scomber scombrus*) originating in FAO 27 (Ireland), rainbow trout muscle (*Oncorhynchus mykiss*) originating in the South Bohemian Region (Czech Republic), wild rabbit muscle (*Oryctolagus cuniculus*) originating from a joint hunt on the grounds of the University facility for breeding and diseases of game, fish, and bees in Rozhanovce (Košice, Slovakia), samples of beef thigh muscle, pork thigh muscle (*Sus scrofa domesticus*) and thigh muscles from chickens (*Gallus gallus domesticus*) were provided from slaughterhouses located in eastern Slovakia. The samples of full-fat bryndza cheese were provided from two dairy farms located in central Slovakia.

Staphylococcus identification

2.1. Detection of the Phenotypic Properties of the Isolates of Isolation of Strains

Free coagulase detection

Staphylococcal colonies were inoculated from the blood–agar surface with sterile bacterial ashes into test tubes with 10 mL of broth containing brain–heart infusion (BHI broth; OXOID, Hampsire, UK). After incubation for 18–24 h at 37 °C, 0.1 mL of multiplied broth culture of the tested strains was added to test tubes with 1 mL of sterile rabbit plasma (StaphyloPK test, IMUNA, Šarišské Michaľany, Slovakia). The inoculated plasma was incubated at 37 °C. The results were read after 1, 2, 3, 6, and 24 h to capture the delayed positive reaction that may occur within 24 h.

Deoxyribonuclease production

DNase agar (OXOID, Hampshire, UK) containing DNA was used to detect nuclease activity. Isolates with positive nuclease activity hydrolyzed the DNA contained in the agar after 24 h incubation at 37 °C. Positive nuclease activity manifested itself on the surface of DNase agar after pouring the surface of the medium with 1 N hydrochloric acid solution as a transillumination zone. If no DNA hydrolysis occurred near the inoculated strains, the soil became cloudy due to HCl.

Other phenotypic properties of isolates

Pigment production as well as hemolysis of individual isolates were evaluated after 24 h of incubation on the agar surface with a 2% addition of lamb blood. Baird–Parker selective diagnostic medium was used to detect the formation of lecithinase and lipase on the surface of which the identified staphylococcal strains were inoculated with sterile bacterial ashes. The inoculated petri dishes were incubated at 37 °C/24 h. After incubation, in the positive case, a precipitation zone around the inoculated colonies in the agar medium in the presence of lecithinase can be detected. This is due to the hydrolysis of the lecithin phospholipid to 1.2-diglyceride and phosphorylcholine [20].

In the presence of the enzyme lipase, a brightening zone was created around the colonies. This is due to the cleavage of triacylglycerides to triacylglycerol and fatty acids, which are present in Baird–Parker agar medium.

2.2. Isolation DNA from Staphylococcal Isolates

Total genomic DNA was isolated according to Sharma et al. [21] from staphylococcal isolates grown in BHI broth. The pellet obtained after centrifugation of 1.5 mL of expanded staphylococcal culture (12,500× *g*/5 min) was resuspended in 100 µL of 0.5% Triton X-100 (Koch-Light Lab., Suffolk, VA, USA) and centrifuged again (12,500× *g*/5 min). Subsequently, the sediment was resuspended in 200 µL of 0.5% Triton X-100. The samples thus prepared were incubated at 95 °C/10 min and centrifuged again (12,500× *g*/5 min). The obtained supernatant was used as a source of DNA in PCR reactions.

2.3. Genotypic Analysis of Staphylococcal Isolates

The identification of individual staphylococcal strains was performed using the PCR method. The identification of individual staphylococcal strains was performed by using specific primers the SwarF (TGTAGCTAACTTAGATAGTGTTCCTTCT) and SwarR (CCGCCACCGTTATTTCTT) synthesized by Amplia s.r.o. (Bratislava, Slovakia) according to Iwas et al. [22]. Commercially produced Hot FIREPol® MasterMix (Amplia s.r.o., Bratislava, Slovakia) was used in PCR reactions. The total reaction mixture volume of 20 μL contained 5 ng/μL of template DNA and each primer at a concentration of 10 pmol. The finished mixture was heated at 95 °C for 12 min at initial denaturation, then heated over 30 amplification cycles: 95 °C/30 s, annealing 60 °C/30 s and 72 °C/2 min, using a thermal cycler (TECHNE TC-512, London, UK) with an extension of 10 min/72 °C. The 65 bp PCR product, which was amplified, was collected for 5 μL for analysis on a 1.5% agarose gel in TBE buffer (Tris-borate-EDTA). GelRed TM (Biotium, Fremont, CA, USA) was added to the agarose gel to visualize the PCR fragments using a Mini Bis Pro® reader (DNR Bio-Imaging Systems Ltd., Neve Yamin, Israel). The obtained PCR products were sequenced at the European Custom Sequencing Center GATCBiotech AG (Cologne, Germany) and compared with the nucleotide sequences of the reference strains in GenBank NCBI. The reference strain *S. warneri* CCM 2730T (Czech Collection of Microorganisms, Brno, Czech Republic) was used as a positive control for PCR strain identification.

2.4. Detection of Genes Encoding Staphylococcal Enterotoxins

To detect genes encoding staphylococcal enterotoxin (SE) production (*sea, seb, sec, sed,* and *see*), a multiplex PCR assay, according to Sharma et al. [21] was performed by using universal forward primer and five specific reverse primers for each staphylococcal enterotoxin to detect each of the five enterotoxins.

The reaction mixture used and the reaction conditions were the same as described above (except for the annealing temperature, which was 45 °C). The sizes of the PCR products are shown in Table 1.

Table 1. Primers used in multiplex PCR to detect staphylococcal enterotoxins [21].

Gene	Primer	Nucleotide Sequence 5′-3′	Product Size (bp)
universal	Sauni-F	TGTATGTATGGAGGTGTAAC	
sea	SAA -R	ATTAACCGAAGGTTCTGT	270
seb	SAB -R	ATAGTGACGAGTTAGGTA	165
sec	SAC -R	AAGTACATTTTGTAAGTTCC	102
sed	SAD -R	TTCGGGAAAATCACCCTTAA	306
see	SAE -R	GCCAAAGCTGTCTGAG	213

bp—base pairs.

The following reference strains of *Staphylococcus aureus* (CCM, Brno, Czech Republic) were used as a positive control for this multiplex PCR: CCM 5756 (*sea* gene), CCM 5757 (*seb* gene), CCM 5984 (*sec* gene), CCM 5973 (*sed* gene), and CCM 5972 (*see* gene).

2.5. Detection of Antimicrobial Resistance

Eighteen isolates of *S. warneri* were analyzed for their antibiotic susceptibility. Minimal inhibitory concentrations (MIC) were determined according to CLSI document [23] and EUCAST document [24], using a Miditech system (Bratislava, Slovakia) with interpretive reading of MIC. The antibiotics used in this study were as follows: ampicillin + sulbactam (SAM), piperacillin + tazobactam (TZP), oxacillin (OXA), erythromycin (ERY), clindamycin (CLI), teicoplanin (TEC), vancomycin (VAN), rifampicin (RIF), gentamicin (GEN), linezolid (LNZ), ciprofloxacin (CIP), moxifloxacin (MFX), tetracycline (TET), tigecycline (TGC), chloramphenenicol (CHL), trimethoprim (TMP), trimethoprim + sulfonamide (COT), and nitrofurantoin (NIT).

In isolates showing phenotypic resistance to oxacillin, the presence of the *mecA* gene was detected by PCR according to Poulsen et al. [25]. The primers used in this PCR reaction were MecA1 (GGGATCATAGCGTCATTATTC) and MecA2 (AACGATTGTGA-CACGATAGCC) (Amplia sro, Bratislava, Slovakia). The PCR reaction conditions were the same as for the species identification of staphylococci mentioned above. The resulting size of the PCR product was 527 bp. The reference isolate *S. aureus* CCM 4750 (Czech Collection of Microorganisms, Brno, Czech Republic) was used as a reference strain for the PCR and for detection of MIC in this study.

3. Results

Typical and atypical colonies were harvested from the surface of the Baird–Parker agar medium after incubation for the initial culture screening of individual samples. A total of 560 staphylococcal isolates were obtained and subjected to further identification. All isolates obtained were subjected to detection of coagulase activity. In the detection of this phenotypic trait, for 200 isolates the result of coagulase assay was negative, showing no coagulase activity. Therefore, these isolates were classified as CNS. Subsequently, in this group of isolates, the detection of other phenotypic characteristics was carried out for a better characterization of the *S. warneri* strains as well as for detection of virulent strains: nuclease activity, pigment, hemolysis, lecithinase, and lipase production (Table 2). In isolates included in the CNS, genotypic identification was performed using the PCR method, where 45 isolates were identified (22% of the CNS group) as *S. warneri*. It was retrospectively evaluated that 8 isolates derived from brown trout, 4 isolates from Atlantic mackerel, 7 isolates from pork thigh muscle, 2 isolates from beef thigh muscle, 14 isolates of *S. warneri* came from wild rabbit, 7 isolates from chicken thigh muscle and, 3 isolates came from bryndza cheese.

Table 2. Virulence phenotypic properties of *S. warneri* isolates.

	Number of Strains	Pigment				Hemolysis	Lecithinase	Lipase	Nuclease
		White (without Pigment)	Gray	Gray-White	Yellow-White				
Oncorhynchus mykiss	8	0	2	6	0	0	3	7	0
Scomber scombrus	4	0	2	1	1	3 (α)	2	3	0
Pork thigh muscle	7	0	0	6	1	0	3	1	0
Beef thigh muscle	2	1	1	0	0	0	1	5	0
Oryctolagus cuniculus	14	0	3	7	4	5 (α)	4	12	0
thigh muscle of chickens	7	1	0	4	2	1 (α)	1	0	0
Bryndza cheese	3	1	1	1	0	1 (α)	3	3	0
Total	45	9	6	25	7	10	17	31	0

(α)—α-hemolysis.

As shown in Table 2, gray pigment was observed most frequently on blood agar. The production of a yellowish-white coloration of *S. warneri* colonies also appeared. The production of α-hemolysis was also confirmed in these isolates. β-hemolysis was not confirmed in any of the isolates. Similarly, nuclease production on DNAse agar was not confirmed in any isolate. However, lipase production was confirmed in a large number of *S. warneri* isolates. Lipase production was manifested by a transillumination zone in Baird–Parker agar medium around the overgrown colony. In contrast, a smaller number of isolates produced the enzyme lecithinase. Its production appeared as a precipitating ring around the overgrown colony on Baird–Parker agar medium. The detection of lecithinase and lipase production confirmed their current production in three isolates of *Oncorhynchus mykiss*, two isolates from *Oryctolagus cuniculus*, and one isolate each from *Scomber scombrus*, pork thigh muscle, and Bryndza cheese.

Subsequently, after the detection of individual phenotypic manifestations, genes encoding the staphylococcal enterotoxins SEA, SEB, SEC, SED, and SEE were detected. As shown in Figure 1, *S. warneri* isolates confirmed the presence of genes encoding the production of SEA and SED enterotoxins. The presence of the *sea* gene was confirmed in three isolates and the *sed* gene in six isolates. Of these, the co-presence of the *sea* and *sed* genes was detected in two isolates. It was an isolate isolated from a sample of *Oryctolagus cuniculus* and a sample of Bryndza cheese. At the same time, these two isolates phenotypically showed α-hemolysis and lecithinase and lipase production.

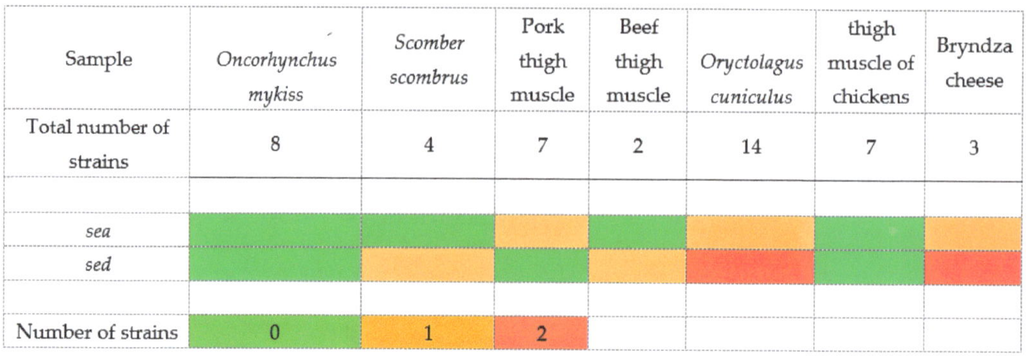

Figure 1. Heatmap demonstration of the presence and absence of genes encoding staphylococcal enterotoxins.

As seen in Figure 2A, we were unable to detect the presence of genes encoding enterotoxin production in *S. warneri* isolates by multiplex PCR, only in the reference strains. For this reason, we approached the detection of individual genes separately using primers from multiplex PCR. As seen in Figure 2B,C, the presence of the *sea* and *sed* genes was subsequently confirmed.

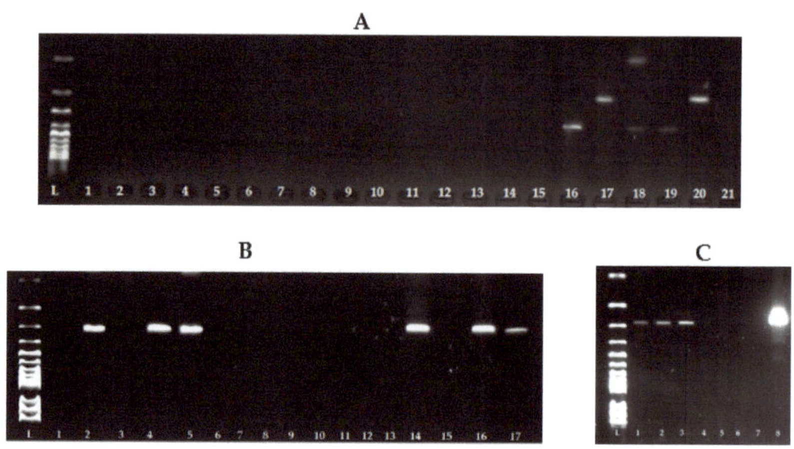

Figure 2. Detection of genes encoding the staphylococcal enterotoxins of *Staphyoccocus warneri* using the PCR method. (**A**) L: 100 bp ladder; lines 1–15: isolates *S. warneri* without genes encoding the staphylococcal enterotoxins; line 16: reference strain for *sea* gene CCM 5756 *S. aureus* (270 bp); line 17: reference strain for *seb* gene CCM 5757 *S. aureus* (165 bp); line 18: reference strain for *sec* gene CCM 5984 *S. aureus* (102 bp); line 19: reference strain for *sed* gene CCM 5973 *S. aureus* (306 bp); line 20: reference strain for *see* gene CCM 5972 *S. aureus* (213 bp); line 21: negative control. (**B**) L: 100 bp ladder; line 1: negative control; line 2: reference strain for *sed* gene CCM 5973 *S. aureus* (306 bp); lines 4, 5, 14, 16, 17: isolates *S. warneri* with *sed* gene (306 bp). (**C**) L: 100 bp ladder; lines 1–3: isolates *S. warneri* with *sea* gene (270 bp); line 7: negative control; line 8: reference strain for *sea* gene CCM 5756 *S. aureus* (270 bp).

After detection of phenotypic characteristics and detection of toxinogenic isolates, MIC detection of antibiotics was performed using the Miditech system. In general, as shown in Table 3, the highest resistance to ciprofloxacin and tetracycline was confirmed. The MIC50 of ciprofloxacin was 2.0 mg/L and the MIC90 was 4.0 mg/L while the MIC of GX was 1.2 mg/L. For tetracycline, the MIC50 was 16.0 mg/L and the MIC90 was 32.0 mg/L. At the same time, the MIC GX of this antibiotic was 8.3 mg/L (Figure 3). At the same time, intermediate sensitivity was confirmed in all 45 isolates against trimethoprim + sulfonamide, and nitrofurantoin. The strains on vancomycin, trimethoprim, trimethoprim + sulfonamide, and nitrofurantoin appeared to be the most sensitive overall (Table 3).

Table 3. Percentage of antimicrobial resistance.

ATB	S	I	R	GX SI	MIC_{50}	MIC_{90}	Total
SAM	86.67%	0.00%	13.33%	0.9	1.00	64.00	45
TZP	86.67%	0.00%	13.33%	0.8	1.00	128.00	45
OXA	86.67%	0.00%	13.33%	0.2	0.25	8.00	45
ERY	75.56%	0.00%	24.44%	0.3	0.50	16.00	45
CLI	89.00%	0.00%	20.00%	0.1	0.25	8.00	45
TEC	97.78%	0.00%	2.22%	1.3	1.00	4.00	45
VAN	100.00%	0.00%	0.00%	0.9	1.00	2.00	45
RIF	42.22%	26.67%	31.11%	0.1	0.50	2.00	45
GEN	64.44%	0.00%	35.56%	0.4	0.50	256.00	45
LNZ	97.78%	0.00%	2.22%	2.0	2.00	4.00	45
CIP	0.00%	26.67%	73.33%	0.2	2.00	4.00	45
MFX	93.33%	0.00%	6.67%	0.1	0.13	0.25	45
TET	26.67%	0.00%	73.33%	0.4	16.00	32.00	45
TGC	100.00%	0.00%	0.00%	0.1	0.06	0.13	45
CHL	44.44%	0.00%	55.56%	5.3	16.00	64.00	45
TMP	0.00%	100.00%	0.00%	1.3	2.00	2.00	45
COT	100.00%	0.00%	0.00%	0.2	0.25	0.50	45
NIT	0.00%	100.00%	0.00%	14.1	16.00	16.00	45

S—sensitive strain, I—intermediate sensitive strain, R—resistant strain; GX SI MIC—Geometric mean MIC of strains that can be treated with a given ATB (S + I). Mean MIC of treatable strains. MIC_{50}—value which expresses the minimum inhibitory concentration of a given antibiotic at which at least 50% of the population is inhibited. MIC_{90}—value which expresses the minimum inhibitory concentration of a given antibiotic at which at least 90% of the population is inhibited.

Specifically, isolates isolated from the *Oncorhynchus mykiss* muscle (eight isolates) were the most frequently confirmed to be resistant to tetracycline (87.50%). Resistance to clindamycin and gentamicin (37.50%) was also among the common resistance detected in these isolates. Isolates isolated from *Scomber scombrus* muscle samples (four isolates) also showed the most common resistance to tetracycline (75.00%) and to gentamicin, ciprofloxacin, and chloramphenicol (50.00%). *S. warneri* strains isolated from the pork thigh muscle (7 isolates) and the *Oryctolagus cuniculus* muscle (14 isolates) showed the most common resistance to tertracycline (100.00%/92.86%). Isolates isolated from chicken thigh muscle (seven isolates) showed the highest resistance to ciprofloxacin (85.71%). In isolates derived from beef thigh muscle (two isolates), 100.00% resistance to clindamycin, rifampicin, and gentamicin was observed. The resistance of all *S. warneri* isolates (three isolates) to rifampicin (100.00%) was confirmed in the last examined bryndza samples.

Based on the antimicrobial resistance using the Miditech system, the various resistance mechanisms, which are shown in Table 4, were confirmed. As can be seen from the table, incomplete fluoroquinolone resistance was generally the most common. Specifically, this resistance mechanism was most commonly detected in isolates isolated from *Oryctolagus cuniculus* samples (13 isolates/92.86%). The presence of methicillin-resistant coagulase-negative staphylococci (MRCNS) in 13.33% of the examined isolates was also phenotypically confirmed.

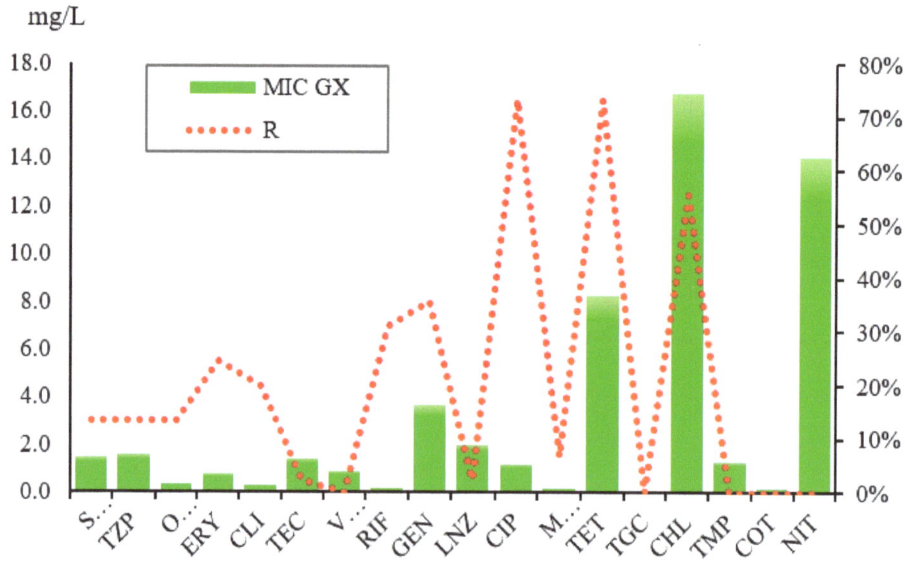

Figure 3. Overview of MIC and MIC GX in resistant *Staphylococcus warneri* isolates. GX MIC—Geometric average MIC—the average MIC value at which the strains are inhibited.

Table 4. Phenotypic detection of various mechanisms of resistance.

Mechanisms of Resistance	Number	%
MRCNS	6	13.33%
Aminogl.PH(2″)-AC(6′)	15	33.33%
Fluoroq.incompl.resistance	32	71.11%
Constitutive MLSB/c	1	2.22%
Inducible MLSB/i	0	0.00%
Multi-resistance	10	22.22%

MRCNS: methicillin-resistant coagulase-negative staphylococci; Amngl.PH (2″)-AC (6′): combined enzymatic resistance to GEN, TOB, and AMI; constitutive MLSB/c: constitutive resistance to macrolides, lincosamides, and streptogramin B; inducible MLSB/i: inducible resistance to macrolides, lincosamides, and streptogramin B; multi-resistance: current resistance in 3 or more unrelated ATB groups; fluoroq.incompl.resistanc: incomplete fluoroquinolone resistance, mutation with incomplete resistance to fluoroquinolones.

Based on the phenotypic expression of these isolates (six isolates), the *mecA* gene, which encodes one of the most common mechanisms of methicillin resistance, was detected. Based on the results of the PCR reaction and subsequent sequencing of the PCR products, the presence of the *mecA* gene was confirmed in four isolates (66.66%). These isolates were isolated from samples of *Scomber scombrus*, *Oryctolagus cuniculus*, thigh muscle of chickens, and Bryndza cheese. The isolates also exhibited virulence phenotypic properties such as α-hemolysis, lipase, and lecithinase production. All isolates carrying the *mecA* gene produced a yellow-white pigment. At the same time, the presence of the *sea* gene (1 isolate), the *sed* gene (1 isolate), and both genes (2 isolates) was confirmed in these isolates.

Based on the detection of resistance of *S. warneri* isolates, multi-drug resistance was also confirmed in 10 isolates (Table 4). Specifically, two isolates isolated from *Oncorhynchus mykiss* samples were confirmed to be resistant to four antimicrobial groups at the same time, namely RIF-GEN-CIP-TET. In isolates derived from *Oryctolagus cuniculus* samples, multi-drug resistance was confirmed in two isolates, namely GEN-CIP-TET. The other two isolates confirmed simultaneous resistance to CIP-TET-CHL-RIF-GEN, of which one isolate also carried the *mecA* gene. In isolates derived from pork thigh muscle, the co-resistance to GEN-CIP-TET was confirmed in three isolates. Multi-resistance was also confirmed in one isolate isolated from Bryndza cheese samples against OXA-ERY-CIP-CHL.

4. Discussion

The isolation of food samples produced 45 isolates that were subsequently tested for phenotypic and genotypic characteristics to confirm that they were *Staphylococcus warneri*, being catalase-positive, oxidase-negative, and coagulase-negative. This species is also a commensal bacterium that occurs as part of the skin microbiota in humans and animals. Although *S. warneri* accounts for less than 1% of the total staphylococcal population, it is responsible for a variety of human diseases, such as immune suppression, skin, eye, and urinary tract infections, nosocomial infections, and weakening the immune systems of patients and neonates [1]. However, in a study by Hoveida et al. [26], *S. warneri* accounted for up to 7.5%, similar to our study, where the incidence rate was up to 22% of the total identified CNS population. In our study, such a higher incidence of *S. warneri* may be due to a different origin of the samples. The samples came from bryndza, game, fish, and thigh muscles of livestock. The microbiota of these commodities can be affected by hygiene levels, where the role of food processors is crucial in determining the hygienic status of the final product, with mishandling leading to an increased likelihood of microbial contamination by humans, including multi-drug-resistant and/or enterotoxigenic staphylococci [27].

In our study, phenotypic and genotypic identification was used to identify this species. In terms of phenotypic characteristics, hemolytic activity was tested, which was demonstrated in 10 isolates in the form of α-hemolysis. This confirms the production of cytolytic alpha-toxin, which is also known as alpha-hemolysin. By binding to the cell surface, it causes necrotic cell death [28]. Alpha-hemolysin production was also confirmed in a study by Noumi et al. [29], in two tested *S. warneri* isolates.

Pigment formation was demonstrated in 36 isolates, where the most frequently produced was a gray-white pigment, to a lesser extent a yellow-white pigment. According to Becker et al. [4], most CNS species are usually unpigmented. Pigmented colonies are characteristic only for the species *S. chromogenes, S. devriesei, S. lugdunensis, S. sciuri, S. vitulinus, S. warneri,* and *S. xylosus*. They form a pigment of gray-white, gray-yellow, yellow, or yellow-orange color.

Other monitored properties were the production of hydrolytic enzymes (lecithinase and lipase). Hydrolytic enzymes are among the CNS virulence factors that contribute to soft-tissue degradation and aid in biofilm formation [30]. Lecithinase was observed in our study in 17 *S. warneri* isolates. It is a type of phospholipase that acts on lecithin [30]. Lipase production was confirmed in 35 *S. warneri* isolates. Staphylococcal lipases are usually released outside the cell, where they perform several functions, some of which are of hygienic importance. Staphylococcal extracellular lipases are overexpressed in connection with pathogenic events. They are also recognized as contributors to human and animal pathogens by improving bacterial nutrition on the skin, disrupting host cell membranes, disrupting the immune response, and host cell signaling [31]. Many studies have been conducted to examine staphylococcal lipase production, and these bacteria have been found to have high levels of lipase activity in meat and dairy production [32,33], as confirmed by our study. Lipolytic bacteria are classified into different genera based on their gene sequences and biochemical properties, and high lipase levels are also a common feature of staphylococci, including *S. warneri* [34]. The presence of the abovementioned hydrolytic properties was confirmed in *S. warneri* isolates by Noumi et al. [29].

Subsequently, the PCR method, which is currently considered to be the most accurate identification method, was used to accurately identify the *S. warneri* isolates obtained. The *16S rRNA* gene is generally used as a target sequence to identify species in the genus Staphylococcus. However, this *16S rRNA* gene in *Staphylococcus epidermidis* is very similar to the sequence of other *16S rRNAs* in other CNS species. To solve this problem, it is possible to use an alternative target sequence that shows a greater sequence divergence than the *16S rRNA* gene. Recently, a highly conserved, ubiquitous *sodA* gene was used that encodes manganese-dependent superoxide dismutase [35] In our study, 45 *S. warneri* isolates from 200 identified CNSs were confirmed using a specific *sodA* gene sequence.

The presence of genes encoding the production of staphylococcal enterotoxins A to E was subsequently detected in the *S. warneri* isolates.

SE production is one of the most notable virulence factors in staphylococci. Staphylococcal enterotoxins (SE) are mainly divided into five classical serological types: SEA, SEB, SEC, SED, and SEE as well as the other recently discovered SEG, SHE, SEI, SER, SES, SET, and the enterotoxin-like proteins, such as SElK, SElN, SE10, SE1P, SE1Q, and SElU [36,37]. Reliable detection of SE genes has a dual function. First, it helps in the genotyping of coagulase-positive staphylococci (CPS) for epidemiological studies. Second, it provides an assessment of the possible occurrence of SE genes in CNS strains which pose a potential risk of staphylococcal enterotoxicosis to consumers [36]. In our study, the presence of the *sea* (6% of isolates) and *sed* (13% of isolates) genes was confirmed. De Freitas Guimarães et al. [38] also confirmed the presence of *sea* and *sed* genes in *S. warneri* isolates isolated from food of animal origin but in higher percentages than those confirmed in our study. In further studies, they confirmed the presence of other genes that encode enterotoxins, namely *sec* and *she* [29,39].

Banaszkiewicz et al. [39] and Hu et al. [40] also suggest the detection of the continuous transfer of elements containing SE genes from staphylococci which contain the stable enterotoxin genes of the CNS originating from wild animals, which was confirmed by our study.

Staphylococcal isolates of phenotypically and genotypically identified isolates were subsequently determined using the Miditech system, where the highest resistance was recorded against CIP and TET. Similarly, frequent resistance to TET was confirmed in the Hoveida et al. [41] study, which accounted for up to 50% of the resistance.

At the same time, our study confirmed the presence of *S. warneri* MRS in 13.33% of the *S. warneri* isolates. On the basis of the detection of the MRSCNS phenotypic expression, the *mecA* gene was detected in these isolates. Methicillin resistance is associated with the presence of the *mecA* gene, which encodes additional binding to a penicillin protein (PBP2A or PBP2′). This protein has a lower affinity for all beta-lactam antibiotics. The *mecA* gene is found on a mobile genetic element called the staphylococcal cassette chromosome mec (SCCmec) [42]. Another mechanism in staphylococcal resistance to beta-lactams is beta-lactamase production, encoded by the *blaZ* gene [43]. In our study, the presence of the *mecA* gene was confirmed in 66.66% of isolates (four isolates) that phenotypically appeared as MRSCNS (six isolates). Humphries et al. [44] confirmed the presence of the *mecA* gene in *S. warneri* isolates in a similarly high percentage. The *mecA*-positive *S. warneri* isolates accounted for up to 41.66% of the total of 48 isolates tested. Similarly, Hoveida et al. [41] confirmed the prevalence of *mecA* genes in *S. warneri* in 30% of the 40 isolates isolated from different types of food.

In addition to the increasing incidence of MRSCNS isolates, resistance to aminoglycosides has recently increased. Aminoglycoside resistance has increased, especially among methicillin-resistant strains carrying the *mecA* gene [45]. Inactivation of antibiotics by aminoglycoside-modifying enzymes (AME), which are encoded by genetic elements, is a major pathway for resistance [46]. The most important of these enzymes is aac $(6')$/aph $(2'')$, which alters aminoglycosides of medical importance, such as tobramycin and gentamicin [47]. This type of resistance to aminoglycosides was also confirmed in our investigated CNS isolates, specifically in 33.33% of the *S. warneri* isolates. Of these, MRSCNS was identified in three strains; they showed resistance to aminoglycosides using combined enzymatic resistance to tobramycin, gentamicin, and amikacin. Consistent with many other studies [47–49], the enzyme aac $(6')$/aph $(2'')$ is the most common, where 85% of AME-positive isolates were found.

Another resistance detected in the isolates we examined was constitutive resistance to macrolides, lincosamides and streptogramin B (cMLSB). However, studies mostly focus on the most common isolated staphylococci, i.e., *S. aureus* and *S. epidermidis* [50–52]. Recently, other staphylococcal species, such as *S. hominis*, *S. haemolyticus*, *S. warneri* and *S. simulans*, have emerged as etiological factors in serious human infections. The phenotypic expression of resistance to MLSB in these staphylococci may be inducible and manifest in clinical resistance to lincosamides and streptogramin B induced by 14- and 15-membered macrolides or constitutive, determining resistance to all MLSB antibiotics [53,54]. In our study, only constitutive macrolide resistance was phenotypically confirmed (2.22%). Similarly, a low percentage (10%) were confirmed cMLSB in *S. warneri* in a study by Szemraj et al. [55].

Incomplete fluoroquinolone resistance was also confirmed in the *S. warneri* isolates in our study, with up to 71.11%; in other words, this is a mutation with incomplete resistance to fluoroquinolones. However, little is known about the mechanisms involved in the development of fluoroquinolone resistance (through exogenous acquisition, de novo mutation, or selection of a minority-resistant mutant) in CNS [56]. However, previous studies have confirmed that methicillin-resistant strains developed resistance to fluoroquinolones more rapidly in *S. aureus* and the CNS than in methicillin-sensitive strains. This difference is partly explained by nosocomial transmission in some environments and thus the potential for co-selection with several antimicrobials (due to the common multi-drug resistance phenotype of MRS isolates [57]. Resistance to more than one antibiotic was also confirmed by Persson-Waller et al. [58], where multi-drug resistance occurred in 9% of the total 56 CoNS isolates. Nunes et al. [59] also confirmed 14 multi-drug-resistant CNS. Our demonstrated multi-drug resistance of CNS isolates, specifically *S. warneri*, is consistent with previous studies on coagulase-negative staphylococci, which found several resistant and multi-drug resistant staphylococcal isolates [60]. Our results are supported by the study of Senga et al. [61], who reported up to 80% resistance to several types of antibiotics in CNS isolates.

5. Conclusions

The results of the paper point to the ever-increasing incidence of resistant and multi-resistant and enterotoxigenic isolates of *S. warneri* in foodstuffs of animal origin, especially game and wild fish. These results point to the risk of the presence of enterotoxigenic and resistant isolates of *S. warneri* in the food industry, as they can serve as vectors for the transfer of resistance determinants into the genomes of bacteria inhabiting the consumer's digestive tract. Therefore, the rational use of antibiotics, preventive measures in environmental hygiene, and the monitoring of antibiotic resistance are important prevention measures to prevent the spread of antimicrobial resistance. At the same time, the presence of enterotoxigenic isolates indicates a potential risk for staphylococcal enterotoxicosis in humans.

Author Contributions: Conceptualization: I.R.; methodology: I.R. and J.V.; software: G.G.; formal analysis: M.P.; investigation and data curation: P.J. and J.B.; writing—original draft preparation: I.R.; writing—review and editing: J.V.; supervision: J.V. and P.J.; project administration and funding acquisition: J.B. and F.Z. All authors have read and agreed to the published version of the manuscript.

Funding: Additional financial support for the implementation of the study comes from Slovak grant KEGA no. 006UVLF-4-2020.

Institutional Review Board Statement: The study was conducted in accordance with the Declaration of Helsinki, and approved by the Institutional Review Board.

Informed Consent Statement: Not applicable.

Data Availability Statement: The data presented in this study are available upon request from the corresponding author.

Conflicts of Interest: The authors declare no conflict of interest.

References

1. Azimi, T.; Mirzadeh, M.; Sabour, S.; Nasser, A.; Fallah, F.; Pourmand, M.R. Coagulase-negative staphylococci (CoNS) meningitis: A narrative review of the literature from 2000 to 2020. *New Microbes New Infect.* **2020**, *37*, 100755. [CrossRef] [PubMed]
2. Bhardwaj, B.; Bhatnagar, U.B.; Conaway, D.G. An Unusual Presentation of Native Valve Endocarditis Caused by *Staphylococcus warneri*. *Rev. Cardiovasc. Med.* **2016**, *17*, 140–143. [CrossRef] [PubMed]

3. Seng, R.; Kitti, T.; Thummeepak, R.; Kongthai, P.; Leungtongkam, U.; Wannalerdsakun, S.; Sitthisak, S. Biofilm formation of methicillin-resistant coagulase negative staphylococci (MR-CoNS) isolated from community and hospital environments. *PLoS ONE* **2017**, *12*, e0184172. [CrossRef] [PubMed]
4. Becker, K.; Heilmann, C.; Peters, G. Coagulase-negative staphylococci. *Clin. Microbiol. Rev.* **2014**, *27*, 870–926. [CrossRef]
5. Fijalkowski, K.; Peitler, D.; Karakulska, J. Staphylococci isolated from ready-toeat meat-identification, antibiotic resistance and toxin gene profile. *Int. J. Food Microbiol.* **2016**, *238*, 113–120. [CrossRef]
6. Bhargava, K.; Zhang, Y. Characterization of methicillin-resistant coagulasenegative staphylococci (MRCoNS) in retail meat. *Food Microbiol.* **2014**, *42*, 56–60. [CrossRef]
7. Incani, R.N.; Hernández, M.; Cortez, J.; González, M.E.; Dorel Salazar, Y. *Staphylococcus warneri* meningitis in a patient with *Strongyloides stercoralis* hyperinfection and lymphoma: First report of a case. *Rev. Inst. Med. Trop. São Paulo* **2010**, *52*, 169–170. [CrossRef]
8. Musharrafieh, R.; Tacchi, L.; Trujeque, J.; LaPatra, S.; Salinas, I. *Staphylococcus warneri*, a resident skin commensal of rainbow trout (*Oncorhynchus mykiss*) with pathobiont characteristics. *Vet. Microbiol.* **2014**, *169*, 80–88. [CrossRef]
9. Wang, H.; Wang, H.; Bai, Y.; Xu, X.; Zhou, G. Pathogenicity and antibiotic resistance of coagulase-negative staphylococci isolated from retailing chicken meat. *LWT* **2018**, *90*, 152–156. [CrossRef]
10. El-Ashker, M.; Gwida, M.; Monecke, S.; Ehricht, R.; Elsayed, M.; El-Gohary, F.; Reißig, A.; Müller, E.; Paul, A.; Igbinosa, E.O.; et al. Microarray-based detection of resistance genes in coagulase-negative staphylococci isolated from cattle and buffalo with mastitis in Egypt. *Trop. Anim. Health Prod.* **2020**, *52*, 3855–3862. [CrossRef]
11. Chajęcka-Wierzchowska, W.; Gajewska, J.; Wiśniewski, P.; Zadernowska, A. Enterotoxigenic potential of coagulase-negative staphylococci from ready-to-eat food. *Pathogens* **2020**, *9*, 734. [CrossRef] [PubMed]
12. Costanzo, N.; Ceniti, C.; Santoro, A.; Clausi, M.T.; Casalinuovo, F. Foodborne pathogen assessment in raw milk cheeses. *Int. J. Food Sci.* **2020**, *2020*, 3616713. [CrossRef] [PubMed]
13. Hleba, L.; Kačániová, M. *Antibiotická Rezistencia Čel'ade Enterobacteriaceae vo Vzt'ahu k Potravinám*; Slovenská Pol'nohospodárska Univerzita v Nitre: Nitra, Slovenska, 2015; p. 93. ISBN 978-80-552-1408-5.
14. Wright, G.D. The antibiotic resistome: The nexus of chemical and genetic diversity. *Nat. Rev. Microbiol.* **2007**, *5*, 175–186. [CrossRef] [PubMed]
15. D'Costa, V.M.; King, C.E.; Kalan, L.; Morar, M.; Sung, W.W.; Schwarz, C.; Froese, D.; Zazula, G.; Calmels, F.; Debruyne, R.; et al. Antibiotic resistance is ancient. *Nature* **2011**, *477*, 457–461. [CrossRef]
16. Schlüter, A.; Szczepanowski, R.; Pühler, A.; Top, E.M. Genomics of IncP-1 antibiotic resistance plasmids isolated from wastewater treatment plants provides evidence for a widely accessible drug resistance gene pool. *FEMS Microbiol. Rev.* **2007**, *31*, 449–477. [CrossRef]
17. Deyno, S.; Fekadu, S.; Seyfe, S. Prevelance and antimicrobial resistance of coagulase staphylococci clinical isolates from Ethiopia: A meta analysis. *BMC Microbiol.* **2018**, *18*, 43. [CrossRef]
18. Szczuka, E.; Krzyminska, S.; Kaznowski, A. Clonality, virulance and the occurrence of genes encoding antibiotic resistance among *Staphylococcus warneri* from bloodstream infections. *J. Med. Microbiol.* **2016**, *65*, 828–836. [CrossRef]
19. Chaves, R.D.; Pradella, F.; Turatti, M.A.; Amaro, E.C.; da Silva, A.R.; dos Santos Farias, A.; Pereira, J.S.; Khaneghah, A.M. Evaluation of *Staphylococcus* spp. in food and kitchen premises of Campinas, Brazil. *Food Control.* **2018**, *84*, 463–470. [CrossRef]
20. Owens, J.J. The egg yolk reaction produced by several species of bacteria. *J. Appl. Bacteriol.* **1974**, *37*, 137–148. [CrossRef]
21. Sharma, N.K.; Rees, C.E.; Dodd, C.E. Development of a single-reaction multiplex PCR toxin typing assay for Staphylococcus aureus strains. *Appl. Environ. Microbiol.* **2000**, *66*, 1347–1353. [CrossRef]
22. Iwase, T.; Seki, K.; Shinji, H.; Mizunoe, Y.; Masuda, S. Development of a real-time PCR assay for the detection and identification of *Staphylococcus capitis*, *Staphylococcus haemolyticus* and *Staphylococcus warneri*. *J. Med. Microbiol.* **2007**, *56*, 1346–1349. [CrossRef] [PubMed]
23. *CLSI Document VET01-S2 Performance Standards for Antimicrobial Disk and Dilution Susceptibility Tests for Bacteria Isolated from Animals*; Clinical and Laboratory Standards Institute: Wayne, PA, USA, 2013; pp. 1–168.
24. *EUCAST Guidelines for Detection of Resistance Mechanisms and Specific Resistances of Clinical and/or Epidemiological Importance. Version 2.0*; Clinical and Laboratory Standards Institute: Wayne, PA, USA, 2017; pp. 1–43.
25. Poulsen, A.B.; Skov, R.; Pallesen, L.V. Detection of methicillin resistance in coagulase-negative staphylococci and in staphylococci directly from simulated blood cultures using the EVIGENE MRSA Detection Kit. *J. Antimicrob. Chemother.* **2003**, *51*, 419–421. [CrossRef] [PubMed]
26. Hoveida, L.; Ataei, B.; Amirmozafari, N.; Noormohammadi, Z. Species diversity and molecular analysis of Staphylococcus in confectioneries of a developing country, Iran. *Infez. Med.* **2018**, *26*, 148–154. [PubMed]
27. Hammad, A.M.; Watanabe, W.; Fujii, T.; Shimamoto, T. Occurrence and characteristics of methicillin-resistant and -susceptible *Staphylococcus aureus* and methicillin-resistant coagulase-negative staphylococci from Japanese retail ready-to-eat raw fish. *Int. J. Food Microbiol.* **2016**, *156*, 286–289. [CrossRef]
28. Vandenesch, F.; Lina, G.; Henry, T. *Staphylococcus aureus* hemolysins, bi-component leukocidins, and cytolytic peptides: A redundant arsenal of membrane-damaging virulence factors. *Front. Cell. Infect. Microbiol.* **2012**, *2*, 12. [CrossRef]

29. Noumi, E.; Merghni, A.; Alreshidi, M.; Del Campo, R.; Adnan, M.; Haddad, O.; De Feo, V.; Snoussi, M. Phenotypic and Genotypic Characterization with MALDI-TOF-MS Based Identification of *Staphylococcus* spp. Isolated from Mobile Phones with their Antibiotic Susceptibility, Biofilm Formation, and Adhesion Properties. *Int. J. Environ. Res. Public Health* **2020**, *17*, 3761. [CrossRef]
30. Fisher, K.; Phillips, C. The ecology, epidemiology and virulence of *Enterococcus*. *Microbiology* **2009**, *155*, 1749–1757. [CrossRef]
31. LEPIDI, A. Chapter 12—Staphylococcal Lipases. In *Pet-to-Man Travelling Staphylococci*; Academic Press: Cambridge, MA, USA, 2018; pp. 147–159. ISBN 978-0-12-813-547-1.
32. Gundogan, N.; Ataol, O.; Torlak, F.O. Determination of Some Virulence Factors in *Staphylococcus aureus*, *Enterococcus faecalis* and *Enterococcus faecium* Isolated from Meat and Milk Products. *J. Food Saf.* **2012**, *33*, 387–393. [CrossRef]
33. Gundogan, N.; Ataol, O. Biofilm, protease and lipase properties and antibiotic resistance profiles of staphylococci isolated from various foods. *Afr. J. Microbiol. Res.* **2013**, *33*, 3582–3588. [CrossRef]
34. Zhao, J.; Ma, M.; Zeng, Z.; Yu, P.; Gong, D.; Deng, S. Production, purification and biochemical characterisation of a novel lipase from a newly identified lipolytic bacterium *Staphylococcus caprae* NCU S6. *J. Enzym. Inhib. Med. Chem.* **2021**, *36*, 248–256. [CrossRef]
35. Poyart, C.; Quesne, G.; Boumaila, C.; Trieu-Cuot, P. Rapid and accurate species-level identification of coagulase-negative staphylococci by using the sodA gene as a target. *J. Clin. Microbiol.* **2001**, *39*, 4296–4301. [CrossRef] [PubMed]
36. Pinchuk, I.V.; Beswick, E.J.; Reyes, V.E. Staphylococcal enterotoxins. *Toxins* **2010**, *2*, 2177–2197. [CrossRef] [PubMed]
37. Sospedra, I.; Soriano, J.M.; Mañes, J. Enterotoxinomics: The omic sciences in the study of staphylococcal toxins analyzed in food matrices. *Food Res. Int.* **2013**, *54*, 1052–1060. [CrossRef]
38. De Freitas Guimarães, F.; Nóbrega, D.B.; Richini-Pereira, V.B.; Marson, P.M.; de Figueiredo Pantoja, J.C.; Langoni, H. Enterotoxin genes in coagulase-negative and coagulase-positive staphylococci isolated from bovine milk. *J. Dairy Sci.* **2013**, *96*, 2866–2872. [CrossRef] [PubMed]
39. Banaszkiewicz, S.; Wałecka-Zacharska, E.; Schubert, J.; Tabiś, A.; Król, J.; Stefaniak, T.; Węsierska, E.; Bania, J. Staphylococcal Enterotoxin Genes in Coagulase-Negative Staphylococci—Stability, Expression, and Genomic Context. *Int. J. Mol. Sci.* **2022**, *23*, 2560. [CrossRef] [PubMed]
40. Hu, D.L.; Li, S.; Fang, R.; Ono, H.K. Update on molecular diversity and multipathogenicity of staphylococcal superantigen toxins. *Anim. Dis.* **2021**, *1*, 7. [CrossRef]
41. Hoveida, L.; Ataei, B.; Amirmozafari, N.; Noormohammadi, Z. Species variety, antibiotic susceptibility patterns and prevalence of enterotoxin genes in Staphylococci isolated from foodstuff in Central Iran. *Iran. J. Public Health* **2020**, *49*, 96. [CrossRef]
42. Saber, H.; Jasni, A.S.; Jamaluddin, T.Z.M.T.; Ibrahim, R. A review of staphylococcal cassette chromosome mec (SCCmec) types in coagulase-negative staphylococci (CoNS) species. Malays. *J. Med. Sci.* **2017**, *24*, 7–18. [CrossRef]
43. Ballhausen, B.; Kriegeskorte, A.; Schleimer, N.; Peters, G.; Becker, K. The *mecA* homolog *mecC* confers resistance against-lactams in *Staphylococcus aureus* irrespective of the genetic strain background. *Antimicrob. Agents Chemother.* **2014**, *58*, 3791–3798. [CrossRef]
44. Humphries, R.M.; Magnano, P.; Burnham, C.A.D.; Dien Bard, J.; Dingle, T.C.; Callan, K.; Westblade, L.F. Evaluation of surrogate tests for the presence of mecA-mediated methicillin resistance in *Staphylococcus capitis*, *Staphylococcus haemolyticus*, *Staphylococcus hominis*, and *Staphylococcus warneri*. *J. Clin. Microbiol.* **2020**, *59*, e02290-20. [CrossRef]
45. Al-Saadi, D.A.A.; Abd Al-Mayahi, F.S. Antibiogram Susceptibility Patterns of *Staphylococcus Aureus* Harboring of *MecA* Gene and Prevalence Aminoglycoside Modifying Enzymes (AMEs) Genes in Iraq. *IOP Conf. Ser. Earth Environ. Sci.* **2021**, *923*, 012049. [CrossRef]
46. Perumal, N.; Murugesan, S.; Krishnan, P. Distribution of genes encoding aminoglycoside-modifying enzymes among clinical isolates of methicillin-resistant staphylococci. *Indian J. Med. Microbiol.* **2016**, *34*, 350–352. [CrossRef] [PubMed]
47. Klingenberg, C.; Sundsfjord, A.; Rønnestad, A.; Mikalsen, J.; Gaustad, P.; Flægstad, T. Phenotypic and genotypic aminoglycoside resistance in blood culture isolates of coagulase-negative staphylococci from a single neonatal intensive care unit, 1989–2000. *J. Antimicrob. Chemother.* **2004**, *54*, 889–896. [CrossRef] [PubMed]
48. Ardic, N.; Sareyyupoglu, B.; Ozyurt, M.; Haznedaroglu, T.; Ilga, U. Investigation of aminoglycoside modifying enzyme genes in methicillin-resistant staphylococci. *Microbiol. Res.* **2006**, *161*, 49–54. [CrossRef]
49. Emaneini, M.; Taherikalani, M.; Eslampour, M.A.; Sedaghat, H.; Aligholi, M.; Jabalameli, F.; Shahsavan, S.; Sotoudeh, N. Phenotypic and genotypic evaluation of aminoglycoside resistance in clinical isolates of staphylococci in Tehran, Iran. *Microb. Drug Resist.* **2009**, *15*, 129–132. [CrossRef]
50. Martineau, F.; Picard, F.J.; Lansac, N.; Ménard, C.; Roy, P.H.; Ouellette, M.; Bergeron, M.G. Correlation between the resistance genotype determined by multiplex PCR assays and the antibiotic susceptibility patterns of *Staphylococcus aureus* and *Staphylococcus epidermidis*. *Antimicrob. Agents Chemother.* **2000**, *44*, 231–238. [CrossRef]
51. Chaudhury, A.; Kumar, A.G. In vitro activity of antimicrobial agents against oxacillin resistant staphylococci with special reference to *Staphylococcus haemolyticus*. *Indian J. Med. Microbiol.* **2007**, *25*, 50–52. [CrossRef]
52. Goudarzi, G.; Tahmasbi, F.; Anbari, K.; Ghafarzadeh, M. Distribution of genes encoding resistance to macrolides among staphylococci isolated from the nasal cavity of hospital employees in Khorramabad, Iran. *Iran. Red Crescent Med. J.* **2016**, *18*, e25701. [CrossRef]
53. Roberts, M.C.; Sutcliffe, J.; Courvalin, P.; Jensen, L.B.; Rood, J.; Seppala, H. Nomenclature for macrolide and macrolide-lincosamide-streptogramin B resistance determinants. *Antimicrob. Agents Chemother.* **1999**, *43*, 2823–2830. [CrossRef]

54. Leclercq, R. Mechanisms of resistance to macrolides and lincosamides: Nature of the resistance elements and their clinical implications. *Clin. Infect. Dis.* **2002**, *34*, 482–492. [CrossRef]
55. Szemraj, M.; Czekaj, T.; Kalisz, J.; Szewczyk, E.M. Differences in distribution of MLS antibiotics resistance genes in clinical isolates of staphylococci belonging to species: *S. epidermidis, S. hominis, S. haemolyticus, S. simulans* and *S. warneri*. *BMC Microbiol.* **2019**, *19*, 124. [CrossRef] [PubMed]
56. Munier, A.L.; de Lastours, V.; Barbier, F.; Chau, F.; Fantin, B.; Ruimy, R. Comparative dynamics of the emergence of fluoroquinolone resistance in staphylococci from the nasal microbiota of patients treated with fluoroquinolones according to their environment. *Int. J. Antimicrob. Agents* **2015**, *46*, 653–659. [CrossRef] [PubMed]
57. Pegues, D.A.; Colby, C.; Hibberd, P.L.; Cohen, L.G.; Ausubel, F.M.; Calderwood, S.B.; Hooper, D.C. The epidemiology of resistance to ofloxacin and oxacillin among clinical coagulase-negative staphylococcal isolates: Analysis of risk factors and strain types. *Clin. Infect. Dis.* **1998**, *26*, 72–79. [CrossRef] [PubMed]
58. Persson-Waller, K.; Aspán, A.; Nyman, A.; Persson, Y.; Grönlund Andersson, U. CNS species and antimicrobial resistance in clinical and subclinical bovine mastitis. *Vet. Microbiol.* **2011**, *152*, 112–116. [CrossRef]
59. Nunes, R.S.C.; Del Aguila, E.M.; Paschoalin, V.M.F. Safety evaluation of the coagulase-negative staphylococci microbiota of salami: Superantigenic Toxin Production and antimicrobial resistance. *BioMed Res. Int.* **2015**, *2015*, 483–548. [CrossRef]
60. Thomas, D.Y.; Jarraud, S.; Lemercier, B.; Cozon, G.; Echasserieau, K.; Etienne, J.; Gougeon, M.L.; Lina, G.; Vandenesch, F. Staphylococcal enterotoxin-like toxins U2 and V, two new staphylococcal superantigens arising from recombination within the enterotoxin gene cluster. *Infect. Immun.* **2006**, *74*, 4724–4734. [CrossRef]
61. Seng, P.; Boushab, B.M.; Romain, F.; Gouriet, F.; Bruder, N.; Martin, C.; Papazian, L. Emerging role of Raoultella ornithinolytica in human infections: A series of cases and review of the literature. *Int. J. Infect. Dis.* **2016**, *45*, 65–71. [CrossRef]

foods

Article

Use of Large-Scale Genomics to Identify the Role of Animals and Foods as Potential Sources of Extraintestinal Pathogenic *Escherichia coli* That Cause Human Illness

Lucas Harrison [1,*], Gregory H. Tyson [1], Errol Strain [1], Rebecca L. Lindsey [2], Nancy Strockbine [2], Olgica Ceric [1], Gamola Z. Fortenberry [3], Beth Harris [4], Sheryl Shaw [3], Glenn Tillman [5], Shaohua Zhao [1] and Uday Dessai [3,*]

1 U.S. Food and Drug Administration, Center for Veterinary Medicine, Laurel, MD 20708, USA; gregory.tyson@fda.hhs.gov (G.H.T.); errol.strain@fda.hhs.gov (E.S.); olgica.ceric@fda.hhs.gov (O.C.); shaohua.zhao@fda.hhs.gov (S.Z.)
2 Centers for Disease Control and Prevention, Atlanta, GA 30333, USA; wmi1@cdc.gov (R.L.L.); nas6@cdc.gov (N.S.)
3 U.S. Department of Agriculture, Food Safety and Inspection Service, Washington, DC 20250, USA; gamola.fortenberry@usda.gov (G.Z.F.); sheryl.shaw@usda.gov (S.S.)
4 U.S. Department of Agriculture, Animal and Plant Health Inspection Service, Ames, IA 50010, USA; beth.n.harris@usda.gov
5 U.S. Department of Agriculture, Food Safety and Inspection Service, Athens, GA 30605, USA; glenn.tillman@usda.gov
* Correspondence: lucas.harrison@fda.hhs.gov (L.H.); uday.dessai@usda.gov (U.D.)

Abstract: Extraintestinal pathogenic *Escherichia coli* (ExPEC) cause urinary tract and potentially life-threatening invasive infections. Unfortunately, the origins of ExPEC are not always clear. We used genomic data of *E. coli* isolates from five U.S. government organizations to evaluate potential sources of ExPEC infections. Virulence gene analysis of 38,032 isolates from human, food animal, retail meat, and companion animals classified the subset of 8142 non-diarrheagenic isolates into 40 virulence groups. Groups were identified as low, medium, and high relative risk of containing ExPEC strains, based on the proportion of isolates recovered from humans. Medium and high relative risk groups showed a greater representation of sequence types associated with human disease, including ST-131. Over 90% of food source isolates belonged to low relative risk groups, while >60% of companion animal isolates belonged to medium or high relative risk groups. Additionally, 18 of the 26 most prevalent antimicrobial resistance determinants were more common in high relative risk groups. The associations between antimicrobial resistance and virulence potentially limit treatment options for human ExPEC infections. This study demonstrates the power of large-scale genomics to assess potential sources of ExPEC strains and highlights the importance of a One Health approach to identify and manage these human pathogens.

Keywords: ExPEC; *Escherichia coli*; virulence factors; foodborne pathogens; companion animals; One Health

Citation: Harrison, L.; Tyson, G.H.; Strain, E.; Lindsey, R.L.; Strockbine, N.; Ceric, O.; Fortenberry, G.Z.; Harris, B.; Shaw, S.; Tillman, G.; et al. Use of Large-Scale Genomics to Identify the Role of Animals and Foods as Potential Sources of Extraintestinal Pathogenic *Escherichia coli* That Cause Human Illness. *Foods* **2022**, *11*, 1975. https://doi.org/10.3390/foods11131975

Academic Editor: Bianca Castiglioni

Received: 28 May 2022
Accepted: 1 July 2022
Published: 3 July 2022

1. Introduction

Escherichia coli is a diverse bacterial species able to adapt to a wide range of environments. *Escherichia* species are part of the normal intestinal microbiota of humans and other warm-blooded animals and can survive in many environmental reservoirs [1]. Most *E. coli* are commensal, living harmlessly within the intestinal tract of their host species, while some are pathogens capable of causing disease within or outside the intestinal tract [2]. Pathogenic strains causing disease inside the intestinal tract are referred to as intestinal pathogenic *E. coli* (IPEC), whereas strains with a propensity for causing disease outside the intestinal tract in otherwise healthy hosts are classified as extraintestinal pathogenic *E. coli* (ExPEC). ExPEC can be further classified into specialized groups, including uropathogenic

E. coli (UPEC), which cause urinary tract infections (UTI), and neonatal meningitis *E. coli* (NMEC), which infect newborns and can cause septicemia or meningitis [3,4]. These strains can also exhibit hybrid pathotypes, encoding for virulence genes of both IPEC and ExPEC pathotypes [5].

Multilocus sequence typing (MLST) has been used to attempt to identify specific sequence types (STs) associated with ExPEC infections [6]. In a review of 217 published ExPEC studies, Manges et. al. noted that the top five clinically relevant sequence types globally (in frequency order) include ST131, ST69, ST10, ST405, and ST38. Of these, ST131 was found in over 90% of studies, and ST69 or ST10 were detected in 50% of studies [7]. However, as ExPEC are genetically heterogeneous, it is difficult to classify *E. coli* as commensals or ExPECs with ease by using traditional MLST typing alone [8]. In addition, individual *E. coli* sequence types can contain multiple pathotypes [9], further complicating associations between MLST and the ExPEC pathotype.

Several efforts have been made to better define virulence genes associated with ExPEC [6]. Recently, the Center for Genomic Epidemiology added 44 ExPEC-associated virulence genes to the *E. coli* VirulenceFinder database for the identification of ExPECs [10]. A database of *E. coli* virulence genes has also been added to AMRFinder, and these genes are automatically identified on public genome sequences [11]. However, there is no single definition for the number and type of virulence genes that designate *E. coli* as ExPEC.

ExPEC infections are important because they are responsible for millions of UTIs in the United States each year [12]. UTIs caused by ExPEC typically result from gut bacteria ascending the urethra [13]. Recent studies have implicated retail meats as a potential source of *E. coli* causing UTIs, and genomic similarities have been identified between UPEC and strains isolated from the chicken gut [14–16]. Prevalence-based work from retail meat sampling has also found that some ExPEC virulence genes are common in *E. coli* from retail meats [17]. Similar associations have been observed in studies evaluating *E. coli* sequence type where human UTI-associated sequence types were recovered from retail meat samples [18]. Further, in a case control study, women with antimicrobial resistant UTIs were reported to be more likely to consume chicken than those with susceptible UTIs [19]. Although no causality was established between retail meat samples and UTIs, the authors demonstrated that common genomic elements existed between these groups. The association between animals, retail meats, environment, and human UTIs is a One Health concern, since the resistance which develops in animals (non-humans) can negatively impact human health [20].

To monitor One Health antimicrobial resistance, the National Antimicrobial Resistance Monitoring System (NARMS) in the United States tracks the prevalence and resistance of foodborne pathogens in food animals, retail meats, and humans [21]. ExPEC strains also affect animal health, as demonstrated by urinary tract infections being common in non-food animals [22]. Collaborative work with the Veterinary Laboratory Investigation and Response Network (Vet-LIRN) and the National Animal Health Laboratory Network sampling includes pathogenic *E. coli* from companion animals, primarily dogs. The increase in antimicrobial resistance determinants among bacteria causing UTIs is a growing threat to both human and non-human animal health [13,23]. Infections caused by antimicrobial resistant strains of *E. coli* impose a greater clinical burden than susceptible strains, and concern is growing that resistant strains from food animals and retail meats are causative agents for a greater portion of extraintestinal infections than previously thought [24,25].

In this study, we use genomics to evaluate the potential of various *E. coli* isolation sources to harbor ExPEC strains. We compare virulence genes among strains isolated from humans to strains isolated from food and companion animals in order to identify strains that may cause ExPEC infections. This study also compares the distribution of antimicrobial resistance determinants to evaluate their potential associations with the ExPEC strains. We discuss similarities and differences in ExPEC from these various sources and how they illustrate the One Health nature of ExPEC and antimicrobial resistance (AMR).

2. Materials and Methods

2.1. Data Collection

The sequences of 41,555 candidate *E. coli* isolates and their associated metadata were collected from humans through the NARMS program by the U.S. Center for Disease Control (CDC) PulseNet (n = 35,621); from food animal cecal samples by the U.S. Department of Agriculture (USDA) Food Safety Inspection Service (FSIS) (n = 2733); from retail meats by the U.S. Food and Drug Administration (FDA) Center for Veterinary Medicine (CVM) (n = 2446); from dog illnesses by the CVM's Vet-LIRN program (n = 663); and from other animal illnesses by the USDA's Animal and Plant Health Inspection Service (APHIS) (n = 92). Publicly available sequences of *E. coli* human isolates from thirty-five academic, research, and government institutions that were not obtained through PulseNet (human non-PulseNet) supplemented our dataset. The addition of these non-PulseNet *E. coli* isolates from the NCBI database (n = 1268) to the NARMS dataset of 41,555 isolates brought the total number of sequences evaluated in this study to 42,823. Metadata collected included isolation source, isolation date, and collection organization, but did not include virulence phenotype or indication as a causative agent of disease. Our final dataset contained *E. coli* strains isolated from humans (n = 36,886); from non-human animal hosts, including cattle (n = 1305), swine (n = 738), dogs (n = 647), chickens (n = 389), turkeys (n = 346), cats (n = 30), horses (n = 29), uncharacterized (n = 6), and sheep (n = 1); and from retail meats of turkey (n = 912), chicken (n = 546), cattle (n = 526), swine (n = 438), and meat products whose isolation source was not characterized (n = 24). The above samples with an uncharacterized isolation source were labeled as untyped meat samples. The sequences collected in this study were obtained from a variety of sources, and the collection criteria for all sources was not reported. As such, trends observed in this dataset may not be representative of trends in the general *E. coli* population.

2.2. Strain Characterization

Strains were characterized by sequence type, phylogenetic group, and virulence type using the following libraries of indicator genes and loci. Sequence type identification was determined using MLST 2.16.1 (https://github.com/tseemann/mlst (accessed on 29 April 2020)). Sequence type for each isolate was assigned using the 7-gene Achtman multilocus sequence typing schema for *E. coli* that assigns clonal complex by the number of alleles common among related sequence types. [26]. Phylogenetic group was assigned using ClermonTyping v1.4.0 [27]. Virulence genes and antimicrobial resistance determinants for each strain were identified using AMRFinderPlus (v3.6.15 National Center for Biotechnology Information, Bethesda, MD, USA) and the VirulenceFinder database updated 05-2021 [11,28]. The results of virulence gene and AMR determinant screening for each strain were combined, and redundant hits were removed. Non-*E. coli* sequences were filtered out from the dataset using Kraken2, MLST v2.16.1 and the presence of the *ipaD* and *ipaH* genes [29,30].

IPEC were characterized with the following criteria: *stx* alleles defined STEC; the combination of *ltcA* and *stb* or *ltcA* and *sta1* defined ETEC; *eae* defined EPEC; and *aggR* defined EAEC [31]. The remaining strains were defined as the non-IPEC population. This non-IPEC population contained all strains not typed by the above criteria and included both extraintestinal pathogenic *E. coli* and commensal *E. coli*.

2.3. Virulence Group

Virulence gene profiles for each strain were analyzed as a presence-absence matrix (PAM) in R v3.6.2 (R Foundation for Statistical Computing, Vienna, Austria, 2021) [32]. A k-modes analysis was used to determine the virulence gene profiles best able to characterize different subpopulations of IPEC *E. coli* [33]. The optimal number of virulence gene profiles was determined by evaluating the sum of within-group differences of a series k-modes analyses allowing for 5–80 allele profiles using the elbow method [34]. Agglomerative clustering of virulence gene profiles determined virulence group relatedness. Virulence groups containing any three of the four genes *chuA*, *fyuA*, *yfcV*, and *vat* were associated with containing UPEC strains [35]. The prevalence of specific virulence gene patterns within *E. coli* isolates from humans was used to determine relative risk group categories for non-IPEC strains. Virulence groups containing >50% *E. coli* strains isolated from humans showed a strong association with a human source and were classified as having a high relative risk of containing strains that cause human disease. Virulence groups with fewer than 25% *E. coli* isolates from humans showed a weaker association with a human source and were classified as a low relative risk to human health. The remaining virulence groups containing 25–50% human isolates showed medium risk to human health relative to the other isolates, and these defined the medium relative risk groups. Due to the broad data collection methods used in this study, the effect of predicting the virulence group from a known isolation source was determined by the Goodman and Kruskal lambda value obtained through the R DescTools package v0.99.44 [36].

3. Results

3.1. Characterization of IPEC and non-IPEC Strains

To identify candidate virulence factors indicative of ExPEC strains, we first needed to identify and separate non-*E. coli* and IPEC sequences in our dataset. Screening by *Escherichia* phylogroup, the presence of *ipaD* or *ipaH*, and membership in sequence types associated with *Shigella* prompted the removal of 4791 strains from our dataset of 42,823 sequences. The resulting dataset of 38,032 strains was subdivided into two groups of 29,890 IPEC and 8142 non-IPEC. A comparison of the IPEC and non-IPEC datasets shows that 99.2% of IPEC were collected from humans through PulseNet, while 85.1% of non-IPEC strains were collected from either food animal, companion animal, or non-PulseNet human sources (Table 1).

Table 1. Distribution of non-IPEC strains by isolation source.

	# of Isolates	IPEC	Non-IPEC
Human (PulseNet)	30,862	29,651	1211
Human (non-PulseNet)	1268	15	1253
Retail meats	2433	31	2402
Food animal cecal	2717	179	2538
Companion animals	752	14	738
Total	38,032	29,890	8142

The # symbol indicates we are describing the quantity, or number of isolates.

We next evaluated the virulence gene composition of the IPEC and non-IPEC isolate collections by querying them against the AMRFinderPlus and VirulenceFinder databases. This evaluation revealed 56 virulence genes that were more than twice as prevalent in the non-IPEC strains than in the IPEC strains (Table 2). Having shown a difference in the rate of occurrence of specific virulence genes between IPEC and non-IPEC strains, we evaluated the combinations of virulence genes found among strains in the non-IPEC population.

Table 2. Virulence genes with >2-fold greater representation in the non-IPEC dataset.

Virulence Gene	IPEC	Non-IPEC	Non-IPEC/IPEC
tsh	2.78%	14.01%	5.046
etsC	5.12%	25.35%	4.954
clbB	1.89%	9.36%	4.945
vat	3.55%	17.56%	4.941
cnf1	1.55%	7.61%	4.911
hlyF	5.61%	27.14%	4.842
f17A	0.75%	3.61%	4.817
iroE	6.98%	33.39%	4.785
sfaF	1.59%	7.58%	4.759

Table 2. *Cont.*

Virulence Gene	IPEC	Non-IPEC	Non-IPEC/IPEC
iroN	7.06%	33.54%	4.751
focI	0.97%	4.58%	4.742
papH	1.59%	7.53%	4.721
tcpC	1.26%	5.96%	4.721
f17G	0.72%	3.39%	4.697
focG	0.88%	4.09%	4.636
papA	4.77%	21.94%	4.594
usp	4.99%	22.92%	4.592
focC	0.77%	3.45%	4.506
papF	3.45%	15.45%	4.478
kpsM	6.37%	28.21%	4.429
papC	5.88%	25.88%	4.402
ibeA	1.95%	8.47%	4.339
sfaD	0.88%	3.75%	4.246
papE	2.09%	8.86%	4.231
sslE	3.64%	15.35%	4.217
cvaC	4.87%	20.40%	4.189
iucD	0.12%	0.49%	4.138
iucB	0.12%	0.50%	4.081
yfcV	6.25%	24.80%	3.967
nfaE	0.50%	1.98%	3.951
sfaS	0.54%	2.08%	3.843
sfaE	0.21%	0.81%	3.785
lngA	0.25%	0.92%	3.733
kpsE	8.46%	31.45%	3.720
air	2.44%	8.90%	3.653
afaE	0.29%	1.04%	3.646
ltcA	0.41%	1.49%	3.586
hlyE	0.22%	0.77%	3.574
eilA	3.36%	11.61%	3.450
eatA	0.50%	1.72%	3.435
mchF	8.09%	27.61%	3.413
sat	3.05%	9.90%	3.241
hra	9.97%	30.19%	3.029
iroD	0.02%	0.06%	2.931
afaB	1.15%	3.33%	2.906
afaA	1.17%	3.34%	2.853
cma	5.63%	14.75%	2.618
afaC	1.01%	2.58%	2.541
[illegible]	0.13%	0.32%	2.407
mchB	3.59%	8.65%	2.406
afaD	1.64%	3.93%	2.398
faeG	0.02%	0.05%	2.345
ireA	3.99%	9.31%	2.333
iroB	0.02%	0.04%	2.261
iroC	0.02%	0.04%	2.261
neuC	4.04%	8.29%	2.054

3.2. Isolation Source Composition of Virulence Groups

Our k-modes analysis of virulence genes in the non-IPEC population defined 40 groups of non-IPEC strains with unique virulence gene profiles (Figure 1). The size of each virulence group ranged from 23 to 1026 isolates, and groups contained 2–34 virulence genes (Supplemental Table S1). Each group was defined by a pattern of virulence genes present in >70% of strains in the group. The proportion of strains isolated from humans within each virulence group was used to characterize the groups as having a low, medium, or high relative risk of containing ExPEC strains (Supplemental Table S2). Twelve virulence groups contained $\geq$50% human isolates and were considered to have a high relative risk of containing ExPEC strains: groups 3, 5, 7, 12, 13, 14, 16, 17, 26, 33, 35, and 36 with a combined population of 1743 isolates. None of the virulence genes were conserved among all the high relative risk virulence groups. Seven virulence genes were exclusive to the high relative risk groups: *sat*, *sfaS*, *iucD*, *iucB*, *sta1*, *capU*, and *nfaE*. Five virulence groups contained from 25% to 50% human isolates, consisted of 845 strains, and were classified as having a medium risk of containing ExPEC strains. The remaining 5554 strains belonged to 23 virulence groups. These 23 virulence groups contained ≤25% strains isolated from humans and were considered to have a low relative risk of containing ExPEC strains.

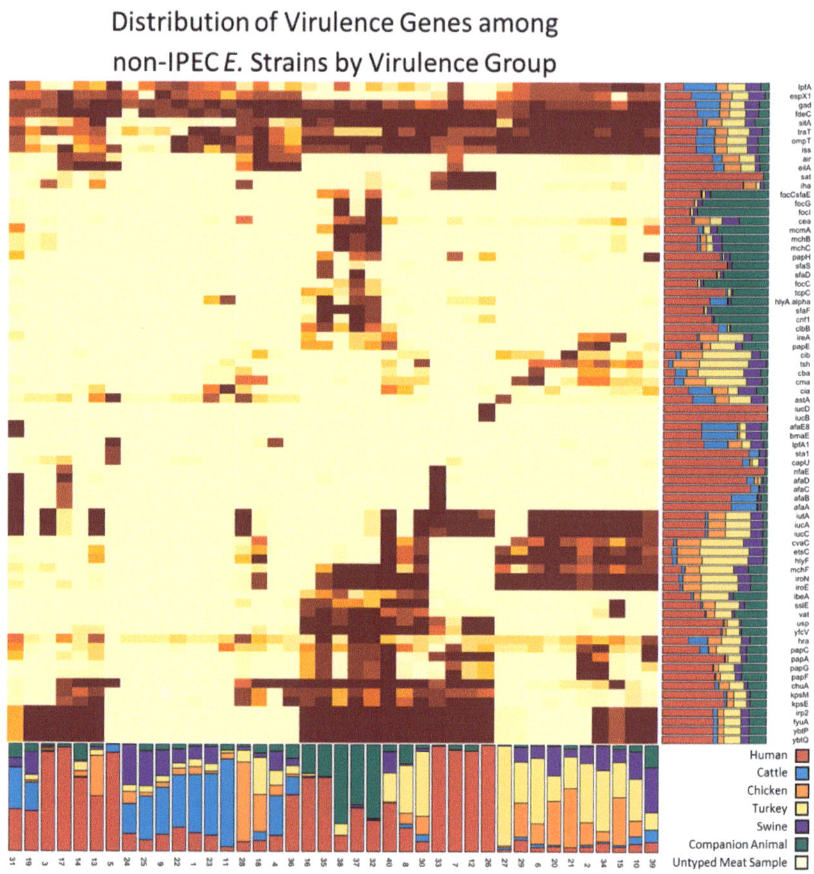

Figure 1. Heatmap of virulence genes organized by virulence groups from the 8142 non-IPEC strains. Stacked bar plots below and to the right of the heatmap show the relative contribution of each isolation source to the virulence groups and virulence genes, respectively. Animal isolation sources are the combined results from all contributing organizations.

Source composition of the non-human fraction of isolates from FSIS-NARMS, CVM-NARMS, APHIS and Vet-LIRN showed that isolates were not found in equal ratios across the virulence relative risk groups (Supplemental Table S3). While strains from all non-human sources were most prevalent in the low relative risk categories, companion animal isolates were more likely to be found in the medium and high relative risk groups compared to other non-human sources. (Figure 2). Evaluation of the dataset by calculating the Goodman and Kruskal lambda returned a lambda value of 0.295 for informing the prediction of relative risk group, given the isolation source.

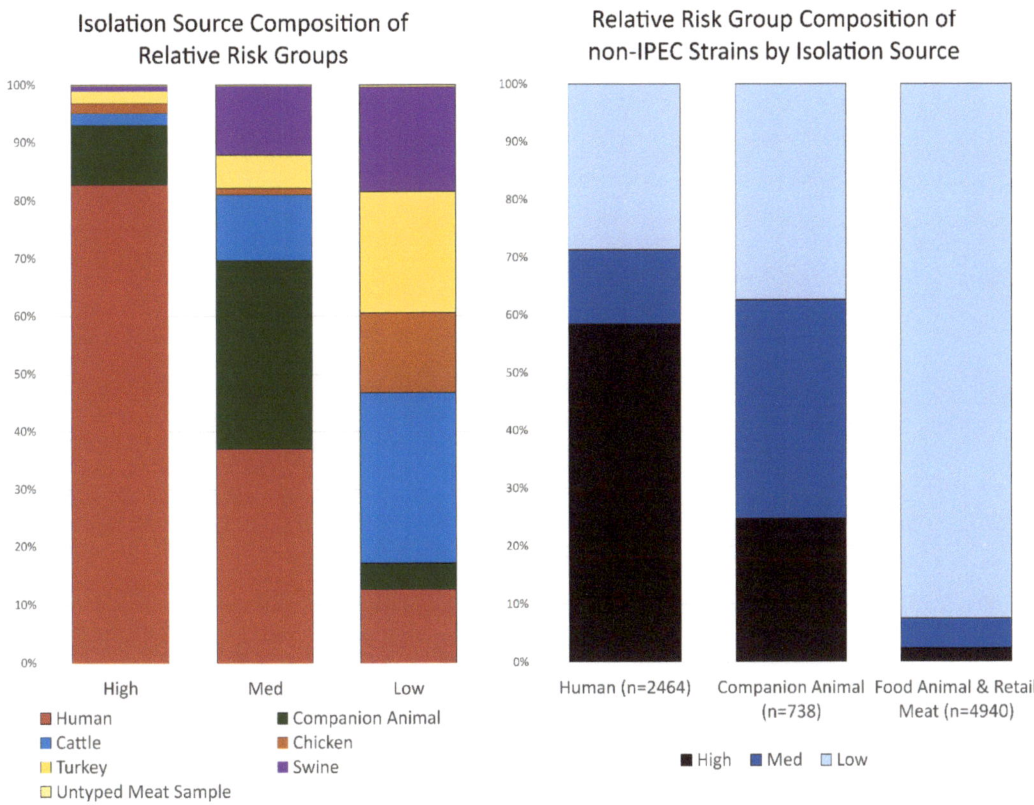

Figure 2. (**left**) Bar graph representation of isolation sources divided into relative risk groups. Isolation sources are the combined results from all contributing organizations. (**right**) Evaluation of the relative risk group composition of strains from humans (PulseNet and non-PulseNet), companion animals (Vet-LIRN and APHIS), and food animals + retail meats (USDA-FSIS and CVM-NARMS).

We then evaluated the composition of the relative risk groups by their *E. coli* phylogenetic groups (Supplemental Table S4). The major phylogenetic groups consisting of at least 20% of the relative risk groups were A, B1, B2, and D (Table 3). In the high relative risk group, 56.2% of the isolates belonged to B2, and 20.4% belonged to the D phylogenetic group. Among the medium relative risk group, 59.3% of isolates belonged to B2, and 20.2% belonged to phylogenetic group A. Phylogenetic group B1 was most common in the low relative risk group, at 48.1%, followed by phylogenetic group A, at 28.6%. In the remaining phylogenetic groups, groups C and G were most common in the low relative risk group, at 3.8% and 3.7%, respectively, while groups E and F were found most often in the high relative risk group, at 4.6% and 4.7%, respectively. A phylogenetic group comparison to virulence group revealed two main clusters of high relative risk virulence groups. In the first cluster, more than 70% of the isolates from high relative risk groups 26, 12, 33, 35, and 16 belonged to phylogenetic group B2. The second cluster of high relative risk groups of 5, 13, 14, 17, and 3 were represented by phylogenetic groups E, D, and A.

Table 3. Distribution of relative risk group by *E. coli* phylogenetic grouping.

	A	B1	B2	C	D	E	E or Clade I	F	G	Unknown
High	10.8	0.9	56.2	0.6	20.4	4.6	0.2	4.7	1.3	0.2
Med	20.2	17	59.3	2.4	0.5	0.4	0	0.1	0.1	0
Low	28.6	48.1	7.6	3.8	4	2.7	0	1.2	3.7	0.3

Values represented as percent of isolates from each relative risk group belonging to the phylogenetic group.

3.3. *Sequence Type and AMR Gene Composition of Virulence Groups*

Seven-gene multi-locus sequence typing analysis identified 1031 sequence types in our dataset of 8142 non-IPEC isolates. Of the 1031 sequence types, 194 sequence types belonged to 43 clonal complexes, accounting for 4139 isolates. The remaining 837 sequence types classified 3710 isolates. The final set of 293 strains did not match any sequence type in the PubMLST database. Sequence types contained multiple virulence gene profiles and were distributed among the 40 virulence groups (Figure 3). While no virulence group was exclusive to a specific sequence type, each of the virulence groups contained a sequence type that represented at least 25% of its strains (Supplemental Table S5). The majority of strains within eight virulence groups belonged to a single sequence type: within virulence group 5, 89.5% of the strains belonged to ST182; 85.7% of strains from virulence group 17 belonged to the ST38 clonal complex; 84.4% of virulence group 26, 80.3% of virulence group 12, and 54.9% of virulence group 33 belonged to the ST131 clonal complex; 82.1% of virulence group 15 belonged to ST117; 68.9% of virulence group 3 belonged to the ST10 clonal complex; and 51.2% of strains from virulence group 39 belonged to the ST23 clonal complex.

Every virulence group contained multiple sequence types, and 39/40 virulence groups were contained in more than one isolation source. We then subdivided the virulence groups by sequence type to determine if the proportion of human isolates was consistent for all sequence types within a virulence group. Of the 572 sequence type/virulence group combinations in our dataset that contained human isolates, 82 sequence type/virulence group combinations had 5 or greater strains isolated from humans and 21 sequence type/virulence group combinations contained only human isolates (Supplemental Figure S1). In 4 of the 21 combinations, the sequence type containing human isolates accounted for all the human isolates in the virulence group. In the remaining 17 of the 21 sequence type-virulence group combinations, the combinations identified sequence types among 6 virulence groups that only contained human isolates.

Twelve virulence groups contained virulence gene patterns associated with UPEC isolates [35]: 8, 12, 15, 16, 26, 30, 32, 33, 35, 37, 38, and 40. The 12 UPEC-associated virulence groups accounted for 25.3% of the non-IPEC isolates, or only 5.4% of our combined IPEC and non-IPEC dataset. Isolates belonging to these UPEC-associated virulence groups represented 24.4% of our non-IPEC PulseNet isolates and 64.05% of non-IPEC human isolates obtained from sources other than PulseNet. The distribution of strains from UPEC-associated virulence groups isolated from non-human sources accounted to 16.2% isolates from retail meats, 3.9% isolates from food animals, and 64.5% isolates from companion animals. Isolates in dogs were common in 5/12 of the UPEC-associated virulence groups and ranged from a low of 26.6% in group 16, up to 74% in group 31. Fewer than 3.8% of the strains isolated from these 5 virulence groups were isolated from food animals or retail meats.

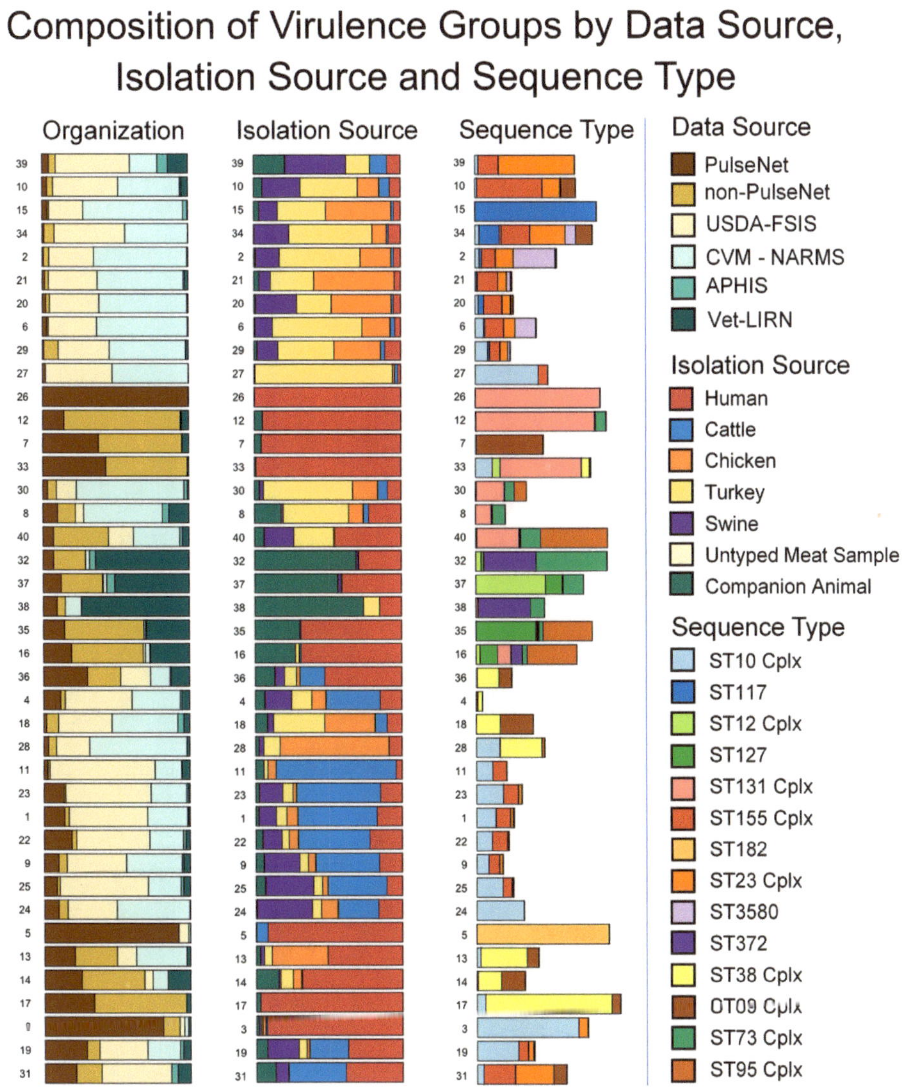

Figure 3. Isolation source composition of, and sequence type distribution among, non-IPEC virulence groups. Cplx designation of sequence type indicates a clonal complex. Virulence group order is consistent with that in Figure 1, to aid visual comparison. Source data can be found in Supplemental Tables S2–S4. Animal isolation sources are the combined results from all contributing organizations.

The AMR profiles of the 40 virulence groups were evaluated by their relative risk group. We found that 18 out of the 26 most common AMR determinants were found more than twice as often in the high relative risk ExPEC virulence groups compared with the other groups (Figure 4). Of these 18 AMR determinants, mutations causing substitutions *gyrA*(S83L), *gyrA*(D87N), *parC*(S80I), *parC*(E84V), *parE*(I529L), and *marR*(S3N) were found in companion animals at a >2-fold higher rate than in the other non-human sources (Table 4). Additionally, the AMR genes *mph(A)*, *dfrA17*, *aadA5*, $bla_{CTX\text{-}M\text{-}15}$, $bla_{OXA\text{-}1}$, and *catB3* were found in companion animals at a >2-fold higher rate than in the other non-human sources. In total, 12/18 AMR determinants more commonly found in the high relative risk virulence groups were isolated from companion animals at a higher rate than from any other non-human source.

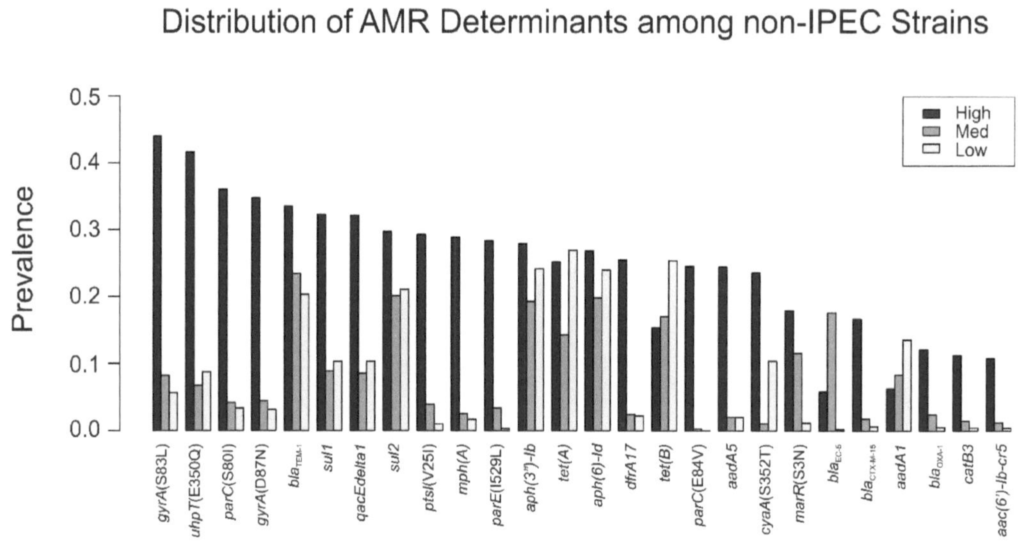

Figure 4. The prevalence of AMR determinants in at least 10% of any of the relative risk groups shows that 18/26 of this set of AMR determinants are present in the high relative risk group at a 2-fold or greater frequency than in either the medium or low relative risk groups.

Table 4. Distribution of AMR determinants associated with high relative risk strains among isolation sources.

Resistance Determinant	Human	Cattle	Chicken	Turkey	Swine	Untyped Meat Sample	Companion Animal
cyaA(S352T)	15.1	2.5	32.6	14.9	5.4	43.5	8.9
gyrA(S83L)	37.2	2.3	2.2	1.7	3.5	4.3	15.5
parC(S80I)	28.9	0.5	0.3	0.8	2.3	0.0	12.4
gyrA(D87N)	27.7	0.4	0.6	0.8	2.1	0.0	12.2
uhpT(E350Q)	29.6	6.9	12.7	13.3	4.1	8.7	13.8
mph(A)	22.1	0.2	0.1	0.2	2.0	0.0	5.8
dfrA17	19.6	0.6	0.3	1.2	2.1	0.0	7.5
parE(I529L)	19.3	0.0	0.1	0.8	1.9	0.0	3.9
ptsI(V25I)	20.1	0.1	1.8	2.7	2.0	4.3	4.0
aadA5	18.5	0.5	0.1	1.3	1.4	0.0	7.4
parC(E84V)	16.8	0.0	0.0	0.0	0.0	0.0	2.4
marR(S3N)	13.3	0.7	1.4	1.0	0.2	4.3	14.1
$bla_{CTX\text{-}M\text{-}15}$	12.8	0.1	0.0	0.1	0.1	0.0	3.1
$bla_{OXA\text{-}1}$	9.6	0.0	0.0	0.4	0.0	0.0	2.6
catB3	8.7	0.1	0.0	0.0	0.0	0.0	2.4
aac(6′)-Ib-cr5	8.2	0.0	0.0	0.0	0.0	0.0	2.7
sul1	28.0	2.8	20.5	14.8	5.8	21.7	7.3
qacEdelta1	27.9	2.7	20.4	14.6	5.8	21.7	7.8

Values represented as percent of isolates from source with AMR determinants.

4. Discussion

ExPEC are a threat to human health, causing millions of urinary tract and other extraintestinal infections in the United States each year [37]. However, the sources of ExPEC are not always known, nor are the precise combinations of genes necessary for pathogenicity. Our study used genomics data from various human and non-human animal sources to initially aid in understanding the markers for ExPEC, and then to assess their relative distributions in retail meat, food animals, companion animals, and humans.

Of note, only a few *E. coli* isolated from food animals and retail meats were in the high and medium relative risk ExPEC categories. Analyzed individually, none of the food animal or retail meat sources had a >15% representation of high and medium relative risk ExPEC strains. This was contradictory to our expectations, as previous work with chicken and human isolates had suggested a strong association between chicken and human ExPEC pathotypes, while our analysis showed less than 5% of non-IPEC isolates from chicken belonged to the high relative risk groups [14]. *E. coli* strains isolated from retail meats, however, were common in three of the UPEC-associated virulence groups. Nevertheless, only 8% of isolates from food animals and retail meats were in the medium and high-relative risk ExPEC groups. Despite the lower relative presence of the higher relative risk groups in food animals and retail meats, their potential as sources of ExPEC risk to humans cannot be completely discounted.

Our approach of grouping ExPEC strains by virulence gene profiles is helpful to further differentiate existing classification schema, such as sequence typing or phylogenetic grouping (Supplemental Figure S2). For example, the high relative risk strains of the broad host range ST10 complex (Cplx) were differentiated from the lower relative risk ST10 Cplx strains by their virulence gene composition. Grouping the dataset by virulence genes showed utility for identifying strains that may pose a threat to animal health, which can be seen in virulence group 15, being made up of ST117 *E. coli*. The ST117 sequence type is known to contain avian pathogenic *E. coli*.

In addition to aiding in the analysis of sequence types, the virulence group subdivision of phylogenetic groups highlighted which strains may be of greater concern than did the classification by phylogenetic grouping alone. While the phylogenetic group B2 is often associated with ExPEC infections, our division by virulence groups classified B2 subsets into low, medium, and high relative risk categories. Our virulence groups also identified a subset of phylogenetic group A strains with a high relative risk of containing ExPEC strains as virulence group 3.

A noteworthy finding was that the proportion of 18 out of 26 resistance determinants was higher among ExPEC in the high relative risk group than in the low relative risk group (Figure 3). This unequal distribution of AMR genes among relative risk groups can be concerning, since ExPEC infections typically require antimicrobial treatment [38]. Given the similarity of human isolates to companion animal isolates, this may warrant further investigation into the directionality of AMR and pathogen transmission between owners and companion animals. Further, since antimicrobial use in food animals can provide selective pressure for AMR, the presence of high relative risk ExPEC strains in food animals can negatively impact human health [39,40]. It is important to note that, in this genomic study, we did not perform any assessment of antimicrobial use and its impact on AMR phenotypes.

Although this study is the largest reported work to date using genomics to assess virulence and AMR in ExPEC, it did have some limitations. For instance, we did not perform virulence assays, so our virulence groups are not validated by phenotypic measures. Thus, the virulence genes found in the high relative risk ExPEC group may not be the most important genes contributing to ExPEC phenotypes. However, even if the genes are only correlated with ExPEC infections, their high rate of occurrence among strains causing clinical illness allows them to be used as indicators for the relative risk of human infection. Moreover, the removal of IPEC isolates was necessary to focus on ExPEC, but we may have eliminated some isolates with IPEC/ExPEC hybrid phenotypes [5]. Another limitation is that the human ExPEC isolates were not collected in a nationally representative and systematic surveillance system, meaning there could be bias in the types and numbers of ExPEC represented in the study. This may have contributed to a greater representation of AMR determinants in isolates of higher-virulence groups, since bacteria with treatment failures may have been more likely to be collected and sequenced. This sampling contrasts with the other isolation sources, which were collected as part of routine surveillance in the United States. An additional limitation of this study is the difference in population size and source composition among the virulence groups and isolation sources. This need for increased data diversity illustrates the importance of collaborative data collection efforts among organizations that represent unique interactions between animal and human health.

This study highlights the power of large-scale genomics and diverse data sources in addressing important One Health questions, particularly those concerning the relationship between antimicrobial resistance and virulence in anthropozoonotic pathogens. We used our large collection of sequencing data from different sources, generated by five federal organizations, to gain an understanding of genes that are linked to extraintestinal infections. Our results show that most *E. coli* from food animals and retail meats are not in the high-risk ExPEC groups. However, a large portion of non-IPEC strains have an increased potential of containing ExPEC strains, as these share virulence genes with isolates causing human or animal illnesses. Further, the contribution of AMR in ExPEC strains can lead to difficult to treat and more serious infections. This is a critical area for future research.

We believe that this unique approach using large-scale genomics on a diverse ExPEC source dataset to arrive at potential isolation sources without the biases of sequence type or multidrug resistance markers was essential to understanding ExPEC in the context of the burden of human illness. With additional datasets and analyses, this approach can further our understanding of ExPEC strains and UPEC pathotypes. This approach can also be applied more broadly to complex and difficult to decipher human-animal disease systems to gain an in-depth understanding of the agents, their roles in disease development, and their risk potential, as well as to predict impacts on human and animal health.

Supplementary Materials: The following supporting information can be downloaded at: https://www.mdpi.com/article/10.3390/foods11131975/s1, Figure S1: Sequence Type Subdivisions of non-IPEC Virulence Groups Containing ≥5 Isolates from Humans; Figure S2: Composition of Virulence Groups by Phylotype, Organization, Isolation Source, and Sequence Type; Table S1: Virulence Group Definitions; Table S2: Isolation Source Composition of Virulence Groups; Table S3: Contributing Data Source Composition of Virulence Groups; Table S4: Phylogenetic Group Composition of Virulence Groups; Table S5: Sequence Type Composition of Virulence Groups.

Author Contributions: Conceptualization, L.H., G.H.T., E.S., R.L.L., N.S., O.C., G.Z.F., B.H., S.S., G.T., S.Z. and U.D.; methodology, L.H., G.H.T. and E.S.; software, L.H. and E.S.; formal analysis, L.H. and E.S.; investigation, L.H., G.H.T., E.S., R.L.L., N.S., O.C., G.Z.F., B.H., S.S., G.T., S.Z. and U.D.; data curation, L.H., G.H.T., E.S., R.L.L., N.S., G.Z.F., B.H., S.S. and G.T.; writing—original draft preparation, L.H., G.H.T. and R.L.L.; writing—review and editing, L.H., G.H.T., E.S., R.L.L., N.S., O.C., G.Z.F., B.H., S.S., G.T., S.Z. and U.D.; visualization, L.H. and S.Z.; supervision, G.H.T., E.S., S.Z. and U.D.; project administration, G.H.T., E.S. and U.D. All authors have read and agreed to the published version of the manuscript.

Funding: The APC was funded by the United States Food and Drug Administration.

Institutional Review Board Statement: Not applicable.

Informed Consent Statement: Not applicable.

Data Availability Statement: The data presented in this study are available in this article and in the supplementary material.

Conflicts of Interest: The authors declare no conflict of interest. The views expressed in this article are those of the authors and do not necessarily reflect the official policy of the Department of Health and Human Services, the Department of Agriculture, the U.S. Food and Drug Administration, or the U.S. government. Reference to any commercial materials, equipment, or process does not in any way constitute approval, endorsement, or recommendation by the Food and Drug Administration or the U.S. government.

References

1. Ishii, S.; Ksoll, W.B.; Hicks, R.E.; Sadowsky, M.J. Presence and Growth of Naturalized *Escherichia coli* in Temperate Soils from Lake Superior Watersheds. *Appl. Environ. Microbiol.* **2006**, *72*, 612–621. [CrossRef] [PubMed]
2. Leimbach, A.; Hacker, J.; Dobrindt, U. *E. coli* as an all-rounder: The thin line between commensalism and pathogenicity. *Curr. Top Microbiol. Immunol.* **2013**, *358*, 3–32. [PubMed]
3. Johnson, J.R.; Russo, T.A. Extraintestinal pathogenic *Escherichia coli*: "the other bad *E. coli*". *J. Lab Clin. Med.* **2002**, *139*, 155–162. [CrossRef] [PubMed]
4. Russo, T.A.; Johnson, J.R. Proposal for a new inclusive designation for extraintestinal pathogenic isolates of *Escherichia coli*: ExPEC. *J. Infect. Dis.* **2000**, *181*, 1753–1754. [CrossRef]
5. Lindstedt, B.A.; Finton, M.D.; Porcellato, D.; Brandal, L.T. High frequency of hybrid *Escherichia coli* strains with combined Intestinal Pathogenic *Escherichia coli* (IPEC) and Extraintestinal Pathogenic *Escherichia coli* (ExPEC) virulence factors isolated from human faecal samples. *BMC Infect Dis.* **2018**, *18*, 544. [CrossRef]
6. Kohler, C.D.; Dobrindt, U. What defines extraintestinal pathogenic *Escherichia coli*? *Int. J. Med. Microbiol.* **2011**, *301*, 642–647. [CrossRef]
7. Manges, A.R.; Geum, H.M.; Guo, A.; Edens, T.J.; Fibke, C.D.; Pitout, J.D.D. Global Extraintestinal Pathogenic *Escherichia coli* (ExPEC) Lineages. *Clin. Microbiol. Rev.* **2019**, *32*, e00135-18. [CrossRef]
8. Salipante, S.J.; Roach, D.J.; Kitzman, J.O.; Snyder, M.W.; Stackhouse, B.; Butler-Wu, S.M.; Lee, C.; Cookson, B.T.; Shendure, J. Large-scale genomic sequencing of extraintestinal pathogenic *Escherichia coli* strains. *Genome Res.* **2015**, *25*, 119–128. [CrossRef]
9. Boll, E.J.; Overballe-Petersen, S.; Hasman, H.; Roer, L.; Ng, K.; Scheutz, F.; Hammerum, A.M.; Dungu, A.; Hansen, F.; Johannesen, T.B. Emergence of enteroaggregative Escherichia coli within the ST131 lineage as a cause of extraintestinal infections. *mBio* **2020**, *11*, e00353-20. [CrossRef]
10. Malberg, T.A.M.; Johnson, J.R.; Johnston, B.D.; Lund, O.; Scheutz, F. In Silico Genotyping of *Escherichia coli* Isolates for Extraintestinal Virulence Genes by Use of Whole-Genome Sequencing Data. *J. Clin. Microbiol.* **2020**, *58*, e01269-20.
11. Feldgarden, M.; Brover, V.; Haft, D.H.; Prasad, A.B.; Slotta, D.J.; Tolstoy, I.; Tyson, G.H.; Zhao, S.; Hsu, C.H.; McDermott, P.F.; et al. Validating the AMRFinder Tool and Resistance Gene Database by Using Antimicrobial Resistance Genotype-Phenotype Correlations in a Collection of Isolates. *Antimicrob Agents Chemother* **2019**, *63*, e00483-19. [CrossRef] [PubMed]
12. Medina, M.; Castillo-Pino, E. An introduction to the epidemiology and burden of urinary tract infections. *Ther. Adv. Urol.* **2019**, *11*, 1756287219832172. [CrossRef] [PubMed]
13. Flores-Mireles, A.L.; Walker, J.N.; Caparon, M.; Hultgren, S.J. Urinary tract infections: Epidemiology, mechanisms of infection and treatment options. *Nat. Rev. Microbiol.* **2015**, *13*, 269–284. [CrossRef] [PubMed]
14. Borges, C.A.; Tarlton, N.J.; Riley, L.W. Escherichia coli from Commercial Broiler and Backyard Chickens Share Sequence Types, Antimicrobial Resistance Profiles, and Resistance Genes with Human Extraintestinal Pathogenic Escherichia coli. *Foodborne Pathog Dis.* **2019**, *16*, 813–822. [CrossRef] [PubMed]
15. Buberg, M.L.; Mo, S.S.; Sekse, C.; Sunde, M.; Wasteson, Y.; Witsø, I.L. Population structure and uropathogenic potential of extended-spectrum cephalosporin-resistant Escherichia coli from retail chicken meat. *BMC Microbiol.* **2021**, *21*, 94. [CrossRef]
16. Díaz-Jiménez, D.; García-Meniño, I.; Fernández, J.; García, V.; Mora, A. Chicken and turkey meat: Consumer exposure to multidrug-resistant Enterobacteriaceae including *mcr*-carriers, uropathogenic *E. coli* and high-risk lineages such as ST131. *Int. J. Food Microbiol.* **2020**, *331*, 108750. [CrossRef]
17. Xia, X.; Meng, J.; Zhao, S.; Bodeis-Jones, S.; Gaines, S.A.; Ayers, S.L.; McDermott, P.F. Identification and antimicrobial resistance of extraintestinal pathogenic Escherichia coli from retail meats. *J. Food Prot.* **2011**, *74*, 38–44. [CrossRef]
18. Yamaji, R.; Friedman, C.R.; Rubin, J.; Suh, J.; Thys, E.; McDermott, P.; Hung-Fan, M.; Riley, L.W. A population-based surveillance study of shared genotypes of Escherichia coli isolates from retail meat and suspected cases of urinary tract infections. *Msphere* **2018**, *3*, e00179-18. [CrossRef]
19. Manges, A.R.; Smith, S.P.; Lau, B.J.; Nuval, C.J.; Eisenberg, J.N.; Dietrich, P.S.; Riley, L.W. Retail meat consumption and the acquisition of antimicrobial resistant Escherichia coli causing urinary tract infections: A case–control study. *Foodborne Pathog. Dis.* **2007**, *4*, 419–431. [CrossRef]
20. Robinson, T.P.; Bu, D.; Carrique-Mas, J.; Fèvre, E.M.; Gilbert, M.; Grace, D.; Hay, S.I.; Jiwakanon, J.; Kakkar, M.; Kariuki, S. Antibiotic resistance is the quintessential One Health issue. *Trans. R. Soc. Trop. Med. Hyg.* **2016**, *110*, 377–380. [CrossRef]
21. FDA. NARMS Now. 2021. Available online: https://www.fda.gov/animal-veterinary/national-antimicrobial-resistance-monitoring-system/narms-now-integrated-data (accessed on 28 January 2021).
22. Kidsley, A.K.; O'Dea, M.; Saputra, S.; Jordan, D.; Johnson, J.R.; Gordon, D.M.; Turni, C.; Djordjevic, S.P.; Abraham, S.; Trott, D.J. Genomic analysis of phylogenetic group B2 extraintestinal pathogenic *E. coli* causing infections in dogs in Australia. *Vet. Microbiol.* **2020**, *248*, 108783. [CrossRef] [PubMed]
23. Paul, R. State of the globe: Rising antimicrobial resistance of pathogens in urinary tract infection. *J. Glob. Infect. Dis.* **2018**, *10*, 117. [CrossRef] [PubMed]
24. Naylor, N.R.; Pouwels, K.B.; Hope, R.; Green, N.; Henderson, K.L.; Knight, G.M.; Atun, R.; Robotham, J.V.; Deeny, S.R. The health and cost burden of antibiotic resistant and susceptible *Escherichia coli* bacteraemia in the English hospital setting: A national retrospective cohort study. *PLoS ONE* **2019**, *14*, e0221944. [CrossRef] [PubMed]

25. Wasiński, B. Extra-intestinal pathogenic *Escherichia coli*–threat connected with food-borne infections. *Ann. Agric. Environ. Med.* **2019**, *26*, 532–537. [CrossRef]
26. Wirth, T.; Falush, D.; Lan, R.; Colles, F.; Mensa, P.; Wieler, L.H.; Karch, H.; Reeves, P.R.; Maiden, M.C.; Ochman, H. Sex and virulence in *Escherichia coli*: An evolutionary perspective. *Mol. Microbiol.* **2006**, *60*, 1136–1151. [CrossRef]
27. Beghain, J.; Bridier-Nahmias, A.; Le Nagard, H.; Denamur, E.; Clermont, O. ClermonTyping: An easy-to-use and accurate in silico method for *Escherichia* genus strain phylotyping. *Microb. Genom.* **2018**, *4*, e000192. [CrossRef]
28. Joensen, K.G.; Scheutz, F.; Lund, O.; Hasman, H.; Kaas, R.S.; Nielsen, E.M.; Aarestrup, F.M. Real-time whole-genome sequencing for routine typing, surveillance, and outbreak detection of verotoxigenic *Escherichia coli*. *J. Clin. Microbiol.* **2014**, *52*, 1501–1510. [CrossRef]
29. Venkatesan, M.M.; Buysse, J.M.; Kopecko, D.J. Use of *Shigella flexneri ipaC* and *ipaH* gene sequences for the general identification of *Shigella* spp. and enteroinvasive *Escherichia coli*. *J. Clin. Microbiol.* **1989**, *27*, 2687–2691. [CrossRef]
30. Wood, D.E.; Lu, J.; Langmead, B. Improved metagenomic analysis with Kraken 2. *Genome Biol.* **2019**, *20*, 257. [CrossRef]
31. Croxen, M.A.; Law, R.J.; Scholz, R.; Keeney, K.M.; Wlodarska, M.; Finlay, B.B. Recent advances in understanding enteric pathogenic *Escherichia coli*. *Clin. Microbiol. Rev.* **2013**, *26*, 822–880. [CrossRef]
32. Team, R.C. R: A Language and Environment for Statistical Computing. R Foundation for Statistical Computing. Vienna, Austria. Available online: https://www.R-Project.org/ (accessed on 28 January 2021).
33. Weihs, C.; Ligges, U.; Luebke, K.; Raabe, N. klaR analyzing German business cycles. In *Data Analysis and Decision Support*; Springer: Berlin/Heidelberg, Germany, 2005; pp. 335–343. [CrossRef]
34. Yuan, C.; Yang, H. Research on K-value selection method of K-means clustering algorithm. *J. Multidiscip. Sci. J.* **2019**, *2*, 226–235. [CrossRef]
35. Spurbeck, R.R.; Dinh, P.C., Jr.; Walk, S.T.; Stapleton, A.E.; Hooton, T.M.; Nolan, L.K.; Kim, K.S.; Johnson, J.R.; Mobley, H.L. *Escherichia coli* isolates that carry vat, fyuA, chuA, and yfcV efficiently colonize the urinary tract. *Infect. Immun.* **2012**, *80*, 4115–4122. [CrossRef] [PubMed]
36. Signorell, A.; Aho, K.; Alfons, A.; Anderegg, N.; Aragon, T.; Arachchige, C.; Arppe, A.; Baddeley, A.; Barton, K.; Bolker, B. DescTools: Tools for Descriptive Statistics. R Package Version 0.99. 44. R Foundation for Statistical Computing, Vienna, Austria. Available online: https://cran.r-project.org/web/packages/DescTools/index.html (accessed on 28 January 2021).
37. Poolman, J.T.; Wacker, M. Extraintestinal pathogenic *Escherichia coli*, a common human pathogen: Challenges for vaccine development and progress in the field. *J. Infect. Dis.* **2016**, *213*, 6–13. [CrossRef] [PubMed]
38. Pitout, J.D. Extraintestinal pathogenic *Escherichia coli*: An update on antimicrobial resistance, laboratory diagnosis and treatment. *Expert Rev. Anti Infect Ther.* **2012**, *10*, 1165–1176. [CrossRef] [PubMed]
39. Hershberger, E.; Oprea, S.F.; Donabedian, S.M.; Perri, M.; Bozigar, P.; Bartlett, P.; Zervos, M.J. Epidemiology of antimicrobial resistance in enterococci of animal origin. *J. Antimicrob Chemother* **2005**, *55*, 127–130. [CrossRef] [PubMed]
40. Witte, W. Selective pressure by antibiotic use in livestock. *Int. J. Antimicrob. Agents* **2000**, *1* (Suppl. 16), S19–S24. [CrossRef]

foods

MDPI

Article

Quantitative Risk Assessment of Susceptible and Ciprofloxacin-Resistant *Salmonella* from Retail Pork in Chiang Mai Province in Northern Thailand

Chaiwat Pulsrikarn [1], Anusak Kedsin [2], Parichart Boueroy [2], Peechanika Chopjitt [2], Rujirat Hatrongjit [3], Piyarat Chansiripornchai [4], Nipattra Suanpairintr [4,5] and Suphachai Nuanualsuwan [5,6,*]

1 National Institute of Health, Department of Medical Science, Ministry of Public Health, Nonthaburi 11000, Thailand
2 Faculty of Public Health, Kasetsart University, Chalermphrakiat Sakon Nakhon Province Campus, Sakon Nakhon 47000, Thailand
3 Faculty of Science and Engineering, Chalermphrakiat Sakon Nakhon Province Campus, Kasetsart University, Sakon Nakhon 47000, Thailand
4 Department of Pharmacology, Faculty of Veterinary Science, Chulalongkorn University, Bangkok 10330, Thailand
5 Center of Excellence for Food and Water Risk Analysis (FAWRA), Department of Veterinary Public Health, Faculty of Veterinary Science, Chulalongkorn University, Bangkok 10330, Thailand
6 Department of Veterinary Public Health, Faculty of Veterinary Sciences, Chulalongkorn University, Bangkok 10330, Thailand
* Correspondence: suphachai.n@chula.ac.th

Abstract: The adverse human health effects as a result of antimicrobial resistance have been recognized worldwide. *Salmonella* is a leading cause of foodborne illnesses while antimicrobial resistant (AMR) *Salmonella* has been isolated from foods of animal origin. The quantitative risk assessment (RA) as part of the guidelines for the risk analysis of foodborne antimicrobial resistance was issued by the Codex Alimentarius Commission more than a decade ago. However, only two risk assessments reported the human health effects of AMR *Salmonella* in dry-cured pork sausage and pork mince. Therefore, the objective of this study was to quantitatively evaluate the adverse health effects attributable to consuming retail pork contaminated with *Salmonella* using risk assessment models. The sampling frame covered pork at the fresh market (n = 100) and modern trade where pork is refrigerated (n = 50) in Chiang Mai province in northern Thailand. The predictive microbiology models were used in the steps where data were lacking. Susceptible and quinolone-resistant (QR) *Salmonella* were determined by antimicrobial susceptibility testing and the presence of AMR genes. The probability of mortality conditional to foodborne illness by susceptible *Salmonella* was modeled as the hazard characterization of susceptible and QR *Salmonella*. For QR *Salmonella*, the probabilistic prevalences from the fresh market and modern trade were 28.4 and 1.9%, respectively; the mean concentrations from the fresh market and modern trade were 346 and 0.02 colony forming units/g, respectively. The probability of illness (P_I) and probability of mortality given illness (P_{MI}) from QR *Salmonella*-contaminated pork at retails in Chiang Mai province were in the range of 2.2×10^{-8}–3.1×10^{-4} and 3.9×10^{-10}–5.4×10^{-6}, respectively, while those from susceptible *Salmonella* contaminated-pork at retails were in the range 1.8×10^{-4}–3.2×10^{-4} and 2.3×10^{-7}–4.2×10^{-7}, respectively. After 1000 iterations of Monte Carlo simulations of the risk assessment models, the annual mortality rates for QR salmonellosis simulated by the risk assessment models were in the range of 0–32, which is in line with the AMR adverse health effects previously reported. Therefore, the risk assessment models used in both exposure assessment and hazard characterization were applicable to evaluate the adverse health effects of AMR *Salmonella* spp. in Thailand.

Keywords: AMR; probabilistic models; quinolone; retail pork; risk assessment; *Salmonella*

Citation: Pulsrikarn, C.; Kedsin, A.; Boueroy, P.; Chopjitt, P.; Hatrongjit, R.; Chansiripornchai, P.; Suanpairintr, N.; Nuanualsuwan, S. Quantitative Risk Assessment of Susceptible and Ciprofloxacin-Resistant *Salmonella* from Retail Pork in Chiang Mai Province in Northern Thailand. *Foods* **2022**, *11*, 2942. https://doi.org/10.3390/foods11192942

Academic Editor: Arun K. Bhunia

Received: 11 July 2022
Accepted: 16 September 2022
Published: 20 September 2022

1. Introduction

The adverse health effects posed by antimicrobial-resistant (AMR) bacteria have been increasing at an alarming rate and have recently been recognized worldwide [1]. Antimicrobial agents are beneficial to a wide variety of sectors, from human and veterinary medicine to animal and plant production. The use of antimicrobial agents in these areas is inevitably connected and this renders a circulating pool of both resistant bacteria and bacteria-borne resistant genes that are eventually delivered to humans [2]. Regardless of the environmental, genetic, or spatial boundary, mobile genetic elements containing resistance determinants can, directly and indirectly, propagate through horizontal transfer among bacteria from foods of animal origin and their environment to humans. Therefore, AMR risk management measures in terms of prevention and control strategy rely heavily on source attribution and risk assessment to evaluate the likelihood and severity of the consequences of AMR bacteria-contaminated foods [3].

The Codex Alimentarius Commission (CAC) has endorsed a systematic framework for foodborne AMR risk analysis. This framework is composed of preliminary risk management activities, risk assessment, and risk management. These three components are connected via risk communication, including the surveillance of AMR and other sources of information. The underlying rationale of the principle of risk analysis is to evaluate the risk to human health from foodborne AMR microorganisms and AMR determinants so that practical risk management measures can be implemented to prevent and control such human health risks [4].

The microbial risk assessment is a scientific process to evaluate the risk of consuming food contaminated with hazards. Hazard identification, hazard characterization, exposure assessment, and risk characterization constitute the four-step risk assessment. Hazard identification is the initial step to examining the risk of the hazards such as foodborne disease viruses, bacteria, protozoa, and parasites; in this study, the hazard is *Salmonella*. Hazard characterization determines the probability of illness upon getting a hazard into the host by a specific dose–response model. The exposure assessment determines the probability of getting hazards through consuming food. The last step is risk characterization, where the risk estimate is derived from the product of probabilities of exposure and illness from the preceding two steps [5,6]. Foodborne AMR risk assessment (AMR RA) is slightly different from the traditional methodology of microbiological risk assessment [5,6] in that hazard characterization is necessary to additionally include the adverse effects of AMR, e.g., antimicrobial treatment failure, prolonged treatment period, more illness severity or virulence, and higher mortality rate [4].

The prevalence of susceptible *Salmonella* spp. from swine manure was in the range of 2–61% and from swine farm swabs it was 95%, whereas that of AMR *Salmonella* spp. isolated from antimicrobial-use swine farms was lowest at 33% against florfenicol and highest at 66% against tetracycline [7]. Likewise, the prevalence of tetracycline-resistant *Salmonella* spp. was even higher at 90% in two independent studies [8,9]. However, isolated *Salmonella* spp. was sensitive to ceftiofur, ceftriaxone, and ciprofloxacin. In addition, fluoroquinolone-resistant *Salmonella* spp. warrant further surveillance by the World Health Organization as a tier group 2 [10].

AMR *Salmonella* spp. in retail pork could be derived either from a farm or abattoir. The prevalence of AMR *Salmonella* spp. from the environment of the abattoir was lowest at 4% against ceftiofur and highest at 86–89% against tetracycline [7,8]. Recently, we investigated a total of 387 non-typhoidal *Salmonella enterica* (NTS) isolated from abattoirs. Approximately 24% of NTS isolates were AMR, while only 6% of NTS isolates were susceptible to all antimicrobial agents tested. However, non-AMR NTS isolates carry extended-spectrum beta-lactamase ($bla_{\text{CTX-M}}$) genes or narrow-spectrum beta-lactamase genes (bla_{TEM} or bla_{SHV}). The rest of the NTS isolates (70%) were susceptible to all fluoroquinolones as well as carbapenems and third-generation cephalosporins [11]. At retail, *Salmonella* spp. isolated from pork was susceptible to ampicillin, norfloxacin, and ciprofloxacin [8,12]. However, the prevalence of AMR *Salmonella* spp. isolated from retail pork were 100% against streptomycin and sulfamethoxazole [12] and 60% against tetracycline [8]. The source attribution of salmonellosis in children from pork was 11% [8].

Scientific evidence demonstrated that AMR *Salmonella* spp. is foodborne-transmitted. On the other hand, AMR *Salmonella* infection is seldom traced from patients in hospitals back through contaminated foods and even further back to animals along the food chain. Some of the implicated commodities in such reports were beef, pork, and milk, where the authors suggested that AMR *Salmonella* spp. in patients was attributable to farm animals [13–15]. While commonly found *Salmonella* serovars with either resistance or multiple resistance to *Salmonella* spp. in foods are Derby, Enteritidis, Hadar, Newport, Paratyphi, Typhimurium, and Virchow [16–18], *Salmonella* Typhimurium is the most prevalent serovar contaminating foods across continents [13–15,17,19,20]. Recently, both cephalosporin-resistant and extended-spectrum beta-lactamase-resistant *Salmonella* spp. have been frequently reported [21,22]. These reports implied that AMR *Salmonella* spp. has been widely circulated regardless of geographical borders, food commodities, serovars, resistance patterns, antimicrobial classes, and host ranges.

Even though 34 AMR RA relating to retail foods have been reported up to 2018, only eight articles investigated the adverse health effects of AMR *Salmonella* spp. Only half of these reports are related to the pork supply chain [23]. Two risk assessments reported the adverse health effects of AMR *Salmonella* in dry-cured pork sausage and pork mince [24,25]. Recently, a farm-to-fork quantitative risk assessment of *Salmonella* Heidelberg resistant to third-generation cephalosporins in broiler chickens was reported [26] while the AMR RA model was developed for anti-*E. coli* drugs [27]. However, a quantitative risk assessment using Monte Carlo simulation of QR *Salmonella* in retail pork has never been reported. In this study, QR *Salmonella*-contaminated pork was the hazard of interest. The sampling frame covered pork at retailers in Chiang Mai province in northern Thailand. The predictive microbiology models were used in the steps where data were lacking. The objective of this study was to comparatively evaluate the adverse health effects attributable to consuming pork contaminated with *Salmonella* susceptible and resistant to quinolone.

2. Materials and Methods

2.1. Pork Samples

The pork samples were collected in Chiang Mai province from both the fresh market and modern trade where pork is refrigerated. Ten pork samples were collected from each retailer. Eleven pork retailers from the fresh market and five butcher shops in the modern trade participated in this study. The sampling unit of pork was at least 100 g. Samples were collected using an aseptic technique to avoid undesirable cross-contamination from environmental fomite and then kept in a leak-proof container between 2 and 8 °C during transportation. The samples arrived at the laboratory and were analyzed within 8–10 h after being collected.

2.2. Enumeration of *Salmonella*

The ten-fold serial dilution of pork samples was achieved using buffered peptone water. For individual dilution, 1 mL of suspension was repeatedly transferred 3 times into 3 separate 9 mL tubes of Rappaport Vassiliades with soya (RVS) broth. Nine tubes of RVS broth for each sample were incubated at 42 °C for 24 h. Only RVS tubes with a turbid appearance and confirmed by xylose lysine desoxycholate agar and then triple sugar iron slant were counted as positive [28]. The concentrations of *Salmonella* in the Most Probable Number unit (MPN) were converted to colony-forming units (cfu) by multiplying by 0.8 since the MPN technique is more sensitive than a standard plate count by 25% [29]. The unit conversion of concentration is necessary to apply for a dose–response model using the dose unit as the cfu [30].

2.3. Antimicrobial Susceptibility Testing (AST)

Susceptibility testing for ampicillin, cefepime, cefotaxime, cefoxitin, chloramphenicol, ciprofloxacin, colistin, gentamicin, imipenem, meropenem, nalidixic acid, streptomycin, sulphamethoxazole, tetracycline, and trimethoprim was performed using a broth microdilution assay to determine the minimum inhibitory concentration (MIC) according to the M07 Clinical and Laboratory Standards Institute (CLSI) guidelines [31]. The results were interpreted according to the 2020 Clinical and Laboratory Standards Institute guidelines for the susceptibility testing of *Salmonella* isolates [32]. *Escherichia coli* ATCC 25922 was used as the control. The broth microdilution assay was performed using two-fold dilution at a concentration in a range of 0.03–64 μg/mL depending on the antimicrobial agents, which are suggested based on the 2020 CLSI.

2.4. Determination of Antimicrobial Resistance Genes

Antimicrobial resistance genes including quinolone, colistin, and carbapenem were conducted using a polymerase chain reaction (PCR). The PCR was carried out to determine the quinolone resistance determining region of *gyrA* and *parC*, and the plasmid-mediated quinolone resistance genes following are described elsewhere [33,34]. The PCR products of the quinolone resistance determining the region from the four genes were purified and subjected to Sanger sequencing (performed by Apical Scientific Sdn Bhd, Selangor, Malaysia) to determine their substitution by comparing with those of wild-type *S.* Typhimurium LT2 [11]. The presence of antibiotic resistance-conferring genes of colistin, including *mcr-1* through *mcr-9*, and carbapenem consisting of bla_{NDM}, $bla_{OXA\text{-}48\text{-}like}$, bla_{IMP}, and bla_{KPC} was investigated using the PCR method described elsewhere [11]. All PCRs performed in this study are described in the Supplementary Materials (Tables S1–S4).

2.5. Risk Assessment Models

2.5.1. Exposure Assessment

1. Probabilistic prevalence variable

The range of prevalence is between zero (0%) and one (100%), inclusively applicable to the range of Beta distribution. The Beta distribution is characterized by 2 parameters, alpha and beta, as shown in Equation (1).

$$P_{PROB} = \text{Beta}\ (\alpha, \beta) \tag{1}$$

To describe the variability of prevalence, the alpha parameter is substituted by $s + \alpha$, and the beta parameter is substituted by $n - s + \beta$ where s is the number of the successful trial (s) in the identical n trials of a binomial process, as shown in Equation (2). In this study, the successful trials were the QR *Salmonella*-contaminated (positive) samples where the identical n trials were the sample size.

$$P_{PROB} = \text{Beta}\ (s + \alpha, n - s + \beta) \tag{2}$$

This study assumes that no prior prevalence of QR *Salmonella* was reported. The uniform probability distribution was assumed, which is equivalent to Beta (1, 1). Therefore, two parameters in Equation (2) were replaced with 1, as shown in Equation (3) [6].

$$P_{PROB} = \text{Beta}\ (s + 1, n - s + 1) \tag{3}$$

2. Thermal inactivation model

The raw pork from retail was subjected to heat treatment before consumption. The cooking temperature and time were 64 °C for 2 min while the decimal reduction time at 64 °C (D_{64}) is 0.48 min [30]. The log reduction of *Salmonella* is shown in Equation (4).

$$LR = \frac{t}{D_{64}} \tag{4}$$

where LR = log reduction (LR) of susceptible or QR *Salmonella* in pork; D_{64} = decimal reduction time of *Salmonella* at 64 °C (min); t = cooking time (min).

3. Concentration variable

If pork samples were all negative, the *Salmonella* concentration was determined by the maximum likelihood estimator (MLE) technique [29,35,36], as shown in Equation (5).

$$\text{log reduction (LR) } C_S = \frac{\sum_{i=1}^{k} N_i}{\sum_{i=1}^{k} V_i} - 10^{LR} \quad (5)$$

where C_S = concentration of susceptible or QR *Salmonella* (g^{-1}); N_i = no. of *Salmonella* detected in retail pork i to k; V_i = analytical unit of pork i to k (g); k = no. of pork retailers; LR = log reduction of *Salmonella* from heat treatment.

4. Consumption variable (C_P)

Food consumption data for Thailand in 2016 from the Agricultural Commodity and Food Standard report showed that the mean and 97.5th percentile consumption of pork among eaters more than 3 years old was 14.12 and 58.28 g/person/day, respectively. The triangular distribution was used to describe the variability of the consumption variable. The three parameters of triangular distribution (minimum, most likely, and maximum) were 0, 14.12, and 58.28 g/person/day, respectively.

5. Dose of *Salmonella* ingested

The dose of *Salmonella* ingested was the product of *Salmonella* concentration after cooking and pork consumption per day. The equation for the dose of *Salmonella* ingested is shown in Equation (6) [6].

$$D = C_S \times C_P \quad (6)$$

where D = dose of susceptible or QR *Salmonella* ingested per day (cfu); C_S = concentration of susceptible or QR *Salmonella* (cfu/g); C_P = pork consumption per day (g).

6. Probability of exposure (P_E)

P_E is the likelihood of experiencing at least one cell of *Salmonella* from pork. Therefore, the input variables to model the P_E are the concentration (C_S) and prevalence (P_{PROB}) of *Salmonella*, including pork consumption (6), as shown in Equation (7).

$$P_E = P_{PROB}\,(1 - \exp - D) \quad (7)$$

2.5.2. Hazard Characterization

1. Probability of illness (P_I)

The dose–response model was used to characterize the probability of illness caused by either residual susceptible or QR *Salmonella*-contaminated pork after cooking, as shown in Equation (8).

$$P_I = 1 - (1 + (D/51.45))^{-0.1324} \quad (8)$$

where P_I = the probability of illness caused by an ingested dose of *Salmonella*; D = dose of susceptible or QR *Salmonella* ingested per day (cfu).

2. Probability of mortality (P_M)

Additional to the conventional hazard characterization of the microbial risk assessment, the adverse effects of AMR such as a higher mortality rate were included [4]. A previous study reported that the mortality rates caused by drug-susceptible and multidrug-resistant non-typhoid *Salmonella* were 0.2 and 3.4%, respectively [13]. Likewise, another study reported that the mortality rates caused by pan-susceptible and AMR *Salmonella* were 0.06 and 0.1%, respectively [18]. Therefore, in this study, the mean mortality rates as P_M caused by susceptible and AMR *Salmonella* were averaged from these two previous reports as 0.13 and 1.75%, respectively.

3. Probability of mortality given illness (P_{MI})

The integration of adverse health effects as the mortality conditional to the foodborne illness is the product of P_I and P_M, as shown in Equation (9).

$$P_{MI} = P_M \times P_I \quad (9)$$

2.5.3. Risk Characterization

In this study, the risk characterization is a two-step linked process of exposure assessment and hazard characterization. The probability of mortality given illness (P_{MI}) is conditional on P_E. Assuming that adverse health effects and hazard exposure are independent, the model for risk estimates in terms of the probability of foodborne mortality (P_{FM}) is the product of P_{MI} and P_E, as shown in Equation (10).

$$P_{FM} = P_{MI} \times P_E \quad (10)$$

The probability of foodborne mortality from at least one day was calculated based on the binomial theorem [36]. The number of annual foodborne mortality cases per 100,000 population is calculated from Equation (11).

$$M_{AFM} = (1 - (1 - P_{FM})^{365}) \times 100{,}000 \quad (11)$$

where M_{AFM} = annual foodborne mortality cases per 100,000 population; P_{FM} = probability of foodborne mortality per day.

Simulations of M_{AFM} were run for 10,000 iterations. The Simulación 4.0 freeware (developed by José Ricardo Varela) was used to run the Monte Carlo simulations.

2.6. Statistical Analysis

The M_{AFM} of susceptible and QR *Salmonella* in pork from the fresh market and modern trade was determined for the statistical difference by one-way analysis of variance (ANOVA) [37]. Tukey's multiple comparison test was followed to determine the pair-wise differences of M_{AFM}. The IBM® SPSS® Statistics version 22 software (SPSS Inc., Chicago, IL, USA) was used to perform statistical analyses.

3. Results

3.1. Exposure Assessment

A total of 150 pork samples collected from pork retailers (fresh market (n = 100) and modern trade (n = 50)) in Chiang Mai province were analyzed for *Salmonella* contamination. The number of *Salmonella*-positive samples is shown in Table 1. All *Salmonella* isolates from positive samples were subject to the AST. We determined antimicrobial-resistant genes in QR isolates for colistin (*mcr-1* through *mcr-9*), carbapenem, and fluoroquinolone including *mcr*, bla_{NDM}, $bla_{OXA\text{-}48\text{-}like}$, bla_{IMP}, bla_{KPC}, plasmid-mediated quinolone resistance, and the quinolone resistance-determining region of *gyrA* and *parC*. No isolates carried the mobile colistin resistance gene (*mcr*) and common carbapenemase genes (bla_{NDM}, $bla_{OXA\text{-}48\text{-}like}$, bla_{IMP}, bla_{KPC}). In the case of fluoroquinolone-resistant genes, among the QR isolates, five isolates carried *qnrS*, there were two substitutions in *parC*, and one isolate carried both *qnrS* and *parC* substitutions. No substitution occurred in *gyrA* in all isolates. P_{PROB} and mean concentrations corresponding to susceptible and QR *Salmonella* contaminated in the pork samples are shown in Table 2. The P_E to susceptible and QR *Salmonella*-contaminated pork at retail in Chiang Mai province was in the range of 2×10^{-7}–0.03 (Table 3).

Table 1. No. of *Salmonella* positive samples collected from retailers in Chiang Mai province.

Retail	No. of *Salmonella*		Total
	Susceptible	QR	
Fresh market	30	28	58 (n =100)
Modern trade	6	0	6 (n = 50)

Table 2. P_{PROB} and mean concentration of contaminants in the pork samples.

Retail	P_{PROB} (%)			Mean Concentration ± SD (log cfu/g)		
	Salmonella spp.		Total	*Salmonella* spp.		Total *
	Susceptible	QR		Susceptible	QR	
Fresh market	30.4	28.4	57.8	1.5 ± 0.8	2.1 ± 0.7	1.8 ± 0.8
Modern trade	13.5	1.9	6.9	1.9 ± 0.9	0	1.9 ± 0.9

* Accounted for only positive samples.

Table 3. Probabilities of exposure (P_E), illness (P_I), and mortality given illness (P_{MI}) from susceptible and QR *Salmonella* spp.

Retail	P_E		P_I		P_{MI}	
	Salmonella spp.		*Salmonella* spp.		*Salmonella*	
	Susceptible	QR	Susceptible	QR	Susceptible	QR
Fresh market	0.020	0.030	1.8×10^{-4}	3.1×10^{-4}	2.3×10^{-7}	5.4×10^{-6}
Modern trade	0.016	2×10^{-7}	3.2×10^{-4}	2.2×10^{-8}	4.2×10^{-7}	3.9×10^{-10}

3.2. Hazard Characterization

The P_I and P_{MI} from QR Salmonella-contaminated pork at retails in Chiang Mai province were in the range of 2.2×10^{-8}–3.1×10^{-4} and 3.9×10^{-10}–5.4×10^{-6}, respectively, while those from susceptible Salmonella-contaminated pork at retails were in the range of 1.8×10^{-4}–3.2×10^{-4} and 2.3×10^{-7}–4.2×10^{-7}, respectively (Table 3).

3.3. Risk Characterization

The descriptive statistics and probability distributions of risk estimates in terms of P_{FM} and M_{AFM} from consuming retail pork contaminated with susceptible and QR *Salmonella* in Chiang Mai province, after performing a Monte Carlo simulation, are shown in Table 4 and Figures 1–3. The mean P_{FM} of susceptible *Salmonella* was lower than that of QR *Salmonella* from the fresh market. On the other hand, in the modern trade, the mean P_{FM} of susceptible *Salmonella* became higher than that of QR *Salmonella*, essentially because the mean concentration of susceptible *Salmonella* was much higher than that of QR *Salmonella*.

Table 4. Descriptive statistics of risk estimate (P_{FM}) and annual mortality rate (M_{AFM}) from consuming pork contaminated with susceptible and AMR *Salmonella* spp.

Retail		Risk Estimate		Annual Cases *	
		Salmonella spp.		*Salmonella* spp.	
		Susceptible	AMR	Susceptible	AMR
Fresh market	min	5.3×10^{-13}	8.8×10^{-11}	<1	<1
	mean	5.7×10^{-9}	2.0×10^{-7}	<1 [a]	7 [b]
	max	2.7×10^{-8}	8.8×10^{-7}	1	32
Modern trade	min	1.5×10^{-12}	4.2×10^{-21}	<1	<1
	mean	7.9×10^{-9}	7.4×10^{-17}	<1 [c]	<1 [d]
	max	4.0×10^{-8}	7.6×10^{-16}	2	<1

* Mean annual cases per 100,000 population (P_{AFM}) with different letters implies that there are statistically significant differences ($p < 0.05$) (letters *a* through *d*).

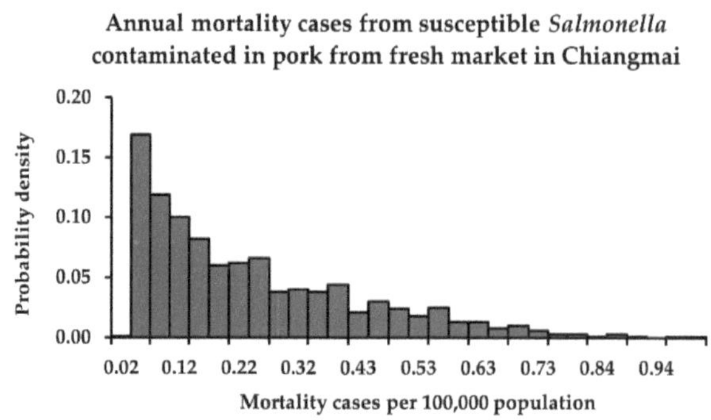

Figure 1. Annual mortality cases from susceptible *Salmonella*-contaminated pork from the fresh market in Chiang Mai.

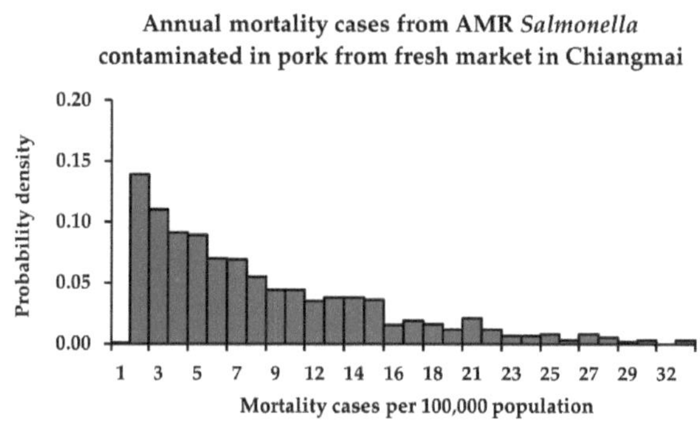

Figure 2. Annual mortality cases from QR *Salmonella*-contaminated pork from the fresh market in Chiang Mai.

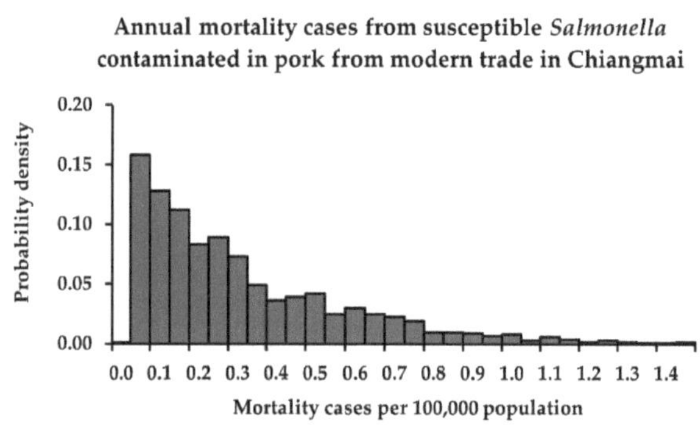

Figure 3. Annual mortality cases from susceptible *Salmonella*-contaminated pork from the modern trade in Chiang Mai.

4. Discussion

Two major approaches to AMR RA were determined by the data characteristics. The qualitative approach requires only a few calculations. The data variable is measured by the ordinal scale, e.g., low, moderate, and high. This could avoid complicated mathematical models and statistics, thus rendering risk assessment more straightforward, prolific, and time-saving. Nevertheless, the major drawback of qualitative AMR RA is the inherent subjectivity. One recommended solution to this dilemma is to transparently state or match the numerical values corresponding to individual descriptive terms for a qualitative variable [38,39]. Even though CAC encourages the quantitative technique to be performed as much as possible, the qualitative technique could not be discounted [4]. For the "quantitative technique", the variables are measured by either interval or ratio scale. Two subcategories of quantitative AMR RA are deterministic and stochastic methods. Variables in the deterministic method possess only one single value, while those in the stochastic method encompass probability density corresponding to all possible values of a variable in the form of probability distribution [40–42]. This technique is more objective than the former technique, while complicated mathematical models are involved in almost every step of AMR RA (from hazard characterization to risk characterization) since the data in this study were allowed to quantitatively evaluate the mortality risk using Monte Carlo simulations. Therefore, the outputs from the mathematical models such as P_E, P_{MI}, and risk estimate are comparable whether between susceptible and QR *Salmonella* or fresh market and modern trade.

To better quantify the risk of exposure to the hazard, the types of hazard should be defined. Hazard, in the context of AMR RA, is either AMR pathogenic bacteria or an AMR determinant. The former hazard or sometimes so-called direct hazard in food is the AMR pathogenic microorganism being capable of colonizing and then infecting a human host. Furthermore, the direct hazard is also derived from handling contaminated food [43], while AMR bacteria harboring resistance genes directly transfer resistance genes to pathogenic bacteria or indirectly transfer to the commensal bacteria. The AMR determinant or resistant genes transferred through the last two mechanisms is a so-called indirect hazard [4]. This study determines the AMR hazard by both phenotypic and genotypic analyses; therefore, the AMR P_{PROB} is more conservative and prevalent than taking into account only the AMR hazard from the genotypic analysis [44].

This study collected pork samples in Chiang Mai province in northern Thailand to investigate the risk of consuming pork contaminated with susceptible and QR *Salmonella*. The P_{PROB} of susceptible and QR *Salmonella* isolated from the fresh market were in the narrow range of 28–30% (Table 2), while the P_{PROB} of susceptible *Salmonella* was about 10 times higher than the P_{PROB} of QR *Salmonella* isolated from the modern trade. The overall P_{PROB} of (both susceptible and QR) *Salmonella* from the fresh market is eight times more than the P_{PROB} of *Salmonella* from the modern trade. Likewise, QR *Salmonella* from the fresh market is almost 15 times more prevalent than susceptible *Salmonella* from the modern trade. In 2014, a similar study collected pork samples to compare the prevalence of susceptible and AMR *Salmonella* from the fresh market and the modern trade in Chiang Mai [45]. Even though 73% of fresh-market pork contaminated with *Salmonella* was more prevalent than only 10% of modern-trade pork contaminated with *Salmonella*, *Salmonella* prevalence from the fresh market in this previous study was slightly higher than the P_{PROB} of *Salmonella* from the fresh market in our study. These compatible findings suggest that the sanitation along the pork supply chain of the fresh market in Chiang Mai province should have been improved.

Even though several *Salmonella* contaminations along the pork supply chain from farms and slaughterhouses to retail were reported in Chiang Mai province in northern Thailand [11,46–50], the magnitude of the contamination of *Salmonella* was reported as a percentage by the detection technique, since the risk assessment approach recommended by the Codex Alimentarius requires both the prevalence and concentration of *Salmonella*, particularly at the point of consumption. Only one previous study in Chiang Mai reported that *Salmonella* prevalence and concentration in pork from the fresh market were 39% (27/70) and 1.31 ± 0.25 log MPN/g, respectively [46]. The mean concentration of *Salmonella* from the previous study was lower than that of *Salmonella* from the fresh market in our study at 1.8 ± 0.8 log cfu/g (Table 2). We assume that MPN/g and cfu/g are compatible units and take into account the standard deviations from these two studies; so far *Salmonella* concentration in pork from the fresh market has never been changed. Note that the Commission Regulation on the microbiological criteria for foodstuffs indicated that *Salmonella* was not detected in the area tested per pig carcass after dressing but before chilling by the EN/ISO 6579 analytical reference method [50].

In this study, P_E as a result of the exposure assessment step was derived from P_{PROB} and the concentration of either susceptible or QR *Salmonella*, including the pork consumption of the Thai population, as shown in Equation (7) [6]. An alternative model to determine human exposure to AMR hazards per person per day requires additional parameters such as cross-contamination, which is dependent upon transfer rates between the food product and the environment [51]. The P_E of QR *Salmonella* from fresh-market pork is considered low at 3×10^{-2}, while the P_E of QR *Salmonella* from modern-trade pork at 2×10^{-7} is considered negligible [52]. These results indicate that the P_E of QR *Salmonella* from fresh market and modern trade followed the magnitude of both P_{PROB} and the concentration of QR *Salmonella*.

In terms of hazard characterization, the consequence of hazard was determined by the dose–response model while AMR RA additionally includes the consequence of AMR [4] as the probability of mortality given illness (P_{MI}) in this study. The P_{MI} of QR *Salmonella* in the fresh market is much higher than P_{MI} in the modern trade (Table 3), primarily because the probability of exposure (P_E) of QR *Salmonella* in the fresh market is higher than the P_E in the modern trade. In general, the P_E model is determined by P_{PROB} and the concentration (C_S) of *Salmonella* (Equation (7)). This indicates that the adverse health effect of QR *Salmonella* from consuming fresh-market pork was higher than that from consuming modern-trade pork in Chiang Mai province.

So far, there have been very few risk assessments evaluating human health effects due to AMR *Salmonella*. One of these studies was the risk assessment of AMR *Salmonella* related to cattle [53,54]. A qualitative approach evaluated the additional risk of QR *Salmonella* recovered from minced pork as high [25]. Another qualitative risk assessment of human health effects from QR *Salmonella* Typhimurium in the EU upon using a (fluoro)quinolone in livestock (not necessarily swine) suggested the risk was low [55]. However, a quantitative risk assessment evaluated the human health effects of multi-resistant *Salmonella* Typhimurium DT104-contaminated Danish pork sausage [24]. The risk of salmonellosis from consuming such dry-cured pork sausages was in the range of 2.5×10^{-8}–1.9×10^{-6}, whereas in our study the mean mortality risks of QR *Salmonella* from modern-trade and fresh-market pork were as low as 7.4×10^{-17} and 2.0×10^{-7}, respectively.

A previous study in Thailand reported that the annual mean mortality rate in 2009 (calculated from an average of the annual mortality cases of four major AMR bacteria (*Acinetobacter baumannii, Staphylococcus aureus, Klebsiella pneumoniae*, and *E. coli*)) was about 14.8 per 100,000 Thai population and was assumed to be the annual mean mortality rate for AMR salmonellosis [3]. In this study, the annual mortality rates for QR salmonellosis simulated by the risk assessment models were in the range of 0–32, which is in line with a previous study. The risk assessment models used in both exposure assessment and hazard characterization were applicable to evaluate the adverse health effects of AMR *Salmonella* in Thailand.

5. Conclusions

As far as we are aware, this is the first study of the quantitative microbial risk assessment of QR *Salmonella* in retail pork using a Monte Carlo simulation to comparatively report the human health adverse effects of susceptible and QR *Salmonella* from consuming retail pork from fresh market and modern trade, particularly in Thailand. The P_{PROB} of both susceptible and QR *Salmonella* from the retail market are higher than the P_{PROB} from modern trade. Likewise, the risk estimate in terms of the annual mortality rate of QR *Salmonella* from the fresh market is higher than that of QR *Salmonella* from modern trade and is also in line with a previous study reporting the mortality rate of AMR pathogens. The risk assessment models used in this study fit for evaluating the adverse health effects of QR *Salmonella* in Thailand and that of other foodborne AMR pathogens.

Supplementary Materials: The following supporting information can be downloaded at: https://www.mdpi.com/article/10.3390/foods11192942/s1, Table S1. Primers for carbapenemase genes; Table S2. Primers for plasmid-mediated colistin resistance genes (*mcr-1-mcr-9*); Table S3. Primers for plasmid-mediated quinolone resistance (PMQR) genes; Table S4. Primers for quinolone resistance-determining region (QRDR) genes. References [56–58] are cited in the Supplementary Materials.

Author Contributions: Conceptualization, S.N.; methodology, C.P., A.K., P.B., P.C. (Peechanika Chopjitt), R.H., P.C. (Piyarat Chansiripornchai) and N.S.; investigation, C.P., A.K., P.B., P.C. (Peechanika Chopjitt), R.H., P.C. (Piyarat Chansiripornchai) and N.S.; formal analysis, S.N.; writing—original draft preparation, C.P. and S.N.; writing—review and editing, C.P., A.K., P.B., P.C. (Peechanika Chopjitt), R.H., P.C. (Piyarat Chansiripornchai) and N.S.; supervision and review, A.K. and S.N.; project administration, S.N.; funding acquisition, A.K., P.B., P.C. (Peechanika Chopjitt) and R.H. All authors have read and agreed to the published version of the manuscript.

Funding: This study was funded by the National Research Council of Thailand (NRCT).

Data Availability Statement: All the data presented within the article is available upon request from the corresponding author.

Conflicts of Interest: The authors declare no conflict of interest.

References

1. Snary, E.; McEwen, S. Antimicrobial Resistance Risk Assessment. In *Guide to Antimicrobial Use in Animals*; Guardabassi, L., Jensen, L.B., Kruse, H., Eds.; Blackwell Publishing, Ltd.: Oxford, UK, 2008; pp. 27–43. [CrossRef]
2. Booton, R.D.; Meeyai, A.; Alhusein, N.; Buller, H.; Feil, E.; Lambert, H.; Mongkolsuk, S.; Pitchforth, E.; Reyher, K.K.; Sakcamduang, W.; et al. One Health drivers of antibacterial resistance: Quantifying the relative impacts of human, animal and environmental use and transmission. *One Health* **2021**, *12*, 8. [CrossRef]
3. Phumart, P.; Phodha, T.; Thamlikitkul, V.; Riewpaiboon, A.; Prakongsai, P.; Limwattananon, S. Health and economic impacts of antimicrobial resistant infections in Thailand: A Preliminary Study. *J. Health Syst. Res.* **2012**, *6*, 360.
4. Codex Alimentarius Commission (CAC). *Guidelines for Risk Analysis of Foodborne Antimicrobial Resistance (CXG 77-2011)*; FAO: Rome, Italy, 2011; p. 28.
5. FAO; WHO. *Principles and Guidelines for the Conduct of Microbiological Risk Assessment (CAC/GL-30)*, 2nd ed.; FAO: Rome, Italy; WHO: Geneva, Switzerland, 1999; p. 10.
6. Khantasup, K.; Tungwongjulaniam, C.; Theerawat, R.; Lamaisri, T.; Piyalikit, K.; Nuengjamnong, C.; Nuanualsuwan, S. Cross-sectional risk assessment of zoonotic *Streptococcus suis* in pork and swine blood in Nakhon Sawan Province in northern Thailand. *Zoonoses Public Health* **2022**, *69*, 625–634. [CrossRef] [PubMed]
7. Hanson, R.; Kaneene, J.B.; Padungtod, P.; Hirokawa, K.; Zeno, C. Prevalence of *Salmonella* and *E. coli*, and their resistance to antimicrobial agents, in farming communities in northern Thailand. *Southeast Asian J. Trop. Med. Public Health* **2002**, *33* (Suppl. S3), 126.
8. Padungtod, P.; Kaneene, J.B. *Salmonella* in food animals and humans in northern Thailand. *Int. J. Food Microbiol.* **2006**, *108*, 354. [CrossRef]
9. Sanpong, P.; Theeragool, G.; Wajjwalku, W.; Amavisit, P. Characterization of multiple antimicrobial resistant *Salmonella* isolated from pig farms in Thailand. *Agric. Nat. Resour.* **2010**, *44*, 651.
10. Tacconelli, E.; Carrara, E.; Savoldi, A.; Harbarth, S.; Mendelson, M.; Monnet, D.L.; Pulcini, C.; Kahlmeter, G.; Kluytmans, J.; Carmeli, Y.; et al. Discovery, research, and development of new antibiotics: The WHO priority list of antibiotic-resistant bacteria and tuberculosis. *Lancet Infect. Dis.* **2018**, *18*, 327. [CrossRef]
11. Poomchuchit, S.; Kerdsin, A.; Chopjitt, P.; Boueroy, P.; Hatrongjit, R.; Akeda, Y.; Tomono, K.; Nuanualsuwan, S.; Hamada, S. Fluoroquinolone resistance in non-typhoidal *Salmonella enterica* isolated from slaughtered pigs in Thailand. *J. Med. Microbiol.* **2021**, *70*, 9. [CrossRef] [PubMed]
12. Angkititrakul, S.; Chomvarin, C.; Chaita, T.; Kanistanon, K.; Waethewutajarn, S. Epidemiology of antimicrobial resistance in *Salmonella* isolated from pork, chicken meat and humans in Thailand. *Southeast Asian J. Trop. Med. Public Health* **2005**, *36*, 1515.
13. Holmberg, S.D.; Solomon, S.L.; Blake, P.A. Health and economic impacts of antimicrobial resistance. *Rev. Infect. Dis.* **1987**, *9*, 1078. [CrossRef]
14. Molbak, K.; Baggesen, D.L.; Aarestrup, F.M.; Ebbesen, J.M.; Engberg, J.; Frydendahl, K.; Gerner-Smidt, P.; Petersen, A.M.; Wegener, H.C. An outbreak of multidrug-resistant, quinolone-resistant *Salmonella enterica* serotype typhimurium DT104. *N. Engl. J. Med.* **1999**, *341*, 1425. [CrossRef] [PubMed]
15. Walker, R.A.; Lawson, A.J.; Lindsay, E.A.; Ward, L.R.; Wright, P.A.; Bolton, F.J.; Wareing, D.R.; Corkish, J.D.; Davies, R.H.; Threlfall, E.J. Decreased susceptibility to ciprofloxacin in outbreak-associated multiresistant *Salmonella typhimurium* DT104. *Vet. Rec.* **2000**, *147*, 395–396. [CrossRef] [PubMed]

16. Meakins, S.; Fisher, I.S.; Berghold, C.; Gerner-Smidt, P.; Tschape, H.; Cormican, M.; Luzzi, I.; Schneider, F.; Wannett, W.; Coia, J.; et al. Antimicrobial drug resistance in human nontyphoidal *Salmonella* isolates in Europe 2000–2004: A report from the Enter-net International Surveillance Network. *Microb. Drug Resist.* **2008**, *14*, 35. [CrossRef] [PubMed]
17. Threlfall, E.J.; Ward, L.R.; Skinner, J.A.; Graham, A. Antimicrobial drug resistance in non-typhoidal *Salmonellas* from humans in England and Wales in 1999: Decrease in multiple resistance in *Salmonella enterica* serotypes Typhimurium, Virchow, and Hadar. *Microb. Drug Resist.* **2000**, *6*, 325. [CrossRef] [PubMed]
18. Varma, J.K.; Greene, K.D.; Ovitt, J.; Barrett, T.J.; Medalla, F.; Angulo, F.J. Hospitalization and antimicrobial resistance in *Salmonella* outbreaks, 1984–2002. *Emerg. Infect. Dis.* **2005**, *11*, 946. [CrossRef] [PubMed]
19. Crook, P.D.; Aguilera, J.F.; Threlfall, E.J.; O'Brien, S.J.; Sigmundsdottir, G.; Wilson, D.; Fisher, I.S.; Ammon, A.; Briem, H.; Cowden, J.M.; et al. A European outbreak of *Salmonella enterica* serotype Typhimurium definitive phage type 204b in 2000. *Clin. Microbiol. Infect.* **2003**, *9*, 845. [CrossRef]
20. Horby, P.W.; O'Brien, S.J.; Adak, G.K.; Graham, C.; Hawker, J.I.; Hunter, P.; Lane, C.; Lawson, A.J.; Mitchell, R.T.; Reacher, M.H.; et al. A national outbreak of multi-resistant *Salmonella* enterica serovar Typhimurium definitive phage type DT 104 associated with consumption of lettuce. *Epidemiol. Infect.* **2003**, *130*, 178. [CrossRef]
21. Bertrand, S.; Weill, F.X.; Cloeckaert, A.; Vrints, M.; Mairiaux, E.; Praud, K.; Dierick, K.; Wildemauve, C.; Godard, C.; Butaye, P.; et al. Clonal emergence of extended-spectrum beta-lactamase (CTX-M-2)-producing *Salmonella enterica* serovar Virchow isolates with reduced susceptibilities to ciprofloxacin among poultry and humans in Belgium and France (2000 to 2003). *J. Clin. Microbiol.* **2006**, *44*, 2903. [CrossRef]
22. Cloeckaert, A.; Praud, K.; Doublet, B.; Bertini, A.; Carattoli, A.; Butaye, P.; Imberechts, H.; Bertrand, S.; Collard, J.M.; Arlet, G.; et al. Dissemination of an extended-spectrum-beta-lactamase *bla*TEM-52 gene-carrying IncI1 plasmid in various *Salmonella enterica* serovars isolated from poultry and humans in Belgium and France between 2001 and 2005. *Antimicrob. Agents Chemother.* **2007**, *51*, 1875. [CrossRef]
23. Caffrey, N.; Invik, J.; Waldner, C.; Ramsay, D.; Checkley, S. Risk assessments evaluating foodborne antimicrobial resistance in humans: A scoping review. *J. Microb. Risk Anal.* **2019**, *11*, 46. [CrossRef]
24. Alban, L.; Olsen, A.M.; Nielsen, B.; Sorensen, R.; Jessen, B. Qualitative and quantitative risk assessment for human salmonellosis due to multi-resistant *Salmonella* Typhimurium DT104 from consumption of Danish dry-cured pork sausages. *Prev. Vet. Med.* **2002**, *52*, 265. [CrossRef]
25. Doménech, E.; Jiménez-Belenguer, A.; Pérez, R.; Ferrús, M.A.; Escriche, I. Risk characterization of antimicrobial resistance of *Salmonella* in meat products. *J. Food Control* **2015**, *57*, 23. [CrossRef]
26. Collineau, L.; Chapman, B.; Bao, X.; Sivapathasundaram, B.; Carson, C.A.; Fazil, A.; Reid-Smith, R.J.; Smith, B.A. A farm-to-fork quantitative risk assessment model for *Salmonella* Heidelberg resistant to third-generation cephalosporins in broiler chickens in Canada. *Int. J. Food Microbiol.* **2020**, *330*, 21. [CrossRef] [PubMed]
27. Li, X.; Liang, B.; Xu, D.; Wu, C.; Li, J.; Zheng, Y. Antimicrobial Resistance Risk Assessment Models and Database System for Animal-Derived Pathogens. *Antibiotics* **2020**, *9*, 829. [CrossRef] [PubMed]
28. Andrews, W.H.; Jacobson, A.; Hammack, T. Chapter 5: Salmonella. In *Bacteriological Analytical Manual (BAM)*; U.S. Food and Drug Administration: Silver Spring, MD, USA, 2018.
29. Haas, C.N.; Rose, J.B.; Gerba, C.P. *Quantitative Microbial Risk Assessment*; John Wiley & Sons, Inc.: Hoboken, NJ, USA, 1999; p. 449.
30. FAO; WHO. *Risk Assessments of Salmonella in Eggs and Broiler Chickens*; FAO: Rome, Italy; WHO: Geneva, Switzerland, 2002; p. 302.
31. *M07-A11*; Methods for Dilution Antimicrobial Susceptibility Tests for Bacteria that Grow Aerobically. 10th ed. Clinical and Laboratory Standards Institute (CLSI): Wayne, PA, USA, 2018.
32. *M100-S30*; Performance Standards for Antimicrobial Susceptibility Testing. 30th ed. Clinical and Laboratory Standards Institute (CLSI): Wayne, PA, USA, 2020.
33. Ciesielczuk, H.; Hornsey, M.; Choi, V.; Woodford, N.; Wareham, D.W. Development and evaluation of a multiplex PCR for eight plasmid-mediated quinolone-resistance determinants. *J. Med. Microbiol.* **2013**, *62*, 1827. [CrossRef] [PubMed]
34. Lu, Y.; Zhao, H.; Liu, Y.; Zhou, X.; Wang, J.; Liu, T.; Beier, R.C.; Hou, X. Characterization of quinolone resistance in *Salmonella enterica* serovar Indiana from chickens in China. *Poult. Sci.* **2015**, *94*, 460. [CrossRef]
35. Alali, W.Q.; Mann, D.A.; Beuchat, L.R. Viability of *Salmonella* and *Listeria monocytogenes* in delicatessen salads and hummus as affected by sodium content and storage temperature. *J. Food Prot.* **2012**, *75*, 1056. [CrossRef]
36. Ruchusatsawat, K.; Nuengjamnong, C.; Tawatsin, A.; Thiemsing, L.; Kawidam, C.; Somboonna, N.; Nuanualsuwan, S. Quantitative Risk Assessments of Hepatitis A Virus and Hepatitis E Virus from Raw Oyster Consumption. *Risk Anal.* **2022**, *42*, 965. [CrossRef]
37. Nuanualsuwan, S.; Songkasupa, T.; Boonpornprasert, P.; Suwankitwat, N.; Lohlamoh, W.; Nuengjamnong, C. Thermal Inactivation of African Swine Fever Virus in Swill. *Front. Vet. Sci.* **2022**, *9*, 8. [CrossRef]
38. Bywater, R.J.; Casewell, M.W. An assessment of the impact of antibiotic resistance in different bacterial species and of the contribution of animal sources to resistance in human infections. *J. Antimicrob. Chemother.* **2000**, *46*, 645. [CrossRef]
39. Presi, P.; Stark, K.D.; Stephan, R.; Breidenbach, E.; Frey, J.; Regula, G. Risk scoring for setting priorities in a monitoring of antimicrobial resistance in meat and meat products. *Int. J. Food Microbiol.* **2009**, *130*, 100. [CrossRef] [PubMed]

40. Berends, B.R.; van den Bogaard, A.E.; Van Knapen, F.; Snijders, J.M. Human health hazards associated with the administration of antimicrobials to slaughter animals. Part II. An assessment of the risks of resistant bacteria in pigs and pork. *Vet. Q.* **2001**, *23*, 21. [CrossRef]
41. Hald, T.; Lo Fo Wong, D.M.; Aarestrup, F.M. The attribution of human infections with antimicrobial resistant *Salmonella* bacteria in Denmark to sources of animal origin. *Foodborne Pathog. Dis.* **2007**, *4*, 326. [CrossRef]
42. Sommer, H.M.; Aabo, S.; Christensen, B.B.; Saadby, P.; Nielsen, N.; Nørrung, B.; Wong, D.L.F. *Risk Assessment of the Impact on Human Health Related to Multiresistant Salmonella Typhimurium DT104 from Slaughter Pigs*; Institute of Food Safety and Nutrition: Mørkhøj, Denmark, 2003; p. 104.
43. European Food Safety Authority (EFSA). Scientific Opinion of the Panel on Biological Hazards on a request from the European Food Safety Authority on foodborne antimicrobial resistance as a biological hazard. *EFSA* **2008**, *765*, 87.
44. Aslam, M.; Checkley, S.; Avery, B.; Chalmers, G.; Bohaychuk, V.; Gensler, G.; Reid-Smith, R.; Boerlin, P. Phenotypic and genetic characterization of antimicrobial resistance in *Salmonella* serovars isolated from retail meats in Alberta, Canada. *Food Microbiol.* **2012**, *32*, 117. [CrossRef]
45. Patchanee, P.; Tansiricharoenkul, K.; Buawiratlert, T.; Wiratsudakul, A.; Angchokchatchawal, K.; Yamsakul, P.; Yano, T.; Boonkhot, P.; Rojanasatien, S.; Tadee, P. *Salmonella* in pork retail outlets and dissemination of its pulsotypes through pig production chain in Chiang Mai and surrounding areas, Thailand. *Prev. Vet. Med.* **2016**, *130*, 105. [CrossRef]
46. Prasertsee, T.; Chokesajjawatee, N.; Santiyanont, P.; Chuammitri, P.; Deeudom, M.; Tadee, P.; Patchanee, P. Quantification and rep-PCR characterization of *Salmonella* spp. in retail meats and hospital patients in Northern Thailand. *Zoonoses Public Health* **2019**, *66*, 309. [CrossRef]
47. Sanguankiat, A.; Pinthong, R.; Padungtod, P.; Baumann, M.P.; Zessin, K.H.; Srikitjakarn, L.; Fries, R. A cross-sectional study of *Salmonella* in pork products in Chiang Mai, Thailand. *Foodborne Pathog. Dis.* **2010**, *7*, 878. [CrossRef]
48. Tadee, P.; Boonkhot, P.; Pornruangwong, S.; Patchanee, P. Comparative phenotypic and genotypic characterization of *Salmonella* spp. in pig farms and slaughterhouses in two provinces in northern Thailand. *PLoS ONE* **2015**, *10*, 11. [CrossRef]
49. Tadee, P.; Kumpapong, K.; Sinthuya, D.; Yamsakul, P.; Chokesajjawatee, N.; Nuanualsuwan, S.; Pornsukarom, S.; Molla, B.Z.; Gebreyes, W.A.; Patchanee, P. Distribution, quantitative load and characterization of *Salmonella* associated with swine farms in upper-northern Thailand. *J. Vet. Sci.* **2014**, *15*, 334. [CrossRef]
50. European Commission. *Microbiological Criteria for Foodstuffs: Commission Regulation (EC) No 2073/2005 on 2005*; European Commission: Brussels, Belgium, 2005; pp. 1–26.
51. Hald, T.; Duarte, A.R.; Stärk, K. Quantifying human exposure to antimicrobial resistance from animals and food. In Proceedings of the International Conference on Animal Health Surveillance, Wellington, New Zealand, 30 April–4 May 2017; p. 2.
52. Biosecurity Australia. *Generic Import Risk Analysis Report for Chicken Meat: Final Report*; Biosecurity Australia: Canberra, Australia, 2008.
53. Hurd, H.S.; Vaughn, M.B.; Holtkamp, D.; Dickson, J.; Warnick, L. Quantitative risk from fluoroquinolone-resistant *Salmonella* and *Campylobacter* due to treatment of dairy heifers with enrofloxacin for bovine respiratory disease. *J. Foodborne Pathog.* **2010**, *7*, 1322. [CrossRef] [PubMed]
54. Singer, R.S.; Ruegg, P.L.; Bauman, D.E. Quantitative Risk Assessment of Antimicrobial-Resistant Foodborne Infections in Humans Due to Recombinant Bovine Somatotropin Usage in Dairy Cows. *J. Food Prot.* **2017**, *80*, 1116. [CrossRef] [PubMed]
55. European Agency for the Evaluation of Medicinal Products. *Antibiotic Resistance in the European Union Associated with Therapeutic Use of Veterinary Medicines: Report and Qualitative Risk Assessment by the Committee for Veterinary Medicinal Products*; European Agency for the Evaluation of Medicinal Products: Amsterdam, The Netherlands, 1999.
56. Poirel, L.; Walsh, T.R.; Cuvillier, V.; Nordmann, P. Multiplex PCR for detection of acquired carbapenemase genes. *Diagn. Microbiol. Infect Dis* **2011**, *70*, 119–123. [CrossRef]
57. Hatrongjit, R.; Kerdsin, A.; Akeda, Y.; Hamada, S. Detection of plasmid-mediated colistin-resistant and carbapenem-resistant genes by multiplex PCR. *MethodsX* **2018**, *5*, 532–536. [CrossRef]
58. Khanawapee, A.; Kerdsin, A.; Chopjitt, P.; Boueroy, P.; Hatrongjit, R.; Akeda, Y.; Tomono, K.; Nuanualsuwan, S.; Hamada, S. Distribution and Molecular Characterization of Escherichia coli Harboring mcr Genes Isolated from Slaughtered Pigs in Thailand. *Microb. Drug Resist.* **2021**, *27*, 971–979. [CrossRef]

foods

Article

Accurate Detection of *Salmonella* Based on Microfluidic Chip to Avoid Aerosol Contamination

Yining Luo [1], Shan Shan [2,3], Shuanglong Wang [4], Jinlin Li [2], Daofeng Liu [3] and Weihua Lai [1,*]

[1] State Key Laboratory of Food Science and Technology, Nanchang University, 235 East Nanjing Road, Nanchang 330047, China
[2] College of Life Science, National R&D Center for Freshwater Fish Processing, Jiangxi Normal University, Nanchang 330022, China
[3] Jiangxi Province Key Laboratory of Diagnosing and Tracing of Foodborne Disease, Jiangxi Province Centre for Disease Control and Prevention, 555 East Beijing Road, Nanchang 330029, China
[4] Jiangxi Key Laboratory for Mass Spectrometry and Instrumentation, East China University of Technology, Nanchang 330013, China
* Correspondence: talktolaiwh@163.com

Abstract: *Salmonella* is a type of common foodborne pathogen of global concern, seriously endangering human health. In molecular biological detection of *Salmonella*, the method of amplifying DNA often faces the problem of aerosol pollution. In this study, a microfluidic chip was developed to integrate loop-mediated isothermal amplification (LAMP) and clustered regularly interspaced short palindromic repeats (CRISPR)/Cas12a system to detect *Salmonella*. The LAMP reaction solution was initially injected into the chamber to amplify at 65 °C for 20 min; the CRISPR/Cas12a reaction solution was subsequently injected to mix with the amplicons for fluorescent signal production at 43 °C for 30 min. Then, the results can be confirmed by naked eyes under 495 nm light or by a fluorescence immunochromatographic reader. The detection limit of this method for *Salmonella* DNA was 118 pg/μL. The sensitivity and specificity of this method was 100%. Furthermore, this method was used to detect *Salmonella* after enrichment for 4 h in salmon and chicken samples spiked with 30 CFU/25 g, and was verified to have a stable detection capability in real samples. The microfluidic chip integrated with the LAMP and CRISPR/Cas12a system not only provides a possibility of highly sensitive endpoint fluorescent visual detection of a foodborne pathogen, but also greatly eliminates the risk of aerosol contamination.

Keywords: *Salmonella*; LAMP; CRISPR/Cas12a; visual detection

Citation: Luo, Y.; Shan, S.; Wang, S.; Li, J.; Liu, D.; Lai, W. Accurate Detection of *Salmonella* Based on Microfluidic Chip to Avoid Aerosol Contamination. *Foods* **2022**, *11*, 3887. https://doi.org/10.3390/foods11233887

Academic Editors: Fernando Pérez-Rodríguez and Arícia Possas

Received: 4 September 2022
Accepted: 28 November 2022
Published: 1 December 2022

1. Introduction

Salmonella is an important pathogen that poses a substantial threat to human health. Approximately 86% of the described salmonellosis cases were caused by foodborne infections [1]. Aquatic products, especially salmon eaten without heat treatment, are easily contaminated with *Salmonella* and more likely cause harm to human health [2]. Human infection by *Salmonella* strains may result in fever, vomiting, abdominal pain, nausea, hemorrhagic enteritis, and other symptoms [3–5]. Therefore, a rapid and accurate method to detect *Salmonella* contamination in food is required.

Loop-mediated isothermal amplification (LAMP) has been developed for its simplicity, high amplification capacity, and rapidity [6–8]. LAMP requires Bst DNA polymerase and a set 4–6 primers, which can be completed at a constant temperature (60–65 °C) without an expensive thermocycler [9]. Furthermore, LAMP can amplify 10^9-fold target sequence copies within 15–60 min [10,11]. The amplification capacity of LAMP is 10–100 times higher than that of traditional polymerase chain reaction (PCR) [12]. On the basis of these advantages, LAMP has been widely applied in the field of foodborne pathogen detection [13].

The clustered regularly interspaced short palindromic repeats (CRISPR)/Cas system can specifically bind and cleave target nucleic acids under the guidance of CRISPR RNA (crRNA), which identifies and captures target DNA by base-complementary pairing [14–16]. The CRISPR/Cas system has shown the potential to improve the specificity and accuracy of genetic detection [17,18]. CRISPR/Cas12a, from the type V-A CRISPR system, exhibits site-specific dsDNA cutting and nonspecific ssDNA trans-cleavage ability, which provides an efficient signal amplification tool for molecular diagnostics [19].

LAMP-binding CRISPR/Cas system has been applied in the field of disease diagnosis and foodborne pathogen detection [20–22]. The reaction process of the LAMP-binding CRISPR/Cas12a system requires two separate steps because the high temperature in the course of LAMP may reduce the activity of Cas12a. However, opening tubes and transferring LAMP solution into CRISPR/Cas reaction solution would lead to aerosol leakage [21,23]. Reports have shown that LAMP amplicons can contaminate reagents in the laboratory by aerosol for several weeks, and they can be promptly used as templates in subsequent reactions, thereby jeopardizing the results of the reactions. When contamination begins, the false-positive rate of the test results would increase sharply, unless all reagents and primers were replaced [24]. Herein, we designed a microfluidic chip as the platform to integrate LAMP and the CRISPR/Cas12a system for the accurate detection of *Salmonella*.

2. Materials and Methods

2.1. Bacterial Strains and DNA Extraction

The standard bacterial strains and clinically isolated strains were from the Jiangxi Province Center for Disease Control and Prevention. Detailed information of all strains is shown in Table 1. All bacterial strains were cultured in trypticase soy broth (TSB, Land Bridge Technology Co., Ltd., Beijing, China) on a shaking incubator at 180 rpm and 37 °C overnight. One milliliter of each bacterial culture in TSB was centrifuged at 8000 rpm/min for 5 min. Then, the pellet was washed with sterilized 1× PBS (Solarbio Co., Ltd., Beijing, China) for three times and resuspended with 1 mL 1× PBS. Resuspension was boiled for lysing bacteria cells, and then centrifuged at 8500 rpm for 5 min. The supernatants were collected to obtain genomic DNA. The concentration of DNA was determined by Micro-Spectrophotometer K5600 (Kaiao Co., Ltd., Beijing, China).

Table 1. Sources and identification numbers of strains.

Bacteria	Source	Identification Number
Salmonella Enteritidis	CMCC	CMCC 50041
Salmonella Enteritidis	Jiangxi CDC	Kalado
Salmonella Enteritidis	Jiangxi CDC	14S39
Salmonella Rissen	Jiangxi CDC	15S2
Salmonella Enteritidis	Jiangxi CDC	15S50
Salmonella Kottbus	Jiangxi CDC	15S59
Salmonella Thompson	Jiangxi CDC	16S24
Salmonella Litchfield	Jiangxi CDC	17S38
Salmonella Newport	Jiangxi CDC	17S40
Salmonella London	Jiangxi CDC	17S68
Salmonella Derby	Jiangxi CDC	18S10
Salmonella Give	Jiangxi CDC	18S49
Salmonella Orion	Jiangxi CDC	18S60
Pseudomonas aeruginosa	ATCC	ATCC 27853
Pseudomonas aeruginosa	CMCC	CMCC(B) 10104
Pseudomonas aeruginosa	BNCC	CGMCC 1.1785
Pseudomonas aeruginosa	BNCC	ATCC 9027
Staphylococcus aureus	CMCC	CMCC 26002
Staphylococcus aureus	Jiangxi CDC	JP-1
Listeria monocytogenes	CMCC	CMCC 54001
Listeria monocytogenes	Jiangxi CDC	DZ-JX-1
Bacillus cereus	CMCC	CMCC 63303
Bacillus cereus	Jiangxi CDC	LY-FC-1
Escherichia coli	CMCC	CMCC 44496
Escherichia coli	CMCC	CMCC 44350

2.2. LAMP Primers, crRNA, and Report DNA Design

invA, a virulence gene, is chromosomally located and conserved in almost all the *Salmonella* serotypes [25,26]. The sequence of *invA* (GenBank No. M90846.1) was used to design LAMP primers. LAMP primers, including the forward inner primer (FIP), the backward inner primer (BIP), the outer forward primer (F3), the outer backward primer (B3), and additional loop primers (FL and BL) were designed via Primer Explorer software Version 5 (http://primerexplorer.jp/lampv5e/index.html accessed on 20 June 2021). The crRNA for specifically recognizing the specific amplicon was designed through the Benchling website (https://www.benchling.com/crispr/ accessed on 3 July 2021). Report DNA was designed as a short ssDNA (5 nt), labeled with fluorophore (6-FAM) and quencher (BHQ1) groups at both ends [27]. The primers, crRNA, and report DNA were synthesized by Sangon Co., Ltd. (Shanghai, China); the detailed sequence information is listed in Table 2.

Table 2. The sequences of primers, crRNA, and report DNA.

Name	Sequence (5′-3′)
F3	GCGAAGCGTACTGGAAAGG
B3	TCAACAATGCGGGGATCTG
FIP	ATGATGCCGGCAATAGCGTCAC-AAAGCCAGCTTTACGGTTCC
BIP	GTGGGGATGACTCGCCATGG-ACCATCACCAATGGTCAGC
LF	AAACTTCATCGCACCGTCAAA
LB	TATGGATTTGTCCTCCGCCCT
crRNA	UAAUUUCUACUAAGUGUAGAUAAUACCGCCAAUAAAGUUCA
Report DNA	6-FAM-TTATT-BHQ1

2.3. LAMP Assay

LAMP solution was prepared according to New England Biolabs (NEB) instruction in a total volume of 25 μL, containing 3.5 μL of 10 mM dNTP mix (Vazyme Co., Ltd., Nanjing, China), 2.5 μL of 10× isothermal amplification buffer, 1.5 μL of 100 mM $MgSO_4$, 1 μL of 800 U/mL Bst 2.0 DNA polymerase (New England Biolabs, Ipswich, MA, USA), 1 μL of 40 μM FIP/BIP primers, 1 μL of 5 μM F3/B3 primers, 1 μL of 10 μM FL/BL primers, and 2 μL DNA template. Then, 13 μL of mixture was incubated in thermostatic metal bath TU-10 (Yiheng Co., Ltd., Shanghai, China) at 65 °C for 5, 10, 15, 20, and 25 min and terminated at 80 °C for 10 min. The product of LAMP was characterized by QIAxcel advanced capillary electrophoresis (Qiagen, Hilden, Germany).

2.4. LAMP–CRISPR/Cas12a Detection System

The LAMP assay was performed according to Section 2.3. The Cas12a-mediated detection system contained 1.95 μL 10× NEBuffer 2.1, 1.3 μL of Cas12a (1, 2, 3, and 4 μM) (New England Biolabs, Ipswich, MA, USA), 1.95 μL of crRNA (0.5, 1.0, 1.5, 2.0, and 2.5 μM), 1.3 μL of report DNA (20, 40, 60, 80, and 100 μM), and 13 μL of LAMP amplicons as an activator. Then, 19.5 μL of mixture was incubated a thermostatic metal bath at 25 °C, 31 °C, 37 °C, 43 °C, and 49 °C for 30 min. The products of LAMP cleaved by Cas12a were characterized by capillary electrophoresis.

2.5. Detection of *Salmonella* by Microfluidic Chip Integrated with LAMP and CRISPR/Cas12a System

The polydimethylsiloxane (PDMS) chip was designed to be a reaction and observation platform, with one observation chamber and two narrow channels for injection of reaction solution. Thirteen microliters of LAMP components were initially injected into the chamber by microinjectors (Anting Co., Ltd., Shanghai, China) with smart syringe pump XMSP-C (Ximai Co., Ltd., Nanjing, China) at the speed of 15 μL/min. Then, the chip was stored at 65 °C for amplification, and subsequently at 80 °C for enzyme inactivation. After amplification, 6.5 μL of CRISPR/Cas12a components were injected into the chamber at the speed of 15 μL/min to mix with LAMP amplicons by another channel, and then the chip was stored at 43 °C. The fluorescence could be observed by the naked eyes under 495 nm light and measured by a fluorescence immunochromatographic reader (Helmen Co., Ltd., Hangzhou, China). Actual devices and the effect diagrams are shown in Figure S1.

2.6. Evaluation of Detection Limit

Typically, the detection limit is determined by setting a threshold, which can be calculated from the average fluorescence intensity and standard deviation of the negative control. The value of the threshold equals the average intensity and three times the standard deviation [28].

Extracted genomic DNA that served as the template was performed a 10-fold gradient dilution from 10^{-2} to 10^{-4} with 1× PBS to evaluate the detection limit of this method. The DNA templates of each concentration were amplified by LAMP. Then, amplicons were employed to activate the CRISPR/Cas12a cleaving system. The fluorescence intensity of these results was measured by a fluorescence immunochromatographic reader.

2.7. Sensitivity and Specificity

One *Salmonella* standard strain, twelve *Salmonella* clinical strains, nine non-*Salmonella* standard strains, and three non-*Salmonella* clinical strains were selected to test the sensitivity and specificity of this method. DNA extracted from these strains was amplified by LAMP. Then, amplicons were employed for activating the CRISPR/Cas12a cleaving system. The fluorescence intensity of these results was measured by a fluorescence immunochromatographic reader.

2.8. Detection of Salmonella in Artificially Contaminated Salmon and Chicken

Salmon and chicken samples were purchased from a local supermarket. Twenty-five grams of salmon and chicken samples, which were verified to be *Salmonella*-free, were made into homogenates and spiked with *Salmonella* at a final concentration of 30 CFU/25 g, whereas un-spiked samples were tested to evaluate the specificity of this method in real samples [29]. Then, samples were placed into TSB media (225 mL of each) and incubated at 37 °C with shaking at 180 rpm. Two milliliters of culture solution were aspirated every hour from 0 to 6 h. Solutions collected were divided into two groups. One group was tested by the LAMP–CRISPR/Cas12a system, and another group was used for colony count by *Salmonella* chromogenic medium (CHRO Magar, Paris, France).

To check the effect of the matrix of real samples on the test efficiency in comparing with the buffer solution, 25 g of salmon and chicken samples were made into homogenates with TSB medium, and all samples were irradiated under a UV lamp (15 W) for 30 min to ensure sterility. Preprocessed samples and the other TSB medium with no sample were spiked with *Salmonella* at final concentrations of 4×10^1, 4×10^2, 4×10^3, 4×10^4 CFU/mL, respectively, and then tested by the LAMP–CRISPR/Cas12a system.

3. Results

3.1. Principle and Operation of the Detection System

The principle and operation of *Salmonella* detection by the microfluidic chip integrated with the LAMP and CRISPR/Cas12a system are shown in Scheme 1. In summary, crude genomic DNA of *Salmonella* extracted using the heated lysis method was rapidly amplified by LAMP. The amplicons could be recognized and captured by crRNA, and then the CRISPR/Cas12a system could cleave LAMP amplicons with site-specific dsDNA cutting ability and report DNA with nonspecific ssDNA trans-cleavage ability. The cleavage of report DNA, which was labeled with fluorophore 6-FAM and quencher BHQ1, resulted in the appearance of fluorescence, as shown in Scheme 1A.

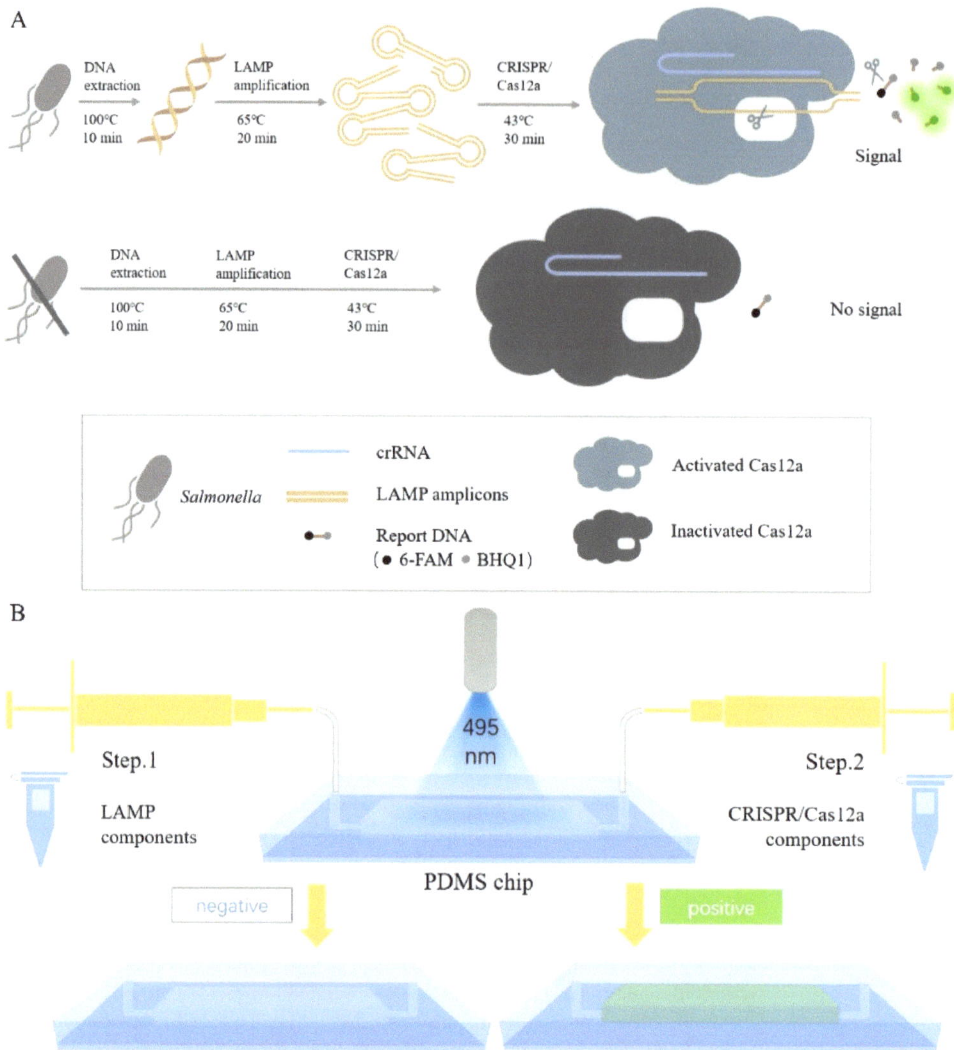

Scheme 1. Schematics of (**A**) principle and (**B**) operation of LAMP–CRISPR/Cas12a detection system on a chip to detect *Salmonella*.

PDMS chip was selected as the detection platform to eliminate the risk of aerosol contamination. LAMP components with or without DNA of *Salmonella* were initially injected into the chamber through one channel, and CRISPR/Cas12a components were injected into the chamber through another channel after LAMP was completed, as shown in Scheme 1B. Then, the result could be confirmed by the naked eyes under 495 nm light or by a fluorescence immunochromatographic reader because of the suitable size of chip pre-designed.

The risk of aerosol leakage was eliminated in this process owing to the leakproofness of the chip, thereby protecting the detection environment and reagents from aerosol pollution. With this chip, the false-positive results are greatly avoided, providing a platform for more accurate detection results in subsequent detection.

3.2. *Analysis of LAMP–CRISPR/Cas12a Detection System*

The extracted genomic DNA of *Salmonella* was amplified by LAMP. The capillary electrophoresis image (Figure S2) showed the bands, which were the LAMP product and the Cas12a-cleaved LAMP product. The difference between the two sets of bands indicated the occurrence of the cleavage.

LAMP amplicons were employed for activating the CRISPR/Cas12a cleaving system. More LAMP amplicons were contributed to obtain a higher fluorescence intensity of the CRISPR/Cas12a detection system. After 20 min, the LAMP amplification system generated ample target molecules for the CRISPR/Cas12a detection system (Figure 1). The optimal reaction time of LAMP to improve the detection efficiency was 20 min.

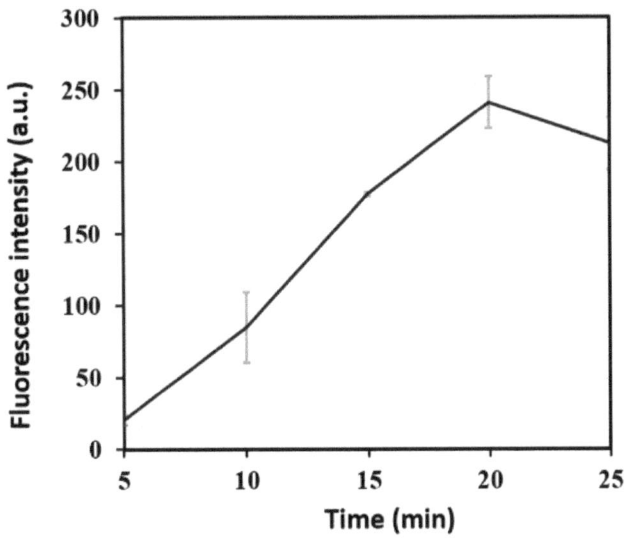

Figure 1. Determination of optimal action time of LAMP. Fluorescence intensity with different reaction time (5, 10, 15, 20, and 25 min) of LAMP.

The experimental factors, including the concentration of CRISPR/Cas12a components and reaction temperature, were investigated to reduce the detection limit. First, the fluorescence intensity increased as the concentration of Cas12a increased, as shown in Figure 2A. When the concentration of Cas12a ranged 3–4 μM, the fluorescence intensity did not increase considerably. Therefore, the optimal concentration of Cas12a was 3 μM. Second, the concentration of crRNA was optimized. As the concentration of crRNA increased, the fluorescence intensity increased significantly initially, and then the increase magnitude gradually decreased, as shown in Figure 2B. Thus, 2 μM was finally selected as the optimal concentration of crRNA. Third, as the concentration of report DNA increased, the fluorescence intensity peaked at 80 μM and then decreased, as shown in Figure 2C. Hence, 80 μM was selected as the optimal concentration. Fourth, five temperatures from 25 °C to 49 °C were selected for the experiment. The fluorescence intensity peaked at 43 °C, and the fluorescence intensity began to decrease as the temperature continued to increase, as shown in Figure 2D. This result might be due to the effect of higher temperature on the activity of the enzyme. Therefore, 43 °C was selected as the optimal reaction temperature.

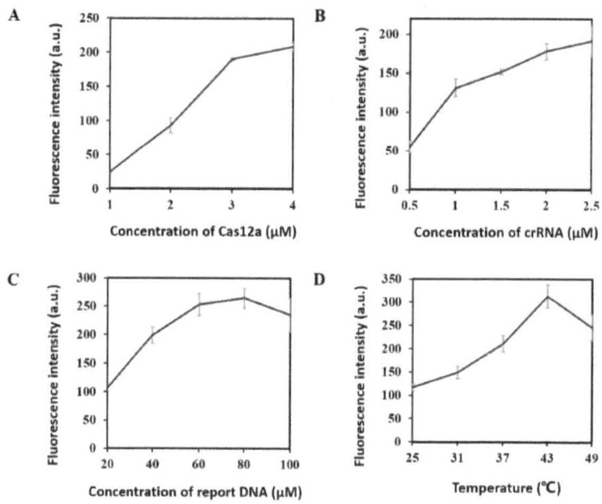

Figure 2. Optimization of LAMP–CRISPR detection system for *Salmonella* detection. (**A**) Fluorescence intensity using different concentrations (1, 2, 3, and 4 μM) of Cas12a; (**B**) Fluorescence intensity using different concentrations (0.5, 1.0, 1.5, 2.0, and 2.5 μM) of crRNA; (**C**) Fluorescence intensity using different concentrations (20, 40, 60, 80, and 100 μM) of report DNA; (**D**) Fluorescence intensity under different temperatures (25 °C, 31 °C, 37 °C, 43 °C, and 49 °C) of CRISPR/Cas12a system.

3.3. Detection Limit of the Proposed Method

After optimizing the conditions, this method was established for *Salmonella* detection. A 10-fold dilution series of extracted genomic DNA from 10^{-2} to 10^{-4} was used to determine the detection limit of this method, and the original concentration of the DNA was 118 ng/μL. The results are shown in Figure 3. The values of fluorescence intensity were 6, 162, and 232 when the concentrations of DNA were 11.8, 118, and 1180 pg/μL, respectively. The threshold was calculated as 6; thus, 118 pg/μL was regarded as the detection limit of the proposed method.

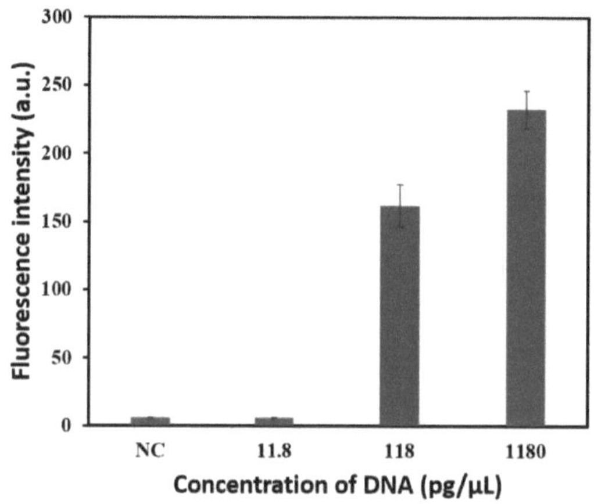

Figure 3. Detection limit test results of LAMP–CRISPR/Cas12a detection system for *Salmonella* detection using 10-fold serial dilutions of crude genomic DNA in ddH_2O (diluted from 1180 pg/μL to 11.8 pg/μL).

3.4. Sensitivity and Specificity

As shown in Figure 4, compared with the threshold (5), all twelve non-*Salmonella* strains had no evident fluorescence signal (5, 5, 5, 4, 5, 5, 4, 4, 5, 4, 5, and 5), thereby confirming that the detection method established in this study had high specificity. Meanwhile, all thirteen *Salmonella* strains had evident fluorescence signals (220, 310, 188, 200, 250, 293, 281, 277, 299, 210, 213, 230, and 259, respectively), confirming that the detection method had high sensitivity.

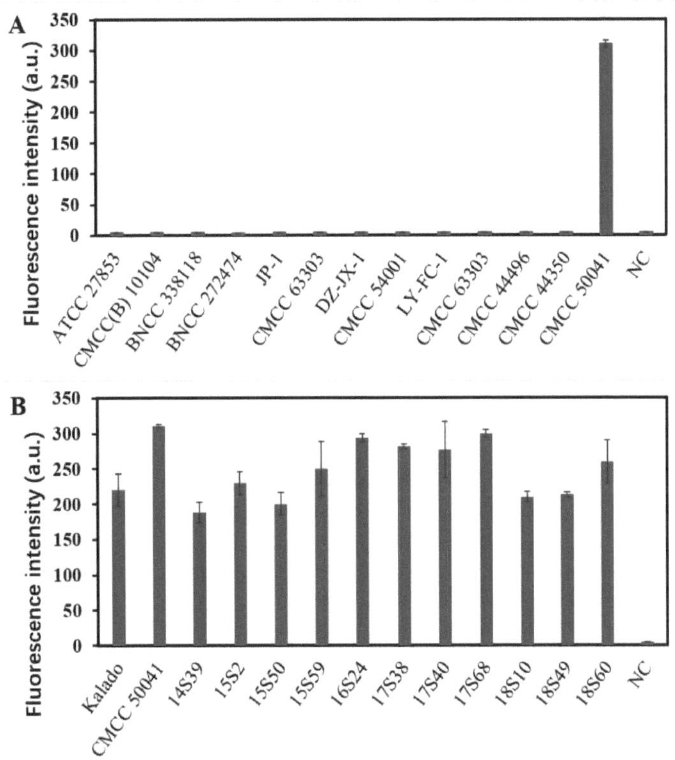

Figure 4. Estimating the specificity and sensitivity of LAMP–CRISPR detection system for *Salmonella* detection. (**A**) Fluorescence intensity using non-*Salmonella* standard strains for specific evaluation. (**B**) Fluorescence intensity using *Salmonella* standard strains for sensitive evaluation.

3.5. Detection of *Salmonella* in Salmon and Chicken Using the Proposed Method

Furthermore, the performance of the LAMP–CRISPR/Cas12a detection system was evaluated by salmon and chicken samples spiked with *Salmonella* (30 CFU/25 g), and un-spiked samples were tested to evaluate the specificity of this method in real samples. DNA extracted from bacteria with different culture times (0, 1, 2, 3, 4, 5, and 6 h) was detected by this method. As Figure 5 shows, *Salmonella* in a spiked salmon sample cultured at 4 h was detected with a fluorescence intensity of 188 when the threshold was 3, and *Salmonella* in a spiked chicken sample cultured at 4 h was detected with a fluorescence intensity of 172. The results indicated that *Salmonella* in both salmon and chicken samples could be detected by this system after only 4 h of enrichment. Figure S3 shows that after 4 h of enrichment, the concentration of *Salmonella* was approximately 5×10^2 CFU/mL in the salmon sample and approximately 4.2×10^2 CFU/mL in the chicken sample (100 μL of culture solution was used for colony count). Meanwhile, the groups of un-spiked samples after 6 h of culture did not show positive results even with large amounts of *Escherichia coli*, indicating that this method had high specificity in real samples.

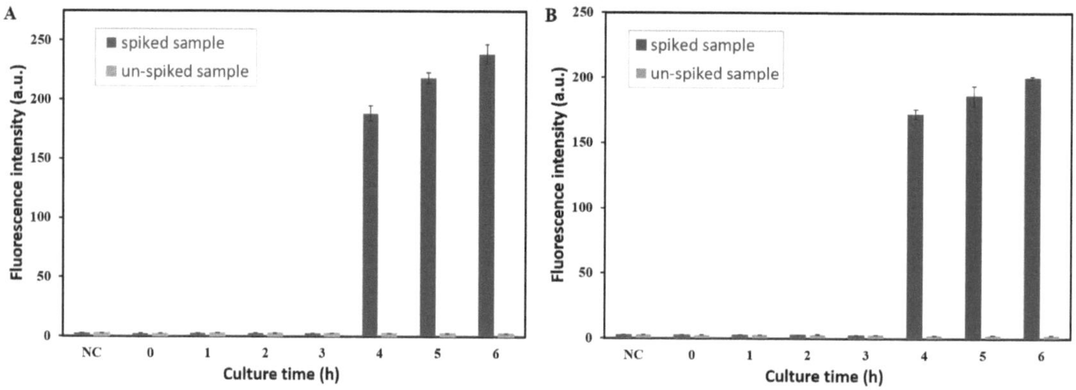

Figure 5. Minimum detection time test results of LAMP–CRISPR detection system for *Salmonella* detection in the (**A**) salmon sample and (**B**) chicken sample spiked and un-spiked.

To check the effect of the matrix of real samples on the test efficiency, TSB medium, salmon sample, and chicken sample were spiked with *Salmonella* (4×10^1, 4×10^2, 4×10^3, and 4×10^4 CFU/mL). As Figure 6 shows, 400 CFU/mL of *Salmonella* could be detected in TSB medium, salmon sample, and chicken sample with fluorescence intensities of 281, 227, and 246, respectively. The results indicated that the matrix of real sample did not affect the detection efficiency of this method, but only slightly weakened the fluorescence intensity. Therefore, this method had a stable detection capability in real samples.

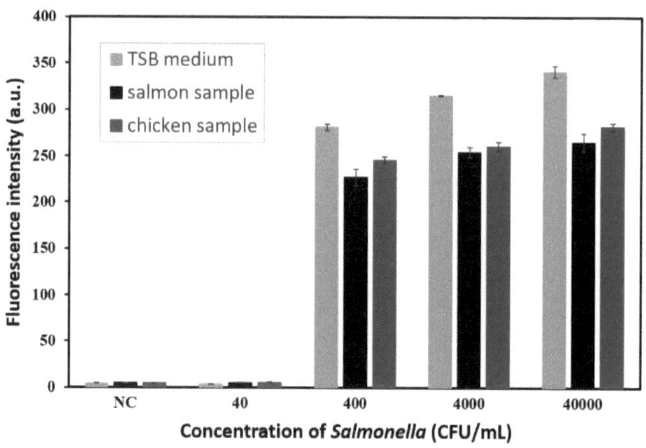

Figure 6. Matrixes test results of LAMP–CRISPR detection system for *Salmonella* detection in TSB medium, salmon sample, and chicken sample spiked with *Salmonella* (4×10^1, 4×10^2, 4×10^3, and 4×10^4 CFU/mL).

3.6. Discussion

Due to inadequate storage temperatures or inadequate cooking, foods can be a source of pathogen infections, causing tremendous harm to human health. Therefore, pathogens including *Salmonella* should be controlled in foods. Mukama et al. [20] reported a simple, inexpensive, and ultrasensitive DNA probe based LFB with CRISPR/Cas and LAMP, which achieved high sensitivity and specificity both in pure and complex samples. However, opening tubes and transferring LAMP products into the CRISPR/Cas reaction solution would lead to aerosol leakage, which would seriously jeopardize the results of the subsequent detection. Therefore, we need a leakproof and observable reaction platform. In an attempt to establish a more rapid and accurate method to detect *Salmonella*, we developed a LAMP combined with CRISPR/Cas12a integrated into a microfluidic chip system.

First, we optimized the reaction conditions of the LAMP and CRISPR/Cas12a system to reduce the detection limit. Then, we evaluated the performance of this method. The results showed this method had high sensitivity and specificity, which were consistent with previous reports [21]. Its excellent performance was due to two sets of pre-designed specific sequences used in the system. Six primers of LAMP made the amplification process have high sensitivity and efficiency. The designed crRNA could specifically recognize the LAMP amplicons for secondary confirmation, which further improved the accuracy of the detection. Furthermore, the performance of the LAMP–CRISPR/Cas12a detection system was evaluated by salmon and chicken samples spiked with *Salmonella*. The results showed the good detection performance for *Salmonella* in real samples, affirming the practical application potential of this method.

Further research aimed at optimizing the design and reducing the cost of the microfluidic chip may enhance the applicability of the assay. This method can also be applied to the detection of other pathogens, providing great potential as a universal platform for pathogen detection. Therefore, exploring the possibility of high-throughput detection on a single chip will also be of great interest.

4. Conclusions

In summary, this study developed a LAMP combined CRISPR/Cas12a integrated into a microfluidic chip system for *Salmonella* detection. Two separate liquid additions were performed through two narrow channels to achieve two-step reactions on one chip, eliminating the risk of aerosol contamination and cross-contamination that could result from opening the cap of a centrifuge tube, which is a typical problem associated with LAMP. The designed chip can be used as a reaction platform as well as a platform for reading the results directly. The size of the chip was designed to be suitable to allow the experimental results to be read by a portable fluorescence immunochromatographic reader or directly interpreted by the naked eyes under the irradiation of a certain wavelength of ultraviolet light. The detection limit of the proposed method could reach 118 pg/μL of crude genomic DNA, and the entire detection process could be completed within 50 minutes. Furthermore, this method was used to detect *Salmonella* after enrichment for 4 h in salmon and chicken samples spiked with 30 CFU/25 g, and was verified to have a stable detection capability in real samples. At the same time, the detection method had high sensitivity and specificity for detecting twelve *Salmonella* strains and thirteen non-*Salmonella* strains. The results showed that the proposed detection system was suitable for on-site rapid *Salmonella* detection with a low detection limit. Its high performance provides great potential as a universal platform for pathogen detection.

Supplementary Materials: The following supporting information can be downloaded at: https://www.mdpi.com/article/10.3390/foods11233887/s1, Figure S1: Devices and the effect diagrams. (A) PDMS chip model marked with actual size. (B) Smart syringe pumps and microinjectors for injection. (C) Fluorescence immunochromatographic reader. (D) Results of naked eyes observation of Salmonella detection on PDMS chip, negative and positive results, respectively. (E) Negative and positive results read by fluorescent immunochromatographic reader; Figure S2: Electropherogram of negative control, LAMP products, and LAMP products cleaved by Cas12a; Figure S3: Culture results of bacteria solution collected from (A) spiked salmon sample, (B) un-spiked salmon sample, (C) spiked chicken sample and (D) un-spiked chicken sample at different times on the *Salmonella* chromogenic medium, in the matrix test. Purple colonies circled were *Salmonella* anthropogenically added, and blue colonies were *Escherichia coli* from sample.

Author Contributions: Investigation, Conceptualization, Data curation, Software, Writing—original draft: Y.L.; Methodology: S.S.; Chip design: S.W.; Resources: D.L. and W.L.; Project administration: J.L.; Funding acquisition: D.L.; Writing—review and editing: W.L and S.S. All authors have read and agreed to the published version of the manuscript.

Funding: This research was funded by National Natural Science Foundation of China (82003467), Key Research and Development Program of Jiangxi Province (20192BBGL70053, 20192BBG70069), General Project of Jiangxi Natural Science Foundation (20202BAB206066), Science and Technology Fund Plan of Jiangxi Provincial Health Commission (202130977).

Institutional Review Board Statement: Not applicable.

Informed Consent Statement: Not applicable.

Data Availability Statement: Data are contained within the article and Supplementary Materials.

Conflicts of Interest: The authors declare no conflict of interest.

References

1. Majowicz, S.E.; Musto, J.; Scallan, E.; Angulo, F.J.; Kirk, M.; O'Brien, S.J.; Jones, T.F.; Fazil, A.; Hoekstra, R.M. The Global Burden of Nontyphoidal *Salmonella* Gastroenteritis. *Clin. Infect. Dis.* **2010**, *50*, 882–889. [CrossRef] [PubMed]
2. Friesema, I.; de Jong, A.; Hofhuis, A.; Heck, M.; Kerkhof, H.V.D.; de Jonge, R.; Hameryck, D.; Nagel, K.; van Vilsteren, G.; van Beek, P.; et al. Large outbreak of *Salmonella* Thompson related to smoked salmon in the Netherlands, August to December 2012. *Eurosurveillance* **2014**, *19*, 20918. [CrossRef]
3. Chiu, C.-H.; Su, L.-H.; Chu, C. *Salmonella enterica* serotype Choleraesuis: Epidemiology, pathogenesis, clinical disease, and treatment. *Clin. Microbiol. Rev.* **2004**, *17*, 311–322. [CrossRef] [PubMed]
4. Qi, X.; Li, P.; Xu, X.; Yuan, Y.; Bu, S.; Lin, D. Epidemiological and Molecular Investigations on *Salmonella* Responsible for Gastrointestinal Infections in the Southwest of Shanghai from 1998 to 2017. *Front. Microbiol.* **2019**, *10*, 2025. [CrossRef] [PubMed]
5. Antonia, K.; André, B.; Ulrich, S.; Burkhard, M.; Madeleine, P.; Amir, A. Establishment and validation of a loop-mediated isothermal amplification (LAMP) assay targeting the ttrRSBCA locus for rapid detection of *Salmonella* spp. in food. *Food Control* **2021**, *126*, 107973. [CrossRef]
6. Shang, Y.; Ye, Q.; Cai, S.; Wu, Q.; Pang, R.; Yang, S.; Xiang, X.; Wang, C.; Zha, F.; Ding, Y.; et al. Loop-mediated isothermal amplification (LAMP) for rapid detection of *Salmonella* in foods based on new molecular targets. *LWT—Food Sci. Technol.* **2021**, *142*, 110999. [CrossRef]
7. Ye, H.; Nowak, C.; Liu, Y.; Li, Y.; Zhang, T.; Bleris, L.; Qin, Z. Plasmonic LAMP: Improving the Detection Specificity and Sensitivity for SARS-CoV-2 by Plasmonic Sensing of Isothermally Amplified Nucleic Acids. *Small* **2022**, *18*, 2107832. [CrossRef]
8. Notomi, T.; Mori, Y.; Tomita, N.; Kanda, H. Loop-mediated isothermal amplification (LAMP): Principle, features, and future prospects. *J. Microbiol.* **2015**, *53*, 1–5. [CrossRef]
9. Heithoff, D.M.; Barnes, L.; Mahan, S.P.; Fox, G.N.; Arn, K.E.; Ettinger, S.J.; Bishop, A.M.; Fitzgibbons, L.N.; Fried, J.C.; Low, D.A.; et al. Assessment of a Smartphone-Based LAMP Assay for Detection of SARS-CoV-2 and Influenza Viruses. *JAMA Netw. Open* **2022**, *5*, e2145669. [CrossRef]
10. Watanabe, S.; Okubo, A.; Miyajima, Y.; Satoh, K.; Makimura, K. Specific detection of *Trichophyton rubrum* and *Trichophyton interdigitale* based on loop-mediated isothermal amplification (LAMP) from onychomycosis specimens. *J. Dermatol.* **2019**, *46*, 1179–1183. [CrossRef] [PubMed]
11. Liao, W.; Long, D.; Huang, Q.; Wei, D.; Liu, X.; Wan, L.; Feng, Y.; Zhang, W.; Liu, Y. Rapid Detection to Differentiate Hypervirulent Klebsiella pneumoniae (hvKp) From Classical *K. pneumoniae* by Identifying peg-344 With Loop-Mediated Isothermal Amplication (LAMP). *Front. Microbiol.* **2020**, *11*, 1189. [CrossRef]
12. Yang, X.; Sun, L.; Sun, H.; Hong, Y.; Xiao, Z.; Pang, X.; Piao, Z.; Feng, J.; Liang, Y. A loop-mediated isothermal DNA amplification (LAMP) assay for detection of the clubroot pathogen *Plasmodiophora brassicae*. *Plant Dis.* **2021**, *106*, 1730–1735. [CrossRef] [PubMed]
13. Sivakumar, R.; Dinh, V.P.; Lee, N.Y. Ultraviolet-induced in situ gold nanoparticles for point-of-care testing of infectious diseases in loop-mediated isothermal amplification. *Lab Chip* **2021**, *21*, 700–709. [CrossRef] [PubMed]
14. Swarts, D.C.; Jinek, M. Mechanistic Insights into the cis- and trans-Acting DNase Activities of Cas12a. *Mol. Cell* **2019**, *73*, 589–600.e4. [CrossRef]
15. Zhu, Z.; Li, R.; Zhang, H.; Wang, J.; Lu, Y.; Zhang, D.; Yang, L. PAM-free loop-mediated isothermal amplification coupled with CRISPR/Cas12a cleavage (Cas-PfLAMP) for rapid detection of rice pathogens. *Biosens. Bioelectron.* **2022**, *204*, 114076. [CrossRef]
16. Liu, P.; Wang, X.; Liang, J.; Dong, Q.; Zhang, J.; Liu, D.; Wang, S.; Bi, J.; Liu, W.; Wang, Z.; et al. A Recombinase Polymerase Amplification-Coupled Cas12a Mutant-Based Module for Efficient Detection of Streptomycin-Resistant Mutations in *Mycobacterium tuberculosis*. *Front. Microbiol.* **2022**, *12*, 796916. [CrossRef] [PubMed]
17. Xu, W.; Jin, T.; Dai, Y.; Liu, C.C. Surpassing the detection limit and accuracy of the electrochemical DNA sensor through the application of CRISPR Cas systems. *Biosens. Bioelectron.* **2020**, *155*, 112100. [CrossRef] [PubMed]

18. Xia, X.; Ma, B.; Zhang, T.; Lu, Y.; Khan, M.R.; Hu, Y.; Lei, C.; Deng, S.; He, Q.; He, G.; et al. G-Quadruplex-Probing CRISPR-Cas12 Assay for Label-Free Analysis of Foodborne Pathogens and Their Colonization In Vivo. *ACS Sens.* **2021**, *6*, 3295–3302. [CrossRef]
19. Zhou, R.; Li, Y.; Dong, T.; Tang, Y.; Li, F. A sequence-specific plasmonic loop-mediated isothermal amplification assay with orthogonal color readouts enabled by CRISPR Cas12a. *Chem. Commun.* **2020**, *56*, 3536–3538. [CrossRef] [PubMed]
20. Mukama, O.; Wu, J.; Li, Z.; Liang, Q.; Yi, Z.; Lu, X.; Liu, Y.; Liu, Y.; Hussain, M.; Makafe, G.G.; et al. An ultrasensitive and specific point-of-care CRISPR/Cas12 based lateral flow biosensor for the rapid detection of nucleic acids. *Biosens. Bioelectron.* **2020**, *159*, 112143. [CrossRef] [PubMed]
21. Mukama, O.; Yuan, T.; He, Z.; Li, Z.; Habimana, J.D.; Hussain, M.; Li, W.; Yi, Z.; Liang, Q.; Zeng, L. A high fidelity CRISPR/Cas12a based Lateral flow biosensor for the detection of HPV16 and HPV18. *Sens. Actuators B Chem.* **2020**, *316*, 128119. [CrossRef]
22. Zhang, T.; Zhao, W.; Zhao, W.; Si, Y.; Chen, N.; Chen, X.; Zhang, X.; Fan, L.; Sui, G. Universally Stable and Precise CRISPR-LAMP Detection Platform for Precise Multiple Respiratory Tract Virus Diagnosis Including Mutant SARS-CoV-2 Spike N501Y. *Anal. Chem.* **2021**, *93*, 16184–16193. [CrossRef] [PubMed]
23. Bai, L.; Wang, L.; Huang, S.; Bai, R.; Lv, X.; Sun, L.; Zhang, F.; Xu, X. Rapid, Visual, and Sequence-Specific Detection of *Salmonella* in Egg Liquid with vis-NEAA, a CRISPR/Cas12 Empowered New Strategy. *J. Agric. Food Chem.* **2022**, *70*, 2401–2409. [CrossRef] [PubMed]
24. Wong, Y.-P.; Othman, S.; Lau, Y.-L.; Radu, S.; Chee, H.-Y. Loop-mediated isothermal amplification (LAMP): A versatile technique for detection of micro-organisms. *J. Appl. Microbiol.* **2018**, *124*, 626–643. [CrossRef] [PubMed]
25. Ye, Y.; Liu, Y.; He, S.; Xu, X.; Cao, X.; Ye, Y.; Zheng, H. Ultrasensitive electrochemical DNA sensor for virulence *invA* gene of *Salmonella* using silver nanoclusters as signal probe. *Sens. Actuators B Chem.* **2018**, *272*, 53–59. [CrossRef]
26. Chaudhary, J.H.; Nayak, J.B.; Brahmbhatt, M.N.; Makwana, P.P. Virulence genes detection of *Salmonella* serovars isolated from pork and slaughterhouse environment in Ahmedabad, Gujarat. *Vet. World* **2015**, *8*, 121–124. [CrossRef] [PubMed]
27. Chen, J.S.; Ma, E.; Harrington, L.B.; Costa, M.D.; Tian, X.; Palefsky, J.M.; Doudna, J.A. CRISPR-Cas12a target binding unleashes indiscriminate single-stranded DNase activity. *Science* **2018**, *360*, 436–439. [CrossRef] [PubMed]
28. Shan, S.; Liu, D.; Guo, Q.; Wu, S.; Chen, R.; Luo, K.; Hu, L.; Xiong, Y.; Lai, W. Sensitive detection of *Escherichia coli* O157:H7 based on cascade signal amplification in ELISA. *J. Dairy Sci.* **2016**, *99*, 7025–7032. [CrossRef] [PubMed]
29. Li, C.; Wang, Y.; Wang, J.; Wang, X. Properties of a Novel *Salmonella* Phage L66 and Its Application Based on Electrochemical Sensor-Combined AuNPs to Detect *Salmonella*. *Foods* **2022**, *11*, 2836. [CrossRef] [PubMed]

foods

MDPI

Article

Surface Hygiene Evaluation Method in Food Trucks as an Important Factor in the Assessment of Microbiological Risks in Mobile Gastronomy

Michał Wiatrowski, Elżbieta Rosiak and Ewa Czarniecka-Skubina *

Department of Food Gastronomy and Food Hygiene, Institute of Human Nutrition Sciences, Warsaw University of Life Sciences (WULS), 166 Nowoursynowska Str., 02-787 Warsaw, Poland
* Correspondence: ewa_czarniecka_skubina@sggw.edu.pl; Tel.: +48-22-59-37063

Abstract: Street food outlets are characterised by poor microbiological quality of the food and poor hygiene practices that pose a risk to consumer health. The aim of the study was to evaluate the hygiene of surfaces in food trucks (FT) using the reference method together with alternatives such as Petrifilm^{TM} and the bioluminescence method. TVC, *S. aureus*, *Enterobacteriaceae*, *E. coli*, *L. monocytogenes*, and *Salmonella* spp. were assessed. The material for the study consisted of swabs and prints taken from five surfaces (refrigeration, knife, cutting board, serving board, and working board) in 20 food trucks in Poland. In 13 food trucks, the visual assessment of hygiene was very good or good, but in 6 FTs, TVC was found to exceed log 3 CFU/100 cm^2 on various surfaces. The assessment of surface hygiene using various methods in the food trucks did not demonstrate the substitutability of culture methods. Petrifilm^{TM} tests were shown to be a convenient and reliable tool for the monitoring of mobile catering hygiene. No correlation was found between the subjective visual method and the measurement of adenosine 5-triphosphate. In order to reduce the risk of food infections caused by bacteria in food trucks, it is important to introduce detailed requirements for the hygiene practices used in food trucks, including techniques for monitoring the cleanliness of surfaces coming into contact with food, in particular cutting boards and work surfaces. Efforts should be focused on introducing mandatory, certified training for food truck personnel in the field of microbiological hazards, appropriate methods of hygienisation, and hygiene monitoring.

Keywords: food trucks; hygiene; Petrifilm^{TM}; reference method; ATP; TVC; pathogenic microorganisms

Citation: Wiatrowski, M.; Rosiak, E.; Czarniecka-Skubina, E. Surface Hygiene Evaluation Method in Food Trucks as an Important Factor in the Assessment of Microbiological Risks in Mobile Gastronomy. *Foods* **2023**, *12*, 772. https://doi.org/10.3390/foods12040772

Academic Editors: Arícia Possas and Fernando Pérez-Rodríguez

Received: 15 January 2023
Revised: 2 February 2023
Accepted: 8 February 2023
Published: 10 February 2023

1. Introduction

Street food refers to ready-to-eat food products, including fruits and vegetables, that are sold in public places, mainly on the streets. Street vending is common in both developing and developed countries. The largest number of such outlets is found in Africa, Asia, and Latin America. It often belongs to the informal food supply sector, which is characterised by unregulated production and hygiene practices [1]. In the last few decades, street food has also become popular in Europe [2,3].

Previous studies on street food establishments focus primarily on consumer choices and frequency of use of this form of facilities [2–8], determining the nutritional value of the meal, the risk of developing diet-related diseases [9–11], assessing hygiene practices, and the risk of health hazards to consumers [7,12–25], which have become important aspects of public health [26,27].

The majority of studies have used visual appraisal to assess the hygiene status of street food outlets. In previous studies [2,24,28–33], it has been stated that vendors are aware of food hygiene and respect good hygiene practices, but street food vendors with elementary education levels should undergo basic training on food hygiene. Thus, the need for health education to improve vendors' knowledge on hygiene practices and food safety becomes apparent [34].

In many parts of the world, street food has been linked to diseases that pose a threat to public health [35,36]. Many authors [17,19,27–45] indicate that, due to the low microbiological quality of street food meals, their consumption may pose a risk of foodborne disease.

Pathogenic microorganisms such as *Escherichia coli, Bacillus cereus, Clostridium perfringens, Staphylococcus aureus,* and *Salmonella* spp. are usually present in street food [7,46]. Among the street-vended products that have been found to be contaminated with aerobic microflora, *S. aureus, Salmonella typhi,* or *E. coli* and the coli group of bacteria are samples of tomato sauce, rice balls, and peanut soups [47] and barbecue chicken [48–53]. *E. coli, S. aureus, Pseudomonas aeruginosa, Klebsiella,* and coliform bacteria have also been found to contaminate ready-to-eat products such as sandwiches, panipuri, momos, chola, samosa, vegetable salads, packaged fried rice, and egg burgers [54–57], as well as chicken meat [58]. The results also indicate poor microbiological quality (TVC, *E. coli*) of meat-based ready-to-eat fast food items (chicken sandwiches, chicken burgers, and hot dogs) sold on the streets [59] and in grilled chicken [60]. Fungi such as *Aspergillus flavus, A. niger, A. candidus, Cladosporium herbarum, Necrospora crassa, Penicillium citrinum, Fusarium, Mucor,* and *Rhizopus* have been identified in various forms of street food [61]. In 118 cooked and uncooked falafels, there was a higher than recommended microbial TVC (10^8 CFU/g), coliform count (3.7×10^3 CFU/g), and mould count (1.3×10^3 CFU/g). Such contamination results in a significant risk of intake of these products [62]. TVC greater than 5 log CFU/g and coliform counts greater than 4 log CFU were considered unsatisfactory and are indicative of poor hygiene standards [60]. The presence of coliforms, including faecal coliforms or *E. coli,* indicates the adoption of poor hygiene practices or unhygienic conditions during food processing [63,64]. The authors have identified the reasons for this to be the limited education of vendors, a lack of training in good hygiene practices, and inadequate food preparation temperatures, all of which result in poor microbiological quality of the final product. Outbreaks of microbiological contamination of street food in many countries have also been linked to poor water quality [65] and poor quality of food ingredients [13,66–68], as well as less attention being paid to hygiene in street food outlets.

Several authors [69–74] identify kitchen utensils and cutlery as a source of serious microbiological risks: spoons, knives, cutting boards, and plates, as well as the hands of employees. These surfaces were found to be contaminated with high numbers of microorganisms, e.g., *B. cereus, E. coli, Shigella sonnei, Clostridium perfringens, Salmonella* spp., and *S. aureus*. Most of the results of microbiological studies of street food outlets have been carried out in developing countries, whereas only a small number of these studies have focused on the assessment of these types of establishments in Europe.

Ensuring appropriate production quality and health safety in mobile catering facilities is determined by the specific nature of the work in this type of establishment. Above all, this type of establishment is characterised by difficult working conditions, often without access to running water; a lack of regularity in the work; a variety of foods offered; a rush to prepare meals and carry out hygiene procedures; often unsatisfactory equipment in the production and serving area; and little attention paid to ensuring food quality and safety. Additionally, the small number of staff recruited, the need for multitasking, the unavailability of employees with adequate professional training in catering technology and food hygiene, the high staff rotation resulting in the constant need to train new employees, the low social status of the employees, and their low salaries all add to the challenges faced [74].

The aim of this study was to assess the status of surface hygiene in mobile food establishments using different analytical methods, traditional and rapid diagnostic methods, and related sampling techniques. The paper addresses the issue of whether and which alternative methods of hygiene assessment might be useful if applied in food trucks, which surfaces in food trucks are best inspected using the available hygiene assessment methods, and with what frequency these surfaces should be inspected with these methods. Our study fills a research gap in the microbiological quality of European food trucks, as well as in comparing and assessing the effectiveness and suitability of traditional and alternative methods of ensuring the health safety of food truck consumers.

2. Materials and Methods

2.1. Materials

The material for the study consisted of swabs and samples taken from 100 different service areas in 20 mobile food catering establishments operating in Warsaw, Poland, in cases in which the owners agreed to provide samples. The food served in the surveyed facilities was prepared in an eat-in and take-out format (Table S1). The material for testing in each establishment was taken from five surfaces: a shelf of a refrigerator, a cutting board, a small utensil such as a knife, a serving surface, and a worktop surface.

Samples were taken before the opening of the establishment and the start of production. According to the owner's declaration, the surfaces were previously cleaned and sanitised. Immediately after collection, the test material was transported to the laboratory under refrigeration [75] and analysed microbiologically. In addition, a subjective visual assessment of the condition and cleanliness of the equipment and work surfaces was conducted by the same person who collected the surface samples.

2.2. Microbiological Methods

Three methods were used to assess microbiological contamination. Two direct culture methods were selected for the detection of microbiological contaminants: the reference swabbing method (plate method) and the alternative contact method (agar swabs) using Petrifilm (3M™Petrifilm™, Kajetany, Poland) plates and the indirect method of bioluminescence measurement of the microbial cell energy metabolite ATP (adenosine 5-triphosphate). A scheme of the tests conducted is shown in Figure 1.

2.2.1. Reference Plate Method

The swabbing method, a modification of the classic stamp method, involves swabbing from a limited template (10×10 cm^2) area with a sterile swab (PROBACT medical, Heywood, Lancashire, UK). The technique of swabbing was standardised with a zig-zag movement of the swab in 4 planes: vertical, horizontal, and two diagonal planes on the template [75]. The swab was transferred to a sterile extender and vortexed (3×5 s) (LP Vortex Mixer, Thermo Scientific, Waltham, MA, USA), and microbiological surface inoculation was performed. The limiting template was sterilised with a burner flame before each material collection. The characteristics of the analyses performed using the reference method are presented in Table 1.

Table 1. Kind and characteristics of reference methods.

Kind of Microorganisms	Medium/Producer	ISO Standard	Typical Growth of Colonies
Total Viable Count (TVC)	NUTRIENT AGAR/ Neogen Co., Heywood, UK	[76]	All colony beside of shape, colour, size
Staphylococcus aureus	BARID PARKER/ BioRad, Watford, UK	[77]	Black-grey with transparent halo
Enterobacteriaceae	VRBG/ Oxoid Ltd., Hampshire, UK	[78]	Pink-violet
Escherichia coli	TBX/Oxoid Ltd., Hampshire, UK	[79]	Blue-green
Listeria monocytogenes	AL OA, PALCAM/ Neogen Co., Heywood, UK	[80]	Blue-green with cloudy halo; olive-grey with black centre
Salmonella spp.	BGA, XLD/Neogen Co., Heywood, UK	[81]	Black, red, pinkish, or white with red halo

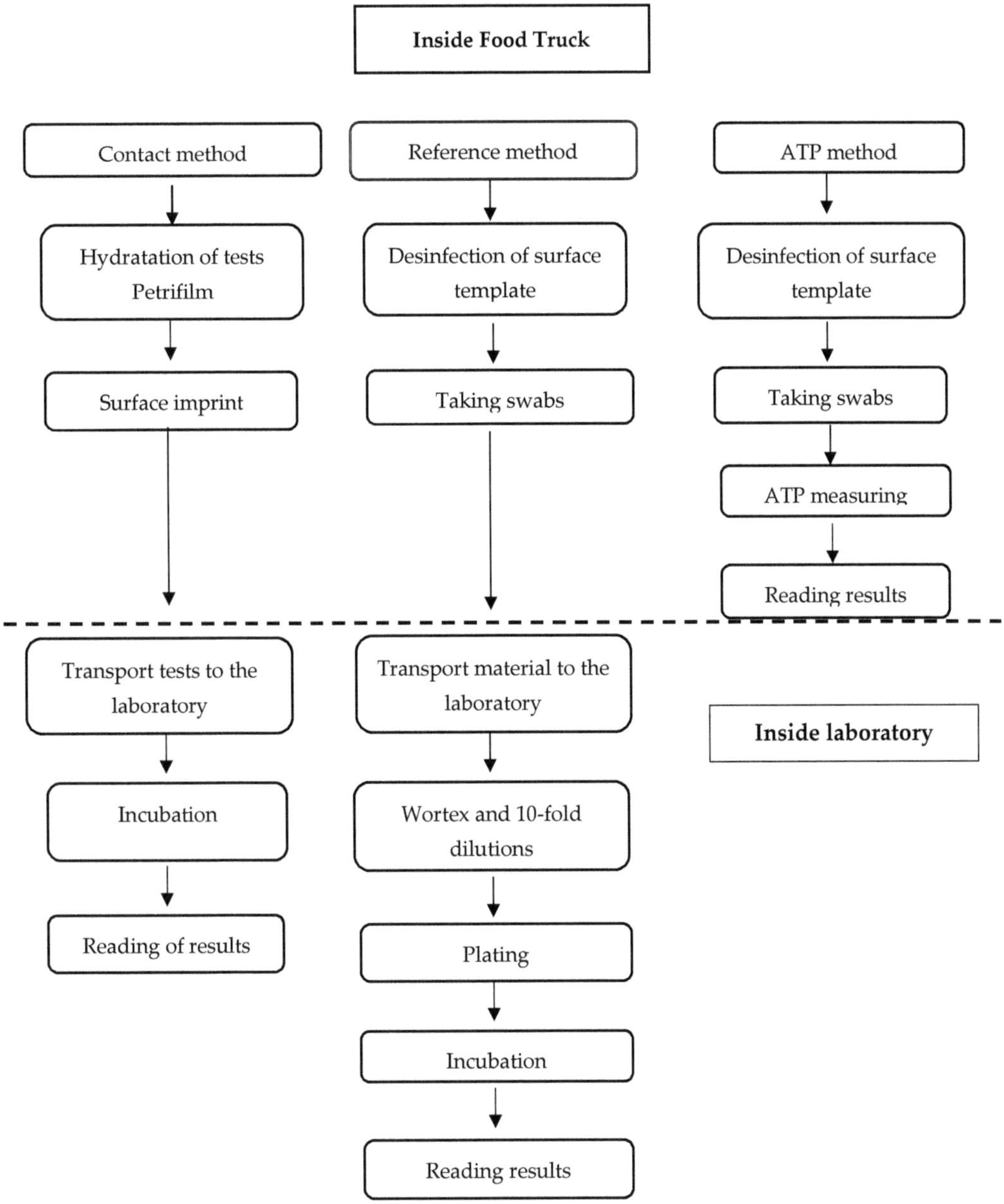

Figure 1. General scheme of analysis of hygiene state of food trucks.

2.2.2. Alternative Contact Method—Petrifilm

Petrifilm tests (3M™Petrifilm™) are ready-to-use plates with dehydrated conditioner dedicated to the detection of specific microorganisms or groups of microorganisms. The Petrifilm plate consists of two parts. The lower part contains the culture medium, which, after rehydration (30 min) with a sterile water conditioner, is transferred to the upper film to make an imprint on the surface to be tested. After collection, the tests were incubated according to the manufacturer's instructions. Analyses were carried out according to the characteristics shown in Table 2.

Table 2. Kinds and characteristics of contact plate method 3M™Petrifilm™.

Type of Petrifilm (Surface cm^2)	Kind of Microorganisms	Medium Composition	Incubation Conditions	Typical Colonies
PAC (20)	TVC	Nutrient agents, tetrazol indicator	30 °C, 72 h	Pink-red
STX (30)	*Staphylococcus aureus*	Baird-Parker medium	35–37 °C, 24 h	Intensive red-purple
EL (40)	*Environmental Listeria* spp.	Selective and nutrient agents, chromogenic indicator	37 °C 26–30 h	Grey-purple
EB (20)	*Enterobacteriaceae*	VRBG medium	37 °C 24 h	Red with gas bubbles or with yellow acid zone or both
EC (20)	*Escherichia coli* and *coliforms*	VRBL medium	35–44 °C, 24–48 h	Blue or red-blue with gas bubble

In order to unify the expression of microbial contamination of surfaces assessed by different methods, microbial contamination was expressed as log CFU/100 cm^2. The obtained results were multiplied by the appropriate coefficient (2.5-EL, 3.3-STX lub 5-PAC, EB, EC). For the Petrifilm STX plates, when uncharacteristic growth was observed, detection of colonies belonging to *S. aureus* was carried out using a disc containing bule-O toluidine, an enzyme produced by *S. aureus*. Readings were taken in accordance with the 3M™Petrifilm™ Interpretation Tables [75].

2.2.3. Interpretation of Culture Methods Results

Results obtained in colony forming units (CFU) per plate area were converted to log CFU/100 cm^2. For the assessment of the surface hygiene with the reference method, the acceptable contamination on surfaces in contact with food in the case of TVC is log 3.0 CFU/100 cm^2; for *S. aureus*, it is log 4.0 CFU/100 cm^2; and any presence of *E. coli*, *Enterobacteriaceae*, *Salmonella* spp., and *L. monocytogenes* is unacceptable [82].

2.2.4. ATP Bioluminescence Measurement

Indirect method tests were performed using a Clean-Trace™ NG (Noack Polen) luminometer Clean-Trace™ NG (3M Health Care, Neuss, Germany) and tests Clean–Trace Surface ATP (ULX100 3M). Swabs were collected from surfaces using a limited metal template (100 cm^2) which was sterilised before each collection.

We use an indirect method to assess hygiene by ascertaining the level of the bioluminescence of adenosine 5-triphosphate present on the surface. ATP is a nucleotide which is the carrier of free energy in every living cell of the sampled biological material. The concentration of ATP in a swab sample is directly proportional to the level of light emitted. A high level of ATP may, therefore, be evidence of the presence of microorganisms or organic remains on the tested surface in a catering establishment.

2.2.5. Interpretation of Bioluminescence Measurement Results

On the basis of the findings of Griffith et al. [83], an ATP value of up to 500 RLU was taken as a realistic limit for clean surfaces.

2.3. Statistical Methods

Statistica 13.3 PL (TIBCO Software Inc., 2017 WA 98109, Seattle, WA, USA) software was used to compare the results and perform cluster analysis. To determine the difference between the samples, one-way ANOVA analysis of variance was used. The significance of differences between individual means was determined using Tukey's post-hoc test (RIR). An α value of 0.05 was used. The cluster analysis method were used to classify the results of microbiological analyses of surfaces [84]. The distance between clusters was measured by Euclidean distance function, whereas the Ward method was used to bind the clusters. The Ward method uses the assumptions of variance analysis and aims to minimise the sum of deviations within clusters. As a result of joining cluster pairs, the pair that gives the cluster with the minimum differentiation is chosen. ESS (Error Sum of Squares) is a measure of the difference from the mean value. The linear coefficient of Pearson to assess the correlation between the used analytical methods was used (previously, the Shapiro–Wilk test was performed). Correlation heatmap was plotted in SRPlot at https://www.bioinformatics.com.cn/en (accessed on 30 December 2022).

3. Results

3.1. Visual Assessment of the Surfaces in Street Food Outlets

The results of the visual assessment of surface cleanliness in street food outlets are presented in Table 3. Only three food trucks received a high visual assessment of the surface (5 points). In 10 FTs, not all surfaces were clean, and they were rated as having good hygiene (4 pts). In turn, surfaces in contact with food in six FTs were rated as having low satisfactory hygiene (3 pts) or unsatisfactory hygiene (2 pts). One of the assessed food trucks was rated as having a poor hygiene state (1 point).

Table 3. The characteristics of the research material.

No. FT	Kind of Offer	Surface Details	Score of Visual Assessments *
FT1	Pizza	The surfaces in the food truck were new, composed of stainless steel and plastic; a round knife designed for cutting pizza was used for evaluation.	5
FT7	Fried chicken	All surfaces were clean and well maintained. The cutting board was composed of hard plastic.	5
FT12	Casseroles	All surfaces were of very good quality and cleanliness; cleaning took place just before testing.	5
FT4	Thai meals	In food truck, the equipment and surfaces were new, but not properly maintained; there were visible traces of dirt from the previous day's work.	4
FT5	Greek meals	The food truck was not new; there were signs of contamination; however, all surfaces and equipment were in good condition—no visible signs of dirt, etc.	4
FT8	Burgers	In the food truck, most of the surfaces were clean and well maintained. The food truck was freshly renovated, but the cutting knives were dirty (there were visible traces of their previous use).	4

Table 3. *Cont.*

No. FT	Kind of Offer	Surface Details	Score of Visual Assessments *
FT9	Burgers	Equipment and surfaces were maintained in good condition and clean. The cutting board was composed of the wrong material (wooden).	4
FT11	Casseroles	The working surfaces, as well as the production equipment, were kept in good condition. The consumer areas (3 tables) were not clean; they had not been cleaned the day before.	4
FT16	Burgers	All surfaces were of average quality; it was evident that they had been intensively used, but there was no visual dirt.	4
FT17	Burgers	All surfaces were of average quality; it was evident that they had been intensively used, but there was no visual dirt.	4
FT18	Burgers	All surfaces were of average quality; it was evident that they had been intensively used, but there was no visual dirt.	4
FT19	Burgers	All surfaces were of average quality; it was evident that they had been intensively used, but there was no visual dirt.	4
FT20	Burgers	All surfaces were of average quality; it was evident that they had been intensively used, but there was no visual dirt.	4
FT10	French fries	The surfaces from which the swab was taken were not of poor quality, whereas the other surfaces were dirty. It was noticed that the "old" frying oil was to be used (there were visible remnants of previous frying, the colour of oil was dark orange).	3
FT13	Ice cream	There were visible marks of raw material remaining after previous work the day before. The condition of the surfaces was average; defects in surface quality and cleanliness were visible.	3
FT15	Burgers	In the food truck, most of the surface was kept in good condition, except for the fridge; dirt and raw material residues were visible.	3
FT2	Israeli meals	The surfaces, which were composed of wood and plastic, had traces of many years of use.	2

Table 3. *Cont.*

No. FT	Kind of Offer	Surface Details	Score of Visual Assessments *
FT3	Burgers	Inside the food truck, it was clear that there had been many years of operation without renovation; the walls were covered with dried fat. Small equipment, i.e., a cutting board and knives, were newly purchased.	2
FT14	Burgers	The equipment used in the street food outlets was not of good quality; the electric grill showed traces of burnt fat, the small appliances were of better quality (there were no visible signs of dirt), and the cutting board was composed of bamboo and was very wet	2
FT6	Ramen	The surfaces were new but not clean; traces of use were visible. The cutting board was composed of compressed bamboo; the worktops were composed of stainless steel.	1

FT—food truck, * subjective, visual assessment of hygiene on a scale 1–5 (1—poor hygiene state; 2—unsatisfactory hygiene state; 3—low satisfactory hygiene state; 4—good hygiene state; 5—very good hygiene state).

3.2. *Presence of Indicator Microorganisms on Surfaces in Street Food Outlets*

3.2.1. Total Viable Count

Table S1 (Supplementary) and Figure 1 show the results of microbiological analyses of the presence of the total number of aerobic mesophilic microorganisms on the tested surfaces. The analyses were conducted using the following methods: imprinting with the use of Petrifilm tests and the reference traditional method. The maximum contamination limit was log 3 CFU/100 cm^2 considering the possibility of pathogenic microorganisms and faecal contamination among the detected microorganisms.

Results obtained from the surfaces of refrigerators (rP, rR), cutting boards (cP, cR), knives (kP, kR), and serving (sP, sR) and working boards (wP, wR) from 20 food truck outlets varied significantly; $p < 0.05$ (Figure 2a,b; Table S1, Supplementary Material).

The results of the analyses provided by the reference method were in the range of 0–3 log CFU/100 cm^2. The exceptions were FT2, FT10, and FT15, in which log 5.22, 5.85, and 6.57 CFU/100 cm^2 were detected on the serving board, knife, and working board, respectively.

Regarding the contamination evaluated using the Petrifilm method, TVC values above the acceptable level were found on the three tested surfaces. On the serving board surface, contamination of log 5.39–6.73 CFU/100 cm^2 was found in FT3, FT4, and FT10, respectively. On the working board surface, contamination of log 5.27–6.93 CFU/100 cm^2 was found in FT2–FT5 and FT10, whereas in FT19, the accepted level was slightly excessive (log 3.32 CFU/100 cm^2). The cutting board surface in FT2, FT5–FT6, and FT10 was contaminated with a TVC of log 4.74–6.73 CFU/100 cm^2.

3.2.2. *Enterobacteriaceae* and *E. coli* Bacteria

Enterobacteriaceae and *E. coli* were detected via growth culture methods in 50% of the tested food truck outlets (Table 4). Using imprint plate methods, *Enterobacteriaceae* contamination was found at levels not exceeding log 1 CFU/100 cm^2 for 15 surfaces in eight food trucks, and contamination was found at levels not exceeding log 2 CFU/100 cm^2 on 5 surfaces in FT1 and FT2. In FT10, *Enterobacteriaceae* were found on all surfaces tested: refrigerator, cutting board, knife, serving board, and working board at levels of log 2.86, 2.90, 3.02, 2.92, and 6.99 CFU/100 cm^2, respectively. By using a conventional swab and the reference analysis method, *Enterobacteriaceae* at levels not exceeding log 1 CFU/100 cm^2 were detected on eight surfaces in FT1–FT5. In food truck outlet FT1, *Enterobacteriaceae* were found at a level not exceeding log 2 CFU/100 cm^2. Contamination with bacteria belonging to the genus *Enterobacteriaceae* indicates the risk of pathogens such as *Salmonella* spp., *Shigella* spp., *Klebsiella* spp., *Cronobacter* spp., *Serratia* spp., and *Citrobacter* spp. Therefore, the hygiene standard for these microorganisms was adopted as "zero tolerance". In food trucks F7, F8, and F13–F20, *Enterobacteriaceae* and *E. coli* were not detected; therefore, they are not included in Table 4.

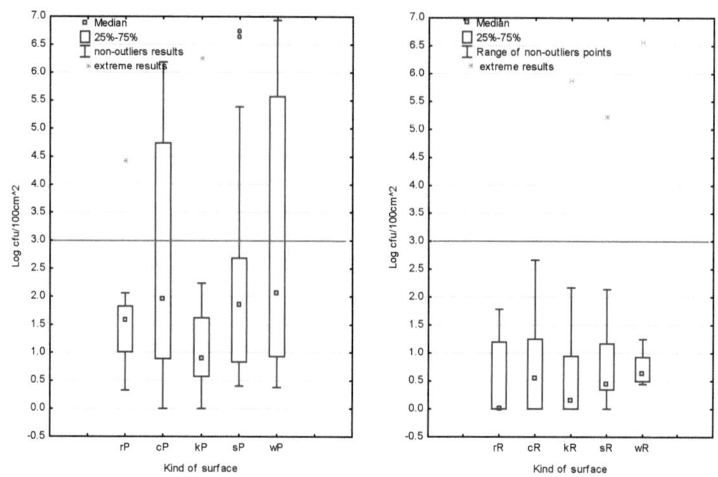

(a) (b)

Figure 2. Mean contamination of TVC on analysed surfaces in food truck outlets (n = 20). (**a**) Petrifilm method (P); (**b**) Reference method (R). rP, rR—refrigerator; cP, cR—cutting board; kP, kR—knife; sP, sR—serving board; wP, wR—working board.

Table 4. Number of surfaces contaminated by faecal bacteria *E. coli* and *Enterobacteriaceae* enumerated in food truck outlets by growth culture methods.

Method/Analysis	Food Truck (FT) *										Sum
	1	2	3	4	5	6	9	10	11	12	
Petrifilm/*Enterobacteriaceae*											
>1 log CFU	2		2	2	2	2	1	-	2	2	15
1–2 log CFU	2	3	-	-	-	-	-	-	-	-	5
<2 log	-	-	-	-	-	-	-	5	-	-	5
Petrifilm/*E. coli*											
>1 log CFU	2	1	-	-	-	-	-	-	-	-	3
1–2 log CFU	2	-	-	-	-	-	-	-	-	-	2
Reference/*Enterobacteriaceae*											
>1 log CFU	2	3	1	1	1	-	-	-	-	-	8
1–2 log CFU	1	-	-	-	-	-	-	-	-	-	1
Reference/*E. coli*											
>1 log CFU	-	-	-	-	-	-	-	-	-	-	0
1–2 log CFU	2	-	-	–	-	-	-	-	-	-	2
Sum of surfaces *Enterobacteriaceae* contaminated P/R	4/3	3/3	2/1	2/1	2/1	2/0	1/0	5/0	2/0	2/0	
Sum of surfaces *E. coli* contaminated P/R	4/2	1/0	0/0	0/0	0/0	0/0	0/0	0/0	0/0	0/0	

P—Petrifilm; R—reference; "-"—not detected; * FT in which the presence of *E. coli* and *Enterobacteriaceae* was found.

The presence of *E. coli* was found by both methods using Petrifilm plates and the reference analysis method on five and two surfaces tested in FT1 and FT2, respectively. No *E. coli* was found in the remaining food truck outlets, that is, FT3–FT20.

3.3. Presence of Pathogenic Microorganisms on Surfaces in Street Food Outlets

Salmonella spp., *L. monocytogenes*, and *Staphylococcus aureus* were detected and quantified using the reference method. Meanwhile, environmental *Listeria* and *S. aureus* were detected using an alternative method. Petrifilm plates are not available for analysis for *Salmonella* spp. The reference method did not identify *Salmonella* spp. or *L. monocytogenes* in any of the food trucks evaluated. An alternative method using Petrifilm plates was used to detect environmental *Listeria* spp. Using this method, *Listeria* spp. were detected in Food Truck No. 5 (refrigeration and serving board), No. 6 (refrigerator), and No. 12 (working board) in numbers not exceeding log 2 CFU/100 cm^2.

Figure 3 and Table S2, Supplementary Material, presents the results of microbiological analyses of the prevalence of *S. aureus* on tested surfaces in the food trucks.

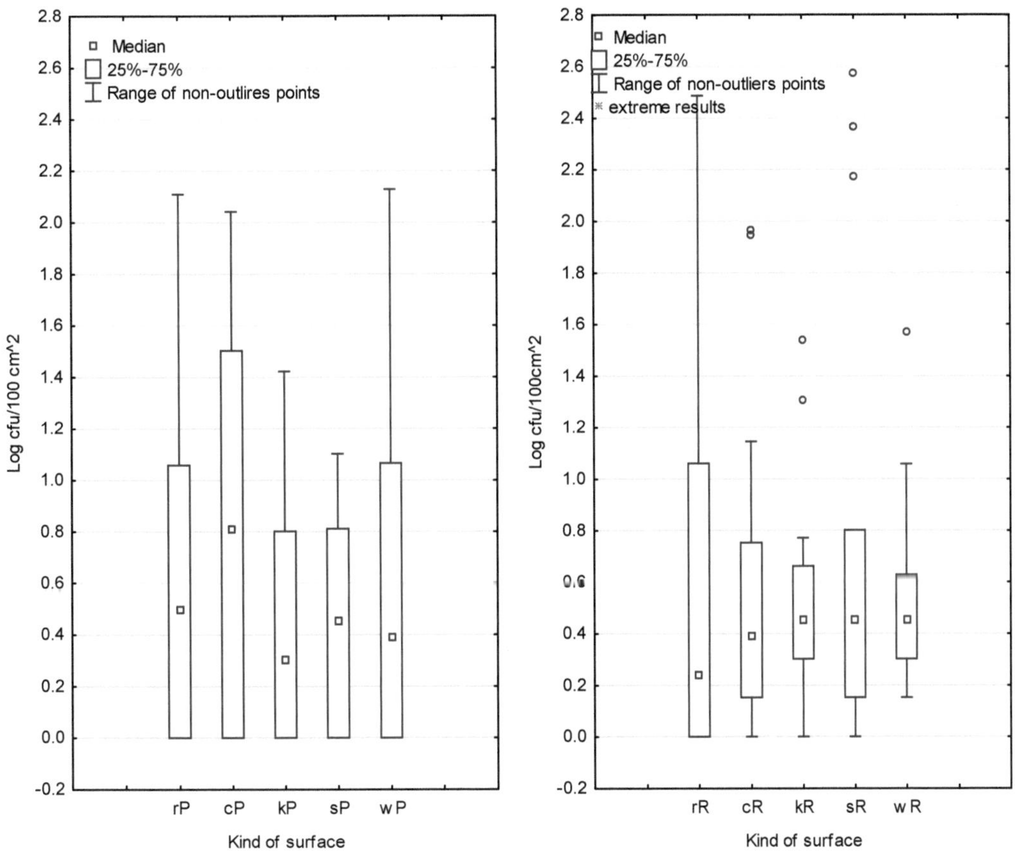

(**a**) (**b**)

Figure 3. Mean contamination of *S. aureus* on analysed surfaces in food truck outlets (n = 20). (**a**) Petrifim method (P). (**b**) Reference method (R). rP, rR—refrigerator; cP, cR—cutting board; kP, kR—knife; sP, sR—serving board; wP, wR—working board.

The presence of vegetative *S. aureus* cells does not represent an immediate threat from the pathogen because what matters is the number of colony-forming units on a given surface. The problem is a concentration above log 4 CFU/1 cm^2 on a given surface or directly in the food (g/mL), at which point *S. aureus* produces an exogenous, thermostable toxin that is a threat to food safety and consumer health. The results of the culture-based analyses (reference and alternative) were statistically different ($p < 0.05$) (Table S2, Supplementary Material). The median value of the results obtained, regardless of the method used, was below log 1 CFU/100 cm^2. The scores in the third quantile as well as the outliers also did not exceed log 3 CFU/100 cm^2. Based on the results obtained, it was observed that the analysed surfaces in the investigated food truck outlets varied in hygienic status, which is evidenced by a significant spread in quantiles 2 and 3 and by outliers.

3.4. Evaluation of Hygiene in Street Food Outlets Using the ATP Method

The indirect method of hygiene evaluation using the measurement of adenosine 5-triphosphate levels is one of the fastest measurement methods compared to culture methods, with results obtained in just a few minutes. Therefore, an experiment was carried out to correlate the results so obtained with those from culture methods. The metabolite that is analysed is generated from the decomposition of a high-energy compound contained in microbial and eukaryotic cells. The result obtained is directly proportional to the level of ATP content of microbial and organic sources. Figure 4 shows a characterisation of the level of ATP occurrence on the food truck outlet surfaces studied. On 25% of the examined surfaces, the RLU level exceeded the tolerance value by more than 500 RLU/100 cm^2. With the exception of FT2, ATP was found on all of the tested surfaces in the range of 0–1000 RLU (Relative Light Unit) (Figure 4a,b). In the case of the exception of food truck outlet FT2, ATP levels in the range of 4000–7000 RLU were found on the refrigerator and on working and cutting board areas, which significantly affected the distribution of values of the results obtained (Figure 4b).

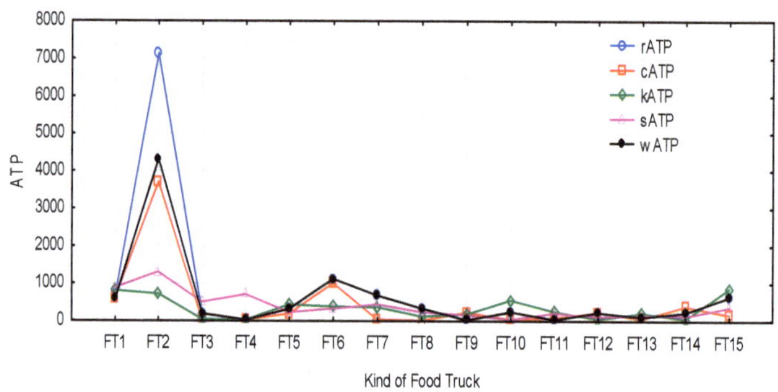

(a)

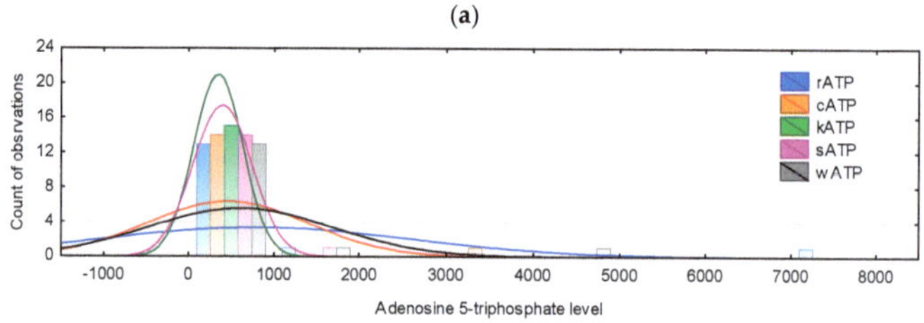

(b)

Figure 4. Level of adenosine 5-triphosphate on surfaces in food truck outlets, frequency and range of ATP results on analysed surfaces: r—refrigerator; c—cutting board; k—knife; s—serving board; w—working board. (**a**) Level of ATP on surfaces in food trucks outlet. (**b**) The frequency and range of ATP results on the analysed surfaces.

3.5. Evaluation of the Suitability of Methods to Measure the State of Hygiene in Street Food Outlets

In the cluster analysis that was conducted, the length of the bond directly represents the level of contamination evaluated on the surface. For TVC analyses (Figure 5, upper) using the Petrifilm method, the surfaces tested were grouped into two clusters. The first included the refrigerator and knife surfaces as surfaces on which the accepted level of log 3 CFU/100 cm^2 was not exceeded, whereas the second included the other surfaces tested, with the cutting board being separated out as a separate cluster, indicating a different microbiological quality to that of the serving and working boards. In contrast, three clusters were identified for the reference method. The surfaces of the refrigerator and the cutting board were found to be a concentration cluster of surfaces for which none of the food truck surfaces was found to exceed the accepted level of contamination. A serving board surface with a contamination level of log 5.22 CFU/100 cm^2 (FT3) was also included in this cluster. Clusters II and III are surfaces on which TVC contamination was found at log 5.87 and 6.57 CFU/100 cm^2, respectively.

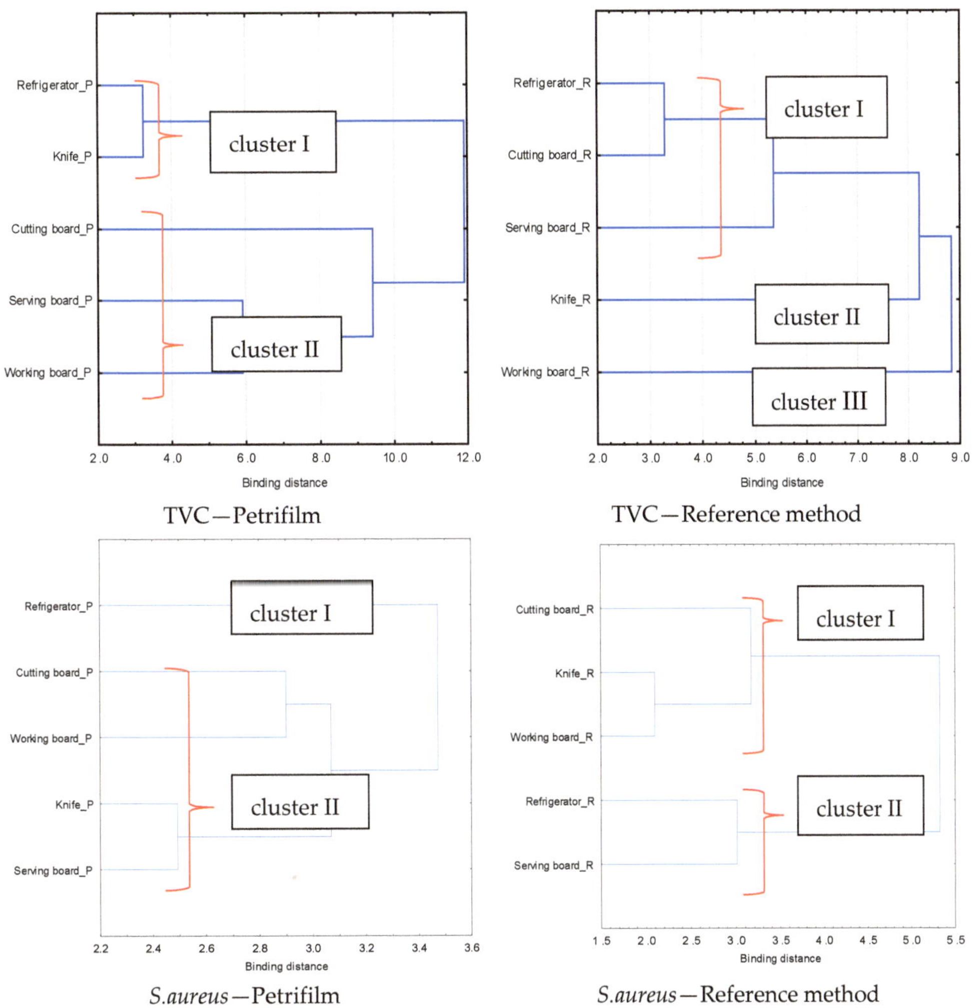

Figure 5. Cluster of surfaces depend on contamination level in analysed food truck outlets.

The cluster analysis that was conducted allows quick identification of clean and contaminated surfaces. Furthermore, it reveals differences in surface contamination depending on the analysis method used. In the case of TVC, the knife and cutting board surfaces show the greatest variation depending on the method. Due to the difficulty of making an imprint with the Petrifilm test on the knife surface, it can be presumed that the swabbing method is better for sampling small equipment, and the result is more accurate. In the case of the cutting board, due to the porosity of the material, the direct agar imprint method from the surface proved to be more effective in sampling.

Analysis of the *S. aureus* surface contamination results identified two clusters for both culture methods (Figure 5, lower). The Petrifilm method classifies the results of *S. aureus* contamination of the refrigerator surface as a separate cluster, similar to the level of contamination of the cutting and working board. In the case of detection of *S. aureus* using the reference method, comparable results of contamination of the tested surfaces were obtained. Only the working board showed lower contamination compared to the Petrifilm.

No significant correlations of $p < 0.05$ were found between the methods used to evaluate the contamination condition of *S. aureus*, whereas for TVC, there was a positive significant correlation between the results from the culture methods: reference and Petrifilm on the knife surface ($r^2 = 0.94$) and between the reference method and adenosine 5-triphosphate values on the serving board ($r^2 = 0.72$) (Figure 6). The results of visual hygiene assessments that were conducted during sampling for analysis in food truck outlets were also analysed. No correlation was found between the results obtained from the visual assessment and the results obtained from the other methods used in the study.

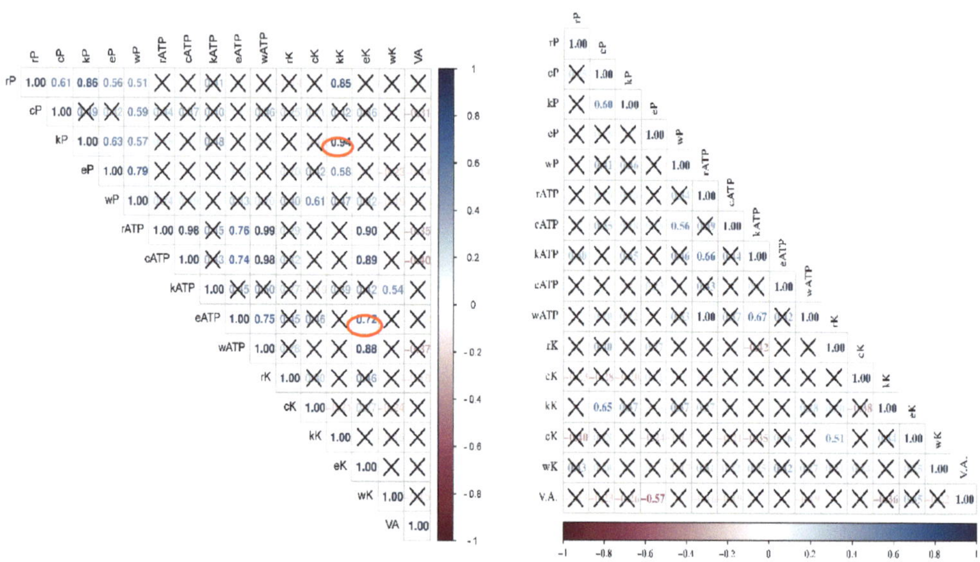

Figure 6. Heat map representing the correlation coefficient between the utilized methods for hygiene assessment in food trucks outlets. Upper triangle—TVC; lower triangle—*S. aureus*; P—Petrifilm method; ATP—adenosine 5-triphosphate; K—reference method; VA—visual assessment method; r—refrigerator; c—cutting board; k—knife; e—serving board; w—working board. X—label represents insignificant *p*-value.

4. Discussion

4.1. Hygienic Status of Working Surfaces in Food Trucks

The hygienic statuses of the evaluated work surfaces in the food trucks varied ($p < 0.05$). The majority of the food trucks fulfilled the surface hygiene requirements; however, in seven of the food trucks, the cutting, serving, and working board surfaces were found to exceed the TVC of log 3 CFU/100 cm^2, indicating high contamination of these surfaces. *Salmonella* spp. and *L. monocytogenes* were not detected in any of the food trucks assessed by the reference method (n = 20) despite the fact that chicken and eggs were used to prepare dishes in the two FT. However, the presence of *Listeria* spp. on the tested surfaces was detected by the alternative Petrifilm method in two food trucks. This does not indicate the presence of the pathogen but is rather an indicator of the existence of positive conditions for its growth. The reference plate method does not provide such information for 30 h (required incubation); therefore, the Petrifilm EL tests can be used to evaluate hygienic conditions in the establishment for the possible growth of pathogenic *L. monocytogenes*.

The risk to consumer health is also evidenced by the detection of *Enterobacteriaceae* and *E. coli* bacteria using culture methods on surfaces in the assessed food trucks. However, only one food truck (F10) showed levels of these bacteria above acceptable levels on all surfaces tested (refrigeration, cutting board, knife, serving board, and working board). High levels of contamination were also visible to the naked eye.

Other authors [13,14,66–68] have also reported problems with ensuring proper hygiene in catering establishments. According to them, poor food quality and inadequate hygiene conditions result in contamination with coliform bacteria, especially *E. coli* [14], which are hygiene indicators of faecal contamination in water and other production related environments [54–57]. The presence of *E. coli*, *S. aureus*, *Pseudomonas aeruginosa*, *Klebesiella*, and coliforms in the finished products shown in these studies is evidence of the poor hygiene and unsanitary practices used in the preparation and packaging of these street foods, as well as hand contamination among employees. These bacteria are indicators of a dirty environment, unhygienic pre- and post-production procedures, and poor water quality. Pathogenic bacteria are also carried by vegetables and street food, which infect workers, food handlers, and consumers in the industry. Previous studies [72,85–87] have also reported that any value greater than 1.0 log10 CFU/cm^2 for total coliforms is not suitable for food preparation.

Contaminated food has a direct effect on human health, but contaminated surfaces (plates, mugs, cutting boards, working tables, and serving tables) are more critical. This is because contaminated surfaces can be one of the factors of food spoilage when RTE (ready-to-eat) food is in direct contact with these surfaces. These surfaces can become re-contaminated after routine cleaning procedures, and, in the case of RTE foods, they will no longer be cooked before being served to consumers. Consequently, equipment, utensils, and areas where food is processed or prepared require attention during cleaning or hygiene tasks so as not to achieve only apparent surface cleaning, which has been found to be the case in small food production facilities, such as food trucks [88]. According to Cooper et al. [86], the reason for inefficient cleaning and consequently higher ATP and TVC levels after cleaning processes may be the spread of microorganisms over the cleaned surface, especially when cleaned with reusable wipes [89–91].

In food service establishments, work areas, cutting boards, sinks, and kitchen taps are identified as key surfaces that can cause cross-contamination of food, particularly if these surfaces are contaminated by mesophilic aerobic bacteria and *Enterobacteriaceae* [71]. Many authors [72–74,92] have also identified significant numbers of TVCs and coliform bacteria taken from cutting boards, knives and spoons, slicers, tabletops, and tables in catering establishments which did not meet the standard for clean surfaces. These contaminations were caused by poor cleaning standards for these surfaces. Microbiological cleanliness for cutting boards depended on the length of time the boards had been in use; only new boards had high cleanliness levels. Boards that have been in use for a long time may have a damaged surface, which will be microbiologically contaminated despite properly conducted cleaning practices. Cutting boards have more irregular surfaces, so proper cleaning and disinfection are more difficult and favour the survival of biofilm-forming microorganisms [93,94]. The relevance of performing the washing operation with care is highlighted by Lee et al. [95]. They demonstrated the effectiveness of hand-washing knives inoculated with *Escherichia* and *Listeria innocua*, thereby obtaining a significant reduction in contamination levels, even at a low temperature and with a low concentration of disinfectant.

In the present study, surface contamination was detected above the acceptable level on refrigeration surfaces in only one case. Similar results are indicated by Czarniecka-Skubina [74], who found satisfactory microbiological quality of samples taken from refrigeration surfaces in 11 stationary catering establishments, finding only one species of coliform bacteria, namely *E. coli*. Contamination in refrigerators is significantly influenced by the type of product stored. The highest values of microorganisms were detected for the storage of raw meat and chicken meat, and the lowest were detected for vegetables and cooked products [69].

In conclusion, there is a strong link between contaminated surfaces in catering establishments and food safety risks.

4.2. Comparison of Control Methods in Food Trucks

Among the important factors for ensuring the safety and hygiene of street food, the knowledge and attitudes of street food vendors and their hygiene practices are crucial. For this reason, finding good control measures for this type of catering activity seems important.

The specifics of catering production make it difficult or even impossible to apply the same control methods in catering establishments as those applied in food industry facilities. In order to make control effective, a systematic approach and the prevention of hazards are essential. As other authors [96,97] point out, even very detailed controls of sanitary inspections are only a fraction of the operations carried out in catering establishments and will not prevent the risk of foodborne diseases. Routine inspections are difficult to conduct in food trucks because of the constant movement of facilities and the frequent change in the range of activities. Traditional microbiological methods for the detection and quantification of microorganisms on surfaces and equipment, which require culture and incubation, are not a good option in this case, as they are time-consuming and do not provide an immediate evaluation of the current state of hygiene in an establishment. In addition, food consumption takes place immediately after production, and it is pointless to obtain a result after this time. In the case of food trucks, preventive measures seem to be a more appropriate solution.

In the evaluated food trucks, a visual assessment was also carried out but was performed before the other methods, as it is inappropriate to use other methods of appraisal when surfaces are visibly dirty. Visual assessment obviously does not replace microbiological analyses. The results of the present study indicate that the visual assessment of the analysed surfaces was not correlated with the results of the microbiological analyses that were carried out using different methods. Visually clean surfaces may still have food residues or microorganisms, which result in food contamination. According to Tebbutt et al. [88], visual assessment underestimates actual surface contamination. When assessing the microbiologically clean surfaces of cutting boards, refrigerator door handles, and microwave oven control buttons, these authors found that they did not meet the conditions of hygienic cleanliness. A periodic visual inspection focusing on hygienic practices and microbiological supervision of surfaces that are at a high risk of cross-contamination provides valuable information for improving the knowledge, attitudes, and practices of food handlers towards food for better food safety [98].

Among the commercially available methods, there are rapid methods for detecting microbial or organic contamination on surfaces that results from improper hygiene processes. These include contact methods, bioluminescent methods, and modifications of plate methods, all of which were used in this study. According to some authors, easy-to-use microbial kits are practical, and the self-check approach in hygiene should be made mandatory or an alternative method for the operator [99].

One study [100] used environmental monitoring controls to look for potential correlations between microbiological indicators and food hygiene and sanitation conditions. It is impossible to completely eliminate pathogenic microorganisms from food production areas, but their growth, spread, and survival can be influenced by regular, thorough cleaning and disinfection of food contact surfaces as well as by monitoring their effectiveness. Surfaces can play a critical role in the development of food poisoning because of the potential for pathogens to grow on them. Then, these surfaces are handled by staff or consumers, and hands can be a medium for the transfer of bacteria and viruses to dishes and vice versa.

The reference method showed statistically significantly lower TVC and *S. aureus* contamination on the surfaces tested than the Petrifilm plate method. *S. aureus* was detected on the surfaces tested at levels not compatible with enterotoxin formation. Among the culture methods, the Petrifilm imprint plate method allowed for more effective recovery of *E. coli* and *Enterobacteriaceae* from the surfaces and better determination of their number compared to the swab method. Petrifilm EL tests provide a convenient tool for the mobile catering industry to assess the growth potential of pathogenic *L. monocytogenes*.

The evaluation of the hygiene status of street food with different methods did not demonstrate the alternate applicability of the culture methods, nor did the correlation between the subjective method and the measurement of adenosine 5-triphosphate.

Opinions vary among researchers on the suitability of alternative hygiene evaluation methods. Larson et al. [101] found no correlation between the bioluminescence method and microbiological reading values. According to these authors, the lower the amount of ATP, the lower the sensitivity of the method. Rosiak et al. [102] obtained a significant correlation between the results that were obtained by the ATP method and the results of TVC evaluation using Petrifilm™ plates on the hand surface of food service personnel (r^2 = 0.63) and work surfaces (r^2 = 0.72). A significant correlation (r^2 = 0.56) between TVC results obtained via the bioluminescence method and the reference method in the case of *E coli* bacteria and a weaker correlation (r^2 = 0.30) on work surfaces between the bioluminescence method and the Petrifilm method in TVC studies were also obtained by Czarniecka-Skubina [74]. The high correlation of hygiene surface results that were obtained via the bioluminescence method in hospital kitchens and the reference method is also indicated by other authors [103]. Petrifilm™ tends to have a lower detection limit than other techniques used to evaluate surface contamination (i.e., swabbing methods) and is widely accepted and approved for microbiological analysis in the food and beverage industry [104].

The results obtained via the ATP method, which indicates the presence of food residues and microorganisms on surfaces, are obtained within 1 min, which is more efficient than surface monitoring using traditional microbiological methods. As highlighted by Aycicek et al. [103], the primary advantage of this method is that it can be used without a laboratory and without specialised personnel. However, it does not reflect quantitative microbiological detection on food contact surfaces. Traditional microbiological methods are cheaper but require more skill and time, and in catering, the result is needed immediately to take corrective action. Despite its many advantages, the ATP method does not quantify microorganisms on food contact surfaces and should be integrated with other techniques that help monitor surface hygiene [103].

Regardless of the level of strict inspection, the hygiene of food trucks in various countries is still unsatisfactory. Trends in a number of countries show that social media, smartphone applications, and online reviews of food trucks provide great opportunities to improve the hygiene practices of street food trucks. Some food standard agencies in several countries such as New Zealand, Australia, and the UK have a social media presence and recommend that customers download one of the food truck apps and look at customer reviews with respect to hygiene [105].

4.3. Limitations

One of limitations of this article is the sample size we considered in this study. We tried to obtain microbiological samples from more food trucks, but private owners refused to provide samples despite knowing that the samples would be obtained from clean surfaces. Food truck owners and employees feared fines.

5. Conclusions

Microbiological analyses that were conducted with two culture and alternative methods to assess the state of surface hygiene in food truck mobile catering establishments showed the presence of pathogenic bacteria *S. aureus* and *E. coli* and a risk of contamination by the pathogenic species *Listeria monocytogenes* and *Salmonella* spp., *Shigella* spp., *Klebsiella* spp., *Cronobacter* spp., *Serratia* spp., and *Citrobacter* spp. In the assessment of microbial contamination with bacteria, statistically higher results were obtained using the Petrifilm^{TM} PAC, STX tests. Furthermore, the recovery of *Enterobacteriaceae* and *E. coli* from the surface using Petrifilm^{TM} EB and EC was better than the swab method.

The results of the research indicate the need to constantly monitor the hygiene of surfaces in food trucks, given the fact that the use of mobile catering is becoming more and more popular due to convenience and price. In order to reduce the risk of foodborne infections caused by bacteria, it is important to introduce specific requirements for monitoring practices for the hygiene of food contact surfaces, in particular cutting boards and work surfaces. These studies did not support an alternative use of the highly convenient measurement of adenosine 5-triphosphate with culture methods and a method of visual assessment of hygiene status. The conducted analyses support the use of Petrifilm tests for routine surface monitoring in food trucks due to better recovery of bacteria from the surface, ease of performance, and interpretation of the results.

Another important cause of microbiological risk in mobile catering establishments is the lack of awareness of the employees. Efforts should focus on introducing mandatory, certified training for food truck personnel in the field of microbiological hazards, methods of hygienisation, and hygiene monitoring.

Supplementary Materials: The following supporting information can be downloaded at: https://www.mdpi.com/article/10.3390/foods12040772/s1, Table S1: Total Viable Count result on five different surfaces in analysed food truck outlets; Table S2: *Staphylococcus aureus* result on five different surfaces in analysed food truck outlets.

Author Contributions: Conceptualisation, E.R., E.C.-S. and M.W.; methodology, E.R.; validation, E.R.; formal analysis, M.W. and E.R.; data curation, M.W. and E.R.; writing—original draft preparation, M.W., E.R. and E.C.-S.; writing—review and editing; M.W., E.R. and E.C.-S.; visualisation, M.W. and E.R.; supervision, E.C.-S. All authors have read and agreed to the published version of the manuscript.

Funding: This research was financed by the Polish Ministry of Science and Higher Education with funds from the Institute of Human Nutrition Sciences, Warsaw University of Life Sciences (WULS).

Institutional Review Board Statement: Not applicable.

Informed Consent Statement: Not applicable.

Data Availability Statement: All related data and methods are presented in this paper. Additional inquiries should be addressed to the corresponding author.

Conflicts of Interest: The authors declare no conflict of interest.

References

1. Akinbode, S.O.; Dipeolu, A.O.; Okuneye, P.A. Willingness to Pay for Street Food Safety in Ogun State, Nigeria. *J. Agric. Food Inf.* **2011**, *12*, 154–166. [CrossRef]
2. Czarniecka-Skubina, E.; Trafiałek, J.; Wiatrowski, M.; Głuchowski, A. An Evaluation of the Hygiene Practices of European Street Food Vendors and a Preliminary Estimation of Food Safety for Consumers, Conducted in Paris. *J. Food Protect.* **2018**, *81*, 1614–1621. [CrossRef] [PubMed]
3. Wiatrowski, M.; Czarniecka-Skubina, E.; Trafiałek, J. Consumer eating behavior and their opinions about the food safety of street food in Poland. *Nutrients* **2021**, *13*, 594. [CrossRef]
4. de Cardoso, C.V.R.; dos Santos, S.M.C.; Silva, E.O. Street food and intervention: Strategies and proposals to the developing world. *Cien. Saude Colet.* **2009**, *14*, 1215–1224. [CrossRef]

5. Choi, J.; Lee, A.; Ok, C. The Effects of Consumers' Perceived Risk and Benefit on Attitude and Behavioral Intention: A Study of Street Food. *J. Travel Tour. Mark.* **2013**, *30*, 222–237. [CrossRef]
6. Calloni, M. Street food on the move: A socio-philosophical approach. *J. Sci. Food Agric.* **2013**, *93*, 3406–3413. [CrossRef]
7. Meher, M.M.; Afrin, M.; Talukder, A.K.; Golam Haider, M.G. Knowledge, attitudes and practices (KAP) of street food vendors on food safety in selected areas of Bangladesh. *Heliyon* **2022**, *8*, e12166. [CrossRef]
8. Isanovic, S.; Constantinides, S.V.; Frongillo, E.A.; Bhandari, S.; Samin, S.; Kenney, E.; Wertheim-Heck, S.; Nordhagen, S.; Holdsworth, M.; Dominguez-Salas, P.; et al. How perspectives on food safety of vendors and consumers translate into food choice behaviors in six African and Asian countries. *Curr. Dev. Nutr.* **2022**, *7*, 100015. [CrossRef]
9. Khairuzzaman, M.; Chowdhury, F.M.; Zaman, S.; Al Mamun, A.; Bari, M.L. Food Safety Challenges towards Safe, Healthy, and Nutritious Street Foods in Bangladesh. *Int. J. Food Sci.* **2014**, *2014*, 483519. [CrossRef]
10. Lucan, S.C.; Maroko, A.R.; Bumol, J.; Varona, M.; Torrens, L.; Schechter, C.B. Mobile food vendors in urban neighborhoods-implications for diet and diet-related health by weather and season. *Health Place* **2014**, *27*, 171–175. [CrossRef]
11. Dal, R.; Cortese, M.; Boro, M.; Feldman, C.; Barletto, S. Food safety and hygiene practices of vendors during the chain of street food production in Florianopolis, Brazil: A cross-sectional study. *Food Control* **2016**, *62*, 178–186. [CrossRef]
12. Chukuezi, O. Food Safety and Hyienic Practices of Street Food Vendors in Owerri, Nigeria. *Stud. Sociol. Sci.* **2010**, *1*, 50–57.
13. Orlando, D.; Suci Hardianti, M.; Sigit Setyabudi, F.M.C. Mounting and effective response an outbreak of viral disease involving street food vendors in Indonesia. In *Case Studies in Food Safety and Authenticity. Lessons from Real-Life Situations*; Woodhead Publishing: Southton, UK, 2012; pp. 161–167.
14. Al Mamun, M.; Rahman, S.M.M.; Turin, T.C. Microbiological quality of selected street food items vended by school-based street food vendors in Dhaka, Bangladesh. *Int. J. Food Microbiol.* **2013**, *166*, 413–418. [CrossRef]
15. Aluko, O.O.; Ojeremi, T.T.; Ajidagba, E.B. Evaluation of food safety and sanitary practices among food vendors at car parks in Ile Ife, southwestern Nigeria. *Food Control* **2014**, *40*, 165–171. [CrossRef]
16. Liu, Z.; Zhang, G.; Zhang, X. Urban street foods in Shijiazhuang city, China: Current status, safety practices and risk mitigating strategies. *Food Control* **2014**, *41*, 212–218. [CrossRef]
17. Campos, J.; Gil, J.; Mourão, J.; Peixe, L.; Antunes, P. Ready-to-eat street-vended food as a potential vehicle of bacterial pathogens and antimicrobial resistance: An exploratory study in Porto region, Portugal. *Int. J. Food Microbiol.* **2015**, *206*, 1–6. [CrossRef] [PubMed]
18. Alimi, B.A. Risk factors in street food practices in developing countries: A review. *Food Sci. Hum. Wellness* **2016**, *5*, 141–148. [CrossRef]
19. Kothe, C.I.; Schild, C.H.; Tondo, E.C.; da Silva Malheiros, P. Microbiological contamination and evaluation of sanitary conditions of hot dog street vendors in Southern Brazil. *Food Control* **2016**, *62*, 346–350. [CrossRef]
20. Trafialek, J.; Drosinos, E.H.; Kolanowski, W. Evaluation of street food vendors' hygienic practices using fast observation questionnaire. *Food Control* **2017**, *80*, 350–359. [CrossRef]
21. Trafialek, J.; Drosinos, E.H.; Laskowski, W.; Jakubowska-Gawlik, K.; Tzamalis, P.; Leksawasdi, M.; Surawang, S.; Kolanowski, W. Street food vendors' hygienic practices in some Asian and EU countries—A survey. *Food Control* **2018**, *85*, 212–222. [CrossRef]
22. Kolanowski, W.J.; Trafialek, E.H.; Drosinos, P.; Tzamalis, P.G. Polish and Greek young adults ' experience of low quality meals when eating out. *Food Control* **2020**, *109*, 106901. [CrossRef]
23. Qureshi, S.; Vaida, N.; Azim, H. Hygiene practices and food safety knowledge among street food vendors in Kashmir. *Int. J. Sci. Res.* **2016**, *4*, 723–726. [CrossRef]
24. Wiatrowski, M.; Czarniecka-Skubina, E.; Trafialek, J.; Rosiak, E. An Evaluation of the Hygiene Practices of Polish Street Food Vendors—On an example of food trucks and stands. *Foods* **2021**, *10*, 2640. [CrossRef]
25. Abid, M.T.; Banna, M.H.A.; Hamiduzzaman, M.; Seidu, A.-A.; Kundu, S.; Rezyona, H.; Disu, T.R.; Akter, N.; Khadleduzzaman, M.; Ahinkorah, B.O.; et al. Assessment of food safety knowledge, attitudes and practices of street food vendors in Chattogram city, Bangladesh: A cross-sectional study. *Public Health Chall.* **2022**, *1*, e16. [CrossRef]
26. Burt, B.M.; Volel, C.; Finkel, M. Safety of vendor-prepared foods: Evaluation of 10 processing mobile food vendors in Manhattan. *Public Health Rep.* **2003**, *118*, 470–476. [CrossRef] [PubMed]
27. WHO. Basic steps to improve safety of street-vended food. *INFOSAN Inf. Note* **2010**, *3*, 1–5. Available online: https://www.yumpu.com/en/document/view/5302144/basic-steps-to-improve-safety-of-street-vended-food (accessed on 12 December 2022).
28. Dun-Dery, E.J.; Addo, H.O. Food hygiene awareness, processing and practice among street food vendors in Ghana. *Food Public Health* **2016**, *6*, 65–74. [CrossRef]
29. McKay, F.H.; Singh, A.; Singh, S.; Good, S.; Osborne, R.H. Street vendors in Patna, India: Understanding the socio-economic profile, livelihood and hygiene practices. *Food Control* **2016**, *70*, 281–285. [CrossRef]
30. Tomar, R.S.; Gupta, M.; Kaushik, S.; Mishra, R.K. Bacteriological Quality of Panipuri in Historical Gwalior City (MP), India. *Asian J. Pharm.* **2018**, *12*, 329. [CrossRef]
31. Abraham, S.; Krishnan, T.A. An Analysis of How Street Food in India can be made Safe Food. *IJCESR* **2017**, *4*, 57–64.
32. Islam, N.; Arefin, S.; Nigar, T.; Haque, S.N.; Haq, K.I.; Emran, T.A.; Nazrul, T. Street food eating habits in Bangladesh: A study on Dhaka city. *Int. J. Manag. Dev. Stud.* **2017**, *6*, 49–57.
33. Rosales, A.P.; Linnemann, A.R.; Luning, P.A. Food safety knowledge, self-reported hygiene practices, and street food vendors' perceptions of current hygiene facilities and services—An Ecuadorean case. *Food Control* **2023**, *144*, 109377. [CrossRef]

34. Azanaw, J.; Engdaw, G.T.; Dejene, H.; Bogale, S.; Degu, S. Food hygiene knowledge, and practices and their associated factors of street food vendors in Gondar city, Northwest Ethiopi. A cross-sectional study. *Heliyon* **2022**, *8*, e11707. [CrossRef] [PubMed]
35. FAO—Food and Agriculture Organization of the United Nations. *Food for the Cities: Street Foods*. 2019. Available online: https://www.fao.org/fcit/food-processing/street-foods/en/ (accessed on 13 May 2022).
36. CDC—Centers for Disease Control and Prevention. Surveillance for Foodborne Disease Outbreaks, United States, 2015, Annual Report. US Department of Health and Human Services. 2017. Available online: https://www.cdc.gov/foodsafety/pdfs/2015 foodborneoutbreaks_508.pdf (accessed on 13 May 2022).
37. Mosupye, F.M.; Von Holy, A. Microbiological hazard identification and exposure assessment of street food vending in Johannesburg, South Africa. *Int. J. Food Microbiol.* **2000**, *61*, 137–145. [CrossRef] [PubMed]
38. Hanashiro, A.; Morita, M.; Matté, G.R.; Matté, M.H.; Torres, E.A.F.S. Microbiological quality of selected street foods from a restricted area of São Paulo City, Brazil. *Food Control* **2005**, *16*, 439–444. [CrossRef]
39. Gadaga, T.H.; Samende, B.K.; Musuna, C.; Chibanda, D. The microbiological quality of informally vended foods in Harare, Zimbabwe. *Food Control* **2008**, *19*, 829–832. [CrossRef]
40. Nyenje, M.E.; Odjadjare, C.E.; Tanih, N.F.; Green, E.; Ndip, R.N. Foodborne pathogens recovered from ready-to-eat foods from roadside cafeterias and retail outlets in alice, eastern cape province, South Africa: Public health implications. *Int. J. Environ. Res. Public Health* **2012**, *9*, 2608–2619. [CrossRef]
41. Adjrah, Y.; Soncy, K.; Anani, K.; Blewussi, K.; Karou, D.S.; Ameyapoh, Y.; de Souza, C.; Gbeassor, M. Socio-economic profile of street food vendors and microbiological quality of ready-to-eat salads in Lomé. *Int. Food Res. J.* **2013**, *20*, 65–70.
42. Nkah, G. *Evaluation de la Qualité Microbiologique Du Maquereau á la Braise Commercialisé Dans Quelques Quartiers de la Ville de Douala*; Mémoire d'Ingénieur, Institut des Sciences Halieutiques de Yabassi, Université de Douala: Douala, Cameroon, 2017; pp. 1–66.
43. Achondi, C. Study on the Factors Affecting the Contamination of Roasted Mackerel within the Town of Douala. Ph.D. Thesis, Institute of Fisheries and Aquatic Sciences of Yabassi, University of Douala, Douala, Cameroon, 2018; pp. 1–74.
44. Nguendo Yongsi, H.B. Eating to live or eating to damage one's health: Microbiological risks associated with street-vended foods in a subtropical urban setting (Yaounde-Cameroon). *Nutr. Food Sci. Int. J.* **2018**, *6*, 555695. [CrossRef]
45. Tchigui Manga Maffouo, S.; Tene Mouafo, H.; Simplice Mouokeu, R.; Manet, L.; Kamgain Tchuenchieu, A.; Noutsa Simo, B.; Tchuitcheu Djeuachi, H.; Nama Medoua, G.; Tchoumbougnang, F. Evaluation of sanitary risks associated with the consumption of street food in the city of Yaoundé (Cameroon): Case of braised fish from Mvog-Ada, Ngoa Ekélé, Simbock, Ahala and Olézoa. *Heliyon* **2021**, *7*, e07780. [CrossRef]
46. Alfred, K.K.; Jean-Paul, B.K.M.; Hermann, C.W.; Mirelle, B.A.; Marcellin, D.K. Assessment of safety risks associated with pork meat sold on the market in Abidjan city (Cote d'Ivoire) using surveys and microbial testing. *Heliyon* **2019**, *5*, e02172. [CrossRef]
47. Aovare, O.P.; Kendie, S.B.; Osei-Kufuorp, P. Safety Assessment of Street Foods in The Bolgatanga Municipality of The Upper East Region, Ghana. *JAFSAT* **2022**, *9*, 21–34.
48. Cardinale, E.; Gros-Claude, J.P.; Tall, F.; Gueye, E.; Salvat, G. Risk factors for contamination of ready-to-eat street-vended poultry dishes in Dakar, Senegal. *Int. J. Food Microbiol.* **2005**, *103*, 157–165. [CrossRef] [PubMed]
49. Ologhobo, A.; Omojola, A.; Ofongo, S.; Moiforay, S.; Jibir, M. Safety of street vended meat products-chicken and beef suya. *Afr. J. Biotechnol.* **2010**, *9*, 4091–4095.
50. Kigigha, L.T.; Ovunda, H.; Izah, S. Microbiological quality assessment of suya sold in Yenegoa Metropolis, Nigeria. *J. Adv. Biol. Basic Res.* **2015**, *1*, 105–109.
51. Dagalea, F.M.S.; Lim, K.M.C.; Vicencio, M.C.G.; Ballicud, J.J.C.; Burac, M.R.B.; Vibar, J.J.B.; Villadolid, V.B.E. Are street foods safe: Detection of *escherichia coli* in street foods sauces. *S. Asian J. Res. Microbiol.* **2021**, *9*, 41–45. [CrossRef]
52. Savvaidis, I.N.; Al Katheeri, A.; Lim, S.-H.E.; Lai, K.-S.; Abushelaibi, A. Traditional Foods, Food Safety Practices, and Food Culture in the Middle East, Chapter 1. In *Food Safety in the Middle East*; Academic Press: Cambridge, MA, USA, 2022; pp. 1–31. [CrossRef]
53. Ali, A.; Ahmad, N.; Liaqat, A.; Farooq, M.A.; Ahsan, S.; Chughtai, M.F.J.; Rahaman, A.; Saeed, K.; Junaid-ur-Rahman, S.; Siddeeg, A. Safety and quality assessment of street-vended barbecue chicken samples from Faisalabad, Pakistan. *Food Sci. Nutr.* **2022**. *online ahead of print*. [CrossRef]
54. Nkere, C.K.; Ibe, N.I.; Iroegbu, C.U. Bacteriological Quality of Foods and Water Sold by Vendors and in Restaurants in Nsukka, Enugu State, Nigeria: A Comparative Study of Three Microbiological Methods. *J. Health Popul. Nutr.* **2011**, *29*, 560–566. [CrossRef] [PubMed]
55. Willis, C.; Elviss, N.; Aird, H.; Fenelon, D.; McLauchlin, J. Evaluation of hygiene practices in catering premises at large-scale events in the UK: Identifying risks for the Olympics 2012. *Public Health* **2012**, *126*, 646–656. [CrossRef] [PubMed]
56. Enoch, D.A.; Mlangeni, D.A.; Ekundayo, J.; Aliyu, M.; Sismey, A.W.; Aliyu, S.H.; Karas, A. Gram negative bacteraemia-are they preventable and what will *E. coli* surveillance add? *J. Infect. Prev.* **2013**, *14*, 55–59. [CrossRef]
57. Martinon, A.; Cronin, U.P.; Quealy, J.; Stapleton, A.; Wilkinson, M.G. Swab sample preparation and viable real-time PCR methodologies for the recovery of *Escherichia coli*, *Staphylococcus aureus* or *Listeria monocytogenes* from artificially contaminated food processing surfaces. *Food Control* **2012**, *24*, 86–94. [CrossRef]
58. Mead, G.C. *Poultry Meat Processing and Quality*; Woodhead Publishing Series in Food Science, Technology and Nutrition: Cambridge, UK, 2004; pp. 232–257.

59. Waliullah, S.; Ahsan, C.R. Assessment of Microbiological Quality of Some Meat-Based Fast Foods Collected From Street Vendors. *J. Innov. Dev. Strategy* **2011**, *5*, 44–46.
60. Manguiat, L.S.; Fang, T.J. Microbiological quality of chicken and pork-based street-vended foods from Taichung, Taiwan, and Laguna Philippines. *Food Microbiol.* **2013**, *36*, 57–62. [CrossRef] [PubMed]
61. Annan-Prah, A.; Amewowor, D.H.A.K.; Osei-Kofi, J.; Amoono, S.E.; Akorli, S.Y.; Saka, E.; Ndadi, H.A. Street foods: Handling, hygiene and client expectations in a World Heritage Site Town, Cape Coast, Ghana. *Afr. J. Micriol. Res.* **2011**, *5*, 1629–1634. [CrossRef]
62. Dabbagh-Zadeh, R.; Hashemi, M.; Massah, M.; Taherian, E.; Sayyahi, N.; Haghparasti, F.; Marhamati, M.; Alavi, L.; Bahrami, N.; Katami, M.; et al. Microbiological and Chemical Properties of Falafel Samples Collected From Street Market in Ahvaz, Iran. *Jundishapur J. Med. Sci.* **2022**, *21*, 68–79. [CrossRef]
63. Jaja, I.F.; Green, E.; Muchenje, V. Aerobic mesophilic, coliform, Escherichia coli, and Staphylococcus aureus counts of raw meat from the formal and informal meat sectors in South Africa. *Int. J. Environ. Res. Public Health* **2018**, *15*, 819. [CrossRef]
64. Odwar, J.A.; Kikuvi, G.; Kariuki, J.N.; Kariuki, S. A cross sectional study on the microbiological quality and safety of raw chicken meats sold in Nairobi, Kenya. *BMC Res. Notes* **2014**, *7*, 627. [CrossRef]
65. Girmay, A.M.; Mengesha, S.D.; Weldetinsae, A.; Alemu, Z.A.; Dinsa, D.A.; Wagari, B.; Weldegebriel, M.G.; Serte, M.G.; Alemayehu, T.A.; Kenea, M.A.; et al. Factors associated with food safety practice and drinking-water quality of food establishments in Bishoftu Town, Ethiopia. *Discov. Food* **2022**, *2*, 35. [CrossRef]
66. Bhattacharjya, H.; Reang, T. Safety of street foods in Agartala, North East India. *Public Health* **2014**, *128*, 746–748. [CrossRef] [PubMed]
67. Parker, J.A.; Kurien, T.T.; Huppatz, C. Hepatitis A outbreak associated with kava drinking. *Commun. Dis. Intell. Q. Rep.* **2014**, *38*, E26–E28.
68. Samapundo, S.; Climat, R.; Xhaferi, R.; Devlieghere, F. Food safety knowledge, attitudes and practices of street food vendors and consumers in Port-au-Prince, Haiti. *Food Control* **2015**, *50*, 457–466. [CrossRef]
69. Evans, J.A.; Russell, S.L.; James, C.; Corry, J.E.L. Microbial contamination of food refrigeration equipment. *J. Food Eng.* **2004**, *62*, 225–232. [CrossRef]
70. Christon, C.A.; Lindsay, D.; Von Holy, A. Microbiological survey of ready- to-eat foods and associated preparation surfaces in retail delicatessens Johannesburg, South Africa. *Food Control* **2008**, *19*, 727–733. [CrossRef]
71. Rodriguez, M.; Valero, A.; Posada-Izquierdo, G.D.; Carrasco, E.; Zurera, G. Evaluation of food handler practices and mirobiological status of ready-to-eat foods in long-term care facilities in Andalusia region of Spain. *J. Food Protect.* **2011**, *74*, 1504–1512. [CrossRef] [PubMed]
72. Cunningham, A.E.; Rajagopal, R.; Lauer, J.; Allwood, P. Assessment of hygienic quality of surfaces in retail food service establishments based on microbial counts and real-time detection of ATP. *J. Food Protect.* **2011**, *74*, 686–690. [CrossRef] [PubMed]
73. Lahou, E.; Jacxsens, L.; Daelman, J.; Van Landeghem, F.; Uyttendaele, M. Microbiological Performance of a Food Safety Management System in a Food Service Operation. *J. Food Protect.* **2012**, *75*, 706–716. [CrossRef]
74. Czarniecka-Skubina, E. *Ocena ryzyka zdrowotnego w zakładach gastronomicznych na wybranych przykładach (Health Risk Assessment in Catering Establishments Based on the Selected Examples)*; Treatises and Monographs; WULS-SGGW: Warsaw, Poland, 2013.
75. *PN-ISO-18593:2018*; Microbiology of food and animal feeding stuffs—Horizontal methods for sampling techniques from surfaces using contact plates and swabs. ISO: Geneva, Switzerland, 2018.
76. *ISO 4833-2:2013/Amd 1:2022*; Microbiology of the food chain—Horizontal method for the enumeration of microorganisms—Part 2: Colony count at 30 °C by the surface plating technique. ISO: Geneva, Switzerland, 2022.
77. *PN-EN ISO 6888-2:2001/A1:2004*; Horizontal method for the enumeration of coagulase-positive staphylococci (*Staphylococcus aureus* and other species)—Part 2: Method using rabbit plasma fibrinogen agar medium. ISO: Geneva, Switzerland, 2004.
78. *PN-EN ISO 21528-2:2017*; Microbiology of the food chain—Horizontal method for the detection and enumeration of *Enterobacteriaceae*—Part 2: Colony-count technique. ISO: Geneva, Switzerland, 2017.
79. *ISO 16649-3:2015*; Microbiology of the food chain—Horizontal method for the enumeration of beta-glucuronidase-positive *Escherichia coli*—Part 3: Detection and most probable number technique using 5-bromo-4-chloro-3-indolyl-ß-D-glucuronide. ISO: Geneva, Switzerland, 2015.
80. *PN-EN ISO 11290-1:2017*; Microbiology of the food chain—Horizontal method for the detection and enumeration of *Listeria monocytogenes* and of *Listeria* spp.—Part 1: Detection method. ISO: Geneva, Switzerland, 2017.
81. *PN-EN ISO 6579-1:2017*; Microbiology of the food chain—Horizontal method for the detection, enumeration and serotyping of *Salmonella*—Part 1: Detection of *Salmonella* spp. ISO: Geneva, Switzerland, 2017.
82. Moore, G.; Griffith, C.J. A comparison of traditional and recently developed methods for monitoring surface hygiene within the food industry: An industry trial. *Int. J. Environ. Health Res.* **2002**, *12*, 317–329. [CrossRef]
83. Griffith, C.J.; Cooper, R.A.; Gilmore, J.; Davies, C.; Lewis, M. An evaluation of hospital regimes and standards. *J. Hosp. Infect.* **2000**, *45*, 19–28. [CrossRef]
84. StatSoft, Inc. STATISTICA (Data Analysis Software System), 2011, Version 13.3. Available online: http://www.statsoft.com (accessed on 4 August 2022).
85. Dancer, S.J. How Do We Assess Hospital Cleaning? A Proposal for Microbiological Standards for Surfaces Hygiene in Hospital. *J. Hosp. Infect.* **2004**, *56*, 10–15. [CrossRef]

86. Cooper, R.A.; Griffith, C.J.; Malik, R.E.; Obee, P.; Looker, N. Monitoring the effectiveness of cleaning in four British hospitals. *Am. J. Infect. Control* **2007**, *35*, 338–341. [CrossRef]
87. Lewis, K.; Arnott, W.P.; Moosmüller, H.; Wold, C.E. Strong spectral variation of biomass smoke light absorption and single scattering albedo observed with a novel dual-wavelength photoacoustic instrument. *J. Geophys. Res.* **2008**, *113*, D16203. [CrossRef]
88. Tebbutt, G.; Bell, V.; Aislabie, J. Verification of cleaning efficiency and its possible role in programmed hygiene inspections of food businesses undertaken by local authority officers. *J. Appl. Microbiol.* **2007**, *102*, 1010–1017. [CrossRef] [PubMed]
89. Griffith, C.J. Food Safety in Catering Establishments. In *Safe Handling of Foods*; Farber, J.M., Todd, E.C.D., Eds.; Marcel Dekker: New York, NY, USA, 2000; pp. 235–256.
90. Yepiz-Gomez, M.S.; Bright, K.R.; Gerba, C.P. Indentity and numbers of bacteria present on tabletops and in dishcloths used to wipe down tabletops in public restaurants and bars. *Food Protec. Trends* **2006**, *26*, 786–792.
91. Bolton, D.J.; Meally, A.; Blair, I.S.; Mcdowell, D.A.; Cowan, C. Food safety knowledge of head chefs and catering managers in Ireland. *Food Control* **2008**, *19*, 291–300. [CrossRef]
92. Legnani, P.; Leoni, E.; Berveglieri, M.; Mirolo, G.; Alvaro, N. Hygienic control of mass catering establishments, microbiological monitoring of food and equipment. *Food Control* **2004**, *15*, 205–211. [CrossRef]
93. Todd, E.C.D.; Greig, J.D.; Bartleson, C.A.; Michaels, B.S. Outbreaks where food workers have been implicated in the spread of foodborne disease. Part 6. Transmission and survival of pathogens in the food processing and preparation environment. *J. Food Protect.* **2009**, *72*, 202–219. [CrossRef]
94. Campdepadros, M.; Stchigel, A.M.; Romeu, M.; Quilez, J.; Sola, R. Effectiveness of two sanitation procedures for decreasing the microbial contamination levels (including *Listeria monocytogenes*) on food contact and non-food contact surfaces in a dessert-processing factory. *Food Control* **2012**, *23*, e31. [CrossRef]
95. Lee, J.; Cartwright, R.; Grueser, T.; Pascall, M.A. Efficiency of manual dishwashing conditions on bacterial survival on eating utensils. *J. Food Eng.* **2007**, *80*, 885–891. [CrossRef]
96. Phlillips, M.L.; Elledge, B.L.; Basara, H.G.; Lynch, R.A.; Boatright, D.T. Recurrent critical violations of the food code in retail food service establishments. *J. Environ. Health* **2006**, *68*, 24–30.
97. Reske, K.A.; Jenkins, T.; Fernandez, C.; Vanamber, D.; Hedberg, C.W. Beneficial effects of implementing an announced restaurant inspection program. *J. Environ. Health* **2007**, *69*, 27–34.
98. Garayoa, R.; Abundancia, C.; Díez-Leturia, M.; Vitas, A.I. Essential tools for food safety surveillance in catering services: On-site inspections and control of high risk cross-contamination surfaces. *Food Control* **2017**, *75*, 48–54. [CrossRef]
99. Saada, M.; Seea, T.P.; Abdullahb, M.F.F.; Nor, N.M. Use of Rapid Microbial Kits for Regular Monitoring of Food-Contact Surfaces towards Hygiene Practices. *Procedia Soc. Behav. Sci.* **2013**, *105*, 273–283. [CrossRef]
100. Tomasevic, I.; Kuzmanovic, J.; Andelkovic, A.; Saračević, M.; Stojanović, M.M.; Djekic, I. The effects of mandatory HACCP implementation on microbiological indicators of process hygiene in meat processing and retail establishments in Serbia. *Meat Sci.* **2016**, *114*, 54–57. [CrossRef] [PubMed]
101. Larson, E.L.; Aiello, A.E.; Gomez-Duarte, C.; Lin, S.X.; Lee, L.; Della-Latta, P.; Lindhart, C. Bioluminescence ATP monitoring as a surrogate marker for microbial load on hands and surfaces in the home. *Food Microbiol.* **2003**, *20*, 735–739. [CrossRef]
102. Rosiak, E.; Czarniecka-Skubina, E.; Wachowicz, I.; Kołożyn-Krajewska, D. Rapid Diagnostic methods for monitoring hygienic conditions in the food production, chapter 6. In *Production Engineering*; Borkowski, S., Ulewicz, R., Eds.; Novosibirsk State Technical University: Novosibirsk, Russia, 2009; pp. 57–66.
103. Aycicek, H.; Oguz, U.; Karci, K. Comparison of results of ATP bioluminescence and traditional hygiene swabbing methods for the determination of surface cleanliness at a hospital kitchen. *Int. J Hyg. Environ. Health* **2006**, *209*, 203–206. [CrossRef] [PubMed]
104. Hooker, E.A.; Allen, S.D.; Gray, L.D. Comparison of Rayon-Tip Swabs and film plates for use in collecting and quantifying bacteria on hospital bed mattresses. *Am. J. Infect. Control* **2011**, *39*, E191–E192. [CrossRef]
105. Okumus, B.; Sonmez, S. An analysis on current food regulations for and inspection challenges of street food: Case of Florida. *J. Culin. Sci. Technol.* **2019**, *17*, 209–223. [CrossRef]

foods

MDPI

Article

Antilisterial and Antimicrobial Effect of *Salvia officinalis* Essential Oil in Beef Sous-Vide Meat during Storage

Robert Gál [1], Natália Čmiková [2], Aneta Prokopová [3] and Miroslava Kačániová [2,4,*]

1 Department of Food Technology, Faculty of Technology, Tomas Bata University in Zlín, Vavrečkova 275, 760 01 Zlín, Czech Republic; gal@ft.utb.cz
2 Institute of Horticulture, Faculty of Horticulture and Landscape Engineering, Slovak University of Agriculture, Tr. A. Hlinku 2, 94976 Nitra, Slovakia
3 Department of Polymer Engineering, Faculty of Technology, Tomas Bata University in Zlín, Vavrečkova 275, 760 01 Zlín, Czech Republic
4 Department of Bioenergetics, Food Analysis and Microbiology, Institute of Food Technology and Nutrition, Rzeszow University, Cwiklinskiej 1, 35-601 Rzeszow, Poland
* Correspondence: miroslava.kacaniova@gmail.com

Abstract: If food is contaminated with pathogens such as *Listeria monocytogenes*, improper cooking during sous-vide preparation can lead to foodborne illnesses. In this study, it was found that *L. monocytogenes* were inactivated with both heat and the essential oil of *Salvia officinalis* (sage EO) in beef tenderloin of the *musculus psoas major* that had undergone sous-vide processing. To determine whether the enhancement of the efficacy of heat treatment is prospective, *L. monocytogenes* and sage EO were mixed. Groups with *L. monocytogenes* alone and sage essential oil combined with *L. monocytogenes* and test groups without EO were established. The samples were vacuum-packed, inoculated with *L. monocytogenes*, and then cooked sous-vide for the predetermined duration at 50, 55, 60, or 65 °C. In both groups with sous-vide beef tenderloin, the total bacterial count, the coliforms bacterial count, and the amount of *L. monocytogenes* were assessed on days 0, 3, 6, 9, and 12. Over these days, the amounts of *L. monocytogenes*, coliform bacteria, and overall bacteria increased. The identification of bacterial strains in various days and categories was performed by MALDI-TOF mass spectrometry. The test group that was exposed to a temperature of 50 °C for 5 min had a higher overall bacterial count for each day that was assessed. *Pseudomonas fragi* and *L. monocytogenes* were the most isolated organisms from the test group and the treated group. To ensure the safety for the consumption of sous-vide beef tenderloin, it was found that the addition of natural antimicrobials could produce effective outcomes.

Keywords: antimicrobial effect; *Listeria monocytogenes*; beef tenderloin (*m. psoas major*); sous-vide; *Salvia officinalis* essential oil

Citation: Gál, R.; Čmiková, N.; Prokopová, A.; Kačániová, M. Antilisterial and Antimicrobial Effect of *Salvia officinalis* Essential Oil in Beef Sous-Vide Meat during Storage. *Foods* **2023**, *12*, 2201. https://doi.org/10.3390/foods12112201

Academic Editor: Baohua Kong

Received: 29 March 2023
Revised: 26 April 2023
Accepted: 28 May 2023
Published: 30 May 2023

1. Introduction

In contrast to conventional cooking techniques, sous-vide involves cooking food in vacuum-sealed containers at precisely controlled temperatures, resulting in a better flavor, texture, and nutritional value as well as a longer shelf life [1]. The evaluation of microbial safety is critical in this cooking technique, so it is essential to understand how this treatment affects microorganisms in order to evaluate the safety of products [2]. The authors of the study [3] found that pathogens present in foods prepared by the sous-vide method at the time of ingestion came from raw ingredients as they had not been destroyed by cooking [3]. The range of temperatures between 30 and 50 °C, where bacterial growth and reproduction are first inhibited, is ideal for the development of most pathogenic bacteria. The temperature of the food during preparation should not be below 54.4 °C to ensure the inactivation of food pathogens, such as *Salmonella* species, *Listeria monocytogenes*, and *Escherichia coli* pathogenic strains [4].

To date, there is not enough evidence to support the antilisterial properties of sage essential oil (EO) in foods. A small number of authors have described how sage affects both Gram-positive and Gram-negative bacteria [5]. *Listeria monocytogenes* can become less resistant to heat treatment and sage chemicals either through a synergistic combination or simply as a result of the cumulative effects of the individual components on microorganisms [6]. The essential oil of *Salvia officinalis* disrupted the cell membrane and altered its permeability, causing the release of several cytoplasmic components, such as macromolecular compounds, ATP, and DNA. Sage essential oil has a broad antibacterial effect that is due not only to a special route mechanism but also to a number of activities on the cell surface and in the cytoplasm. To fully understand the antibacterial mechanism of essential sage oil, further research is needed [7]. Natural food preservation methods have recently attracted a lot of attention from both consumers and food technologists [8]. According to research by Korczak et al. [9], the inclusion of sage significantly reduced the development of unpleasant flavors and odors in precooked meat products while they were stored at 4 °C. Madsen et al. [10] claim that sage can be used successfully to prevent the emergence of a warmed-over flavor and thus improve the sensory quality of heat-treated meat products. Gas chromatography/mass spectrometry (GC/MS) and gas chromatography (GC-FID) were used to identify the major compounds in *S. officinalis* EO, including α-thujone (24.6%), camphor (20.6%), 1,8-cineole (12.1%), and α-humulene (5.8%) [11].

The impact of sage EO during meat preservation is a subject that has not yet been thoroughly investigated. Sage could be added to beef patties, according to Zhang et al. [12], without having a negative impact on the sensory qualities of the burger. Natural dried sage powder, even in high concentrations, has been shown to be effective in preserving the sensory quality of cooked beef burgers, according to Mizi et al. [13]. Turkey meatballs with sage extract exhibited superior sensory quality attributes compared to test samples, according to Karpinska-Tymoszczyk [14]. According to data, sage EO may help limit the development of *L. monocytogenes*. Sage may be used as a natural preservative, but to successfully limit microbial growth it must be combined with other substances [15].

Fresh beef vacuum packaging has been shown to be effective in extending shelf life and preserving the sensory characteristics of the product for a long period of time. By limiting oxidation and the development of aerobic microorganisms during cooling, the vacuum increases the shelf life of the meat. The vacuum packaging method has been used more frequently in the institutional market for the distribution of whole pieces of beef [16].

The purpose of this experiment was to study the behavior of *Listeria monocytogenes* that had been inoculated into beef meat with the addition of sage essential oil and to observe the effect of vacuum packing, various heat treatments, and 12 days of storage time. To simulate the storage at a temperature commonly used by consumers in the refrigerator, a temperature of 6 °C was used. The total number of microorganisms and their identification were studied.

2. Materials and Methods

2.1. Sample Preparation

In this experiment, beef meat samples from the thigh (*m. psoas major*) were used. According to the information on the label produced in the Czech Republic, the meat sample was obtained from Charolais breading that was purchased from an authorized retailer. The meat samples were delivered to the microbiological facility in a clean refrigerator under hygienic conditions, where they were kept at 6 °C until the analyses were performed. Within 120 min, the samples were moved from the approved store to the laboratory. The meat was diced and samples weighing 5 g were treated with 1% SOEO solutions (Hanus, Nitra, Slovakia), dissolved in sunflower oil, and vacuum-packaged using a vacuum packer (Concept, Choceň, Czech Republic). Good-quality sunflower oil was purchased from an authorized dealer. A total of 480 different beef samples were examined. The samples studied were treated in the following manner:

BM—fresh beef meat was vacuum-packaged in polyethylene bags, stored anaerobically at 6 °C, and treated at 50–65 °C for 5–25 min.

BMLMEO—fresh beef meat treated with *L. monocytogenes* and 1% EO sage was vacuum-packaged in polyethylene bags, stored anaerobically at 6 °C, and treated at 50–65 °C for 5–25 min.

The control samples were prepared from uncooked raw meat on day zero. Essential oils were added to the samples and maceration was performed for 24 h. The samples were placed in the CASO SV1000 sous-vide device. *L. monocytogenes* CCM 4699 was prepared at 1.5×10^8 CFU and added to the sample at a volume of 100 µL.

2.2. Samples Cultivation

Microbiological tests were performed at 6 °C on days 0, 1, 3, 6, and 12. Samples of five grams were diluted in an Erlenmeyer beaker with 45 mL of a 0.1% sterile saline solution. The samples were homogenized for 30 min in the GFL 3031 shaking incubator of Burgwedel, Germany. The microbial communities were examined: Violet Red Bile Lactose Agar (VRBL, Oxoid, Basingstoke, UK) was used for coliform bacteria culture and incubated at 37 °C for 24 to 48 h. Total viable counts (TVCs) were grown on Plate Count Agar (PCA, Oxoid, Basingstoke, UK), which was incubated at 30 °C for 48 to 72. Then, the total viable counts in this medium were calculated. A 0.1 mL sample was used to inoculate Oxford Agar with an Oxford supplement (Oxoid, Basingstoke, UK) for *L. monocytogenes count*. Incubation took place at 37 °C for 24 h.

2.3. Identification of Microorganisms by MALDI-TOF MS

Using the MALDI-TOF (Matrix-Assisted Laser Desorption/Ionization Time of Flight) MS Bio-typer (Bruker, Daltonics, Bremen, Germany) and reference libraries, microorganisms isolated from beef meat samples were identified.

As an organic substance, a stock solution was created. The standard solution contained 2.5% trifluoroacetic acid, 47.5% water, and 50% acetonitrile. Amounts of 500 mL of pure 100% acetonitrile, 475 mL of purified water, and 25 mL of pure 100% trifluoroacetic acid were combined to create 1 mL of stock solution. The organic solvent was made and combined with "HCCA matrix portioned" in a 250 L Eppendorf flask. All of the substances used to prepare the matrix were purchased from Lambda Life (Bratislava, Slovakia).

The samples were prepared according to the previous instructions [17]. Eight colonies per Petri dish were briefly examined. In an Eppendorf flask, biological material was transferred from a Petri plate along with 300 µL of distilled water, then it was mixed and 900 µL of ethanol was added. The mixture was then centrifuged for two minutes at 10,000× *g* (ROTOFIX 32A, Ites, Vranov, Slovakia). The precipitate was removed from the Eppendorf tube after the supernatant was removed and left to dry at room temperature 20 °C. The particle was then treated with 30 L of 70% formic acid and 30 µL of acetonitrile. The mixture was then centrifuged for 2 min at 10,000× *g*. A MALDI plate was coated with 1 µL of liquid, which was then followed by the addition of 1 µL of a MALDI matrix solution. The samples were dried before being processed for microorganism identification on a MALDI-TOF mass spectrometer (Bruker, Daltonics, Bremen, Germany). The LT MALDI-TOF microflex mass spectrometer (Bruker Daltonics, Bremen, Germany) was used to create mass spectra automatically and was set to work in a linear positive mode with a mass range of 2000–20,000 Da. The device was calibrated using the Bruker bacterial test standard. The results of the mass spectra were examined using MALDI Bio-typer 3.0 software (Bremer, Germany-based Bruker Dal-tonics). The following were the identification criteria: scores between 2.300 and 3.000 denoted a highly probable species identification; scores between 2.000 and 2.299 secured a genus identification with a probable species identification; scores between 1.700 and 1.999 denoted a probable genus identification; and a score less than 1700 was considered an unreliable identification.

2.4. Statistical Analysis

Triplicates of each test and analysis were performed. The mean and standard deviation (SD) of the microbial numbers were calculated using Microsoft Excel. Using Prism 8.0.1 (GraphPad Software, San Diego, CA, USA), one-way analysis of variance (ANOVA) was carried out prior to a Tukey's test with a significance level of 0.05. Data analysis was carried out using SAS®® software version 8.

3. Results

3.1. Total Count of Bacteria

In our investigation, the total numbers of bacteria in the control group and the group treated with sage EO and *L. monocytogenes* were assessed on day 0. The number of total bacteria counts ranged in the control group from 2.20 ± 0.07 (50 °C, 20 min) to 2.50 ± 0.13 log CFU/g (50 °C, 5 min) and in the treated group from 1.91 ± 0.07 (50 °C, 20 min) to 2.24 ± 0.07 log CFU/g (50 °C, 5 min) (Table 1).

Table 1. The total count of bacteria in the control group and the group treated with sage EO and *L. monocytogenes* (log CFU/g) in 0 days.

Treatment	Temperature (°C)	Time (min)	Average	SD	*p* Value
BM	50	5	2.50	0.13	3.699×10^{-2}
BMLMEO	50	5	2.24	0.07	
BM	50	10	2.42	0.06	5.546×10^{-3}
BMLMEO	50	10	2.13	0.07	
BM	50	15	2.28	0.06	6.336×10^{-3}
BMLMEO	50	15	2.03	0.06	
BM	50	20	2.20	0.07	6.139×10^{-3}
BMLMEO	50	20	1.91	0.07	

BM: fresh beef meat was vacuum-packaged in polyethylene bags, stored anaerobically at 6 °C, and treated at 50–65 °C for 5–25 min; BMLMEO: fresh beef meat treated with *L. monocytogenes* and 1% of sage EO was vacuum-packaged in polyethylene bags, stored anaerobically at 6 °C, and treated at 50–65 °C for 5–25 min.

There was no count of microorganisms in sous-vide beef meat that had been heated to higher temperatures. The total number of bacteria decreased on day 0 in counts from the time used for treatment. The number of coliform bacteria on day 0 was zero. The number of *L. monocytogenes* decreased in the treated group with time used at a temperature of 50 °C (Figure 1).

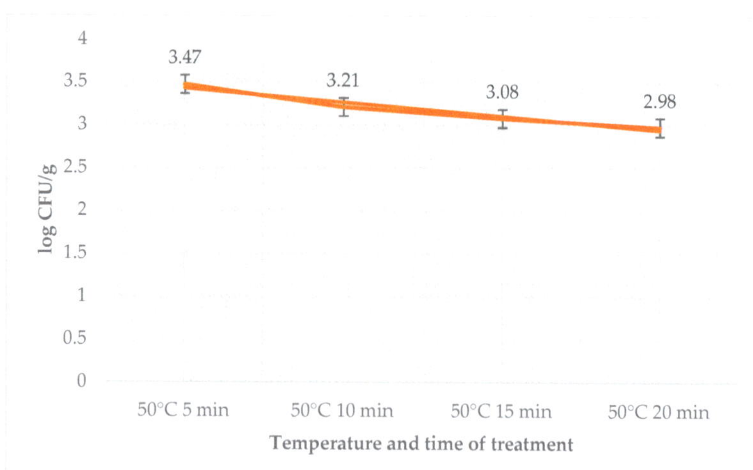

Figure 1. Number of *L. monocytogenes* (log CFU/g) on day 0 in the sage EO treatment group.

Table 2 shows the impact of sage EO for each temperature treatment on day 3 in sous-vide beef samples. Average counts obtained in samples with or without sage EO over time and in accordance with heat treatment are shown in this table. The highest number in the test group was found in samples treated at 50 °C for 5 min. The number of coliforms bacteria on day 3 was zero. The number of *L. monocytogenes* in the treated groups ranged from 3.30 ± 0.10 log CFU/g (50 °C, 20 min) till 3.57 ± 0.12 log CFU/g (50 °C, 5 min) (Figure 2).

Table 2. Total count of bacteria in the control group and group treated with sage EO and *L. monocytogenes* (log CFU/g) on day 3.

Treatment	Temperature (°C)	Time (min)	Average	SD	*p* Value
BM	50	5	2.63	0.06	8.301×10^{-2} *
BMLMEO	50	5	2.52	0.06	
BM	50	10	2.56	0.08	9.114×10^{-2} *
BMLMEO	50	10	2.45	0.03	
BM	50	15	2.47	0.05	1.890×10^{-1} *
BMLMEO	50	15	2.40	0.06	
BM	50	20	2.30	0.20	3.730×10^{-1} *
BMLMEO	50	20	2.44	0.12	
BM	55	5	2.36	0.02	3.290×10^{-2}
BMLMEO	55	5	2.31	0.02	
BM	55	10	2.30	0.10	2.894×10^{-1} *
BMLMEO	55	10	2.23	0.08	
BM	55	15	2.25	0.07	1.721×10^{-1} *
BMLMEO	55	15	2.18	0.02	
BM	55	20	2.17	0.05	1.432×10^{-1} *
BMLMEO	55	20	2.09	0.05	

BM: fresh beef meat was vacuum-packaged in polyethylene bags, stored anaerobically at 6 °C and treated at 50–65 °C for 5–25 min; BMLMEO: fresh beef meat treated with *L. monocytogenes* and 1% of sage EO was vacuum-packaged in polyethylene bags, stored anaerobically at 6 °C and treated at 50–65 °C for 5–25 min. * The data are not statistically significant at the 95% significance level.

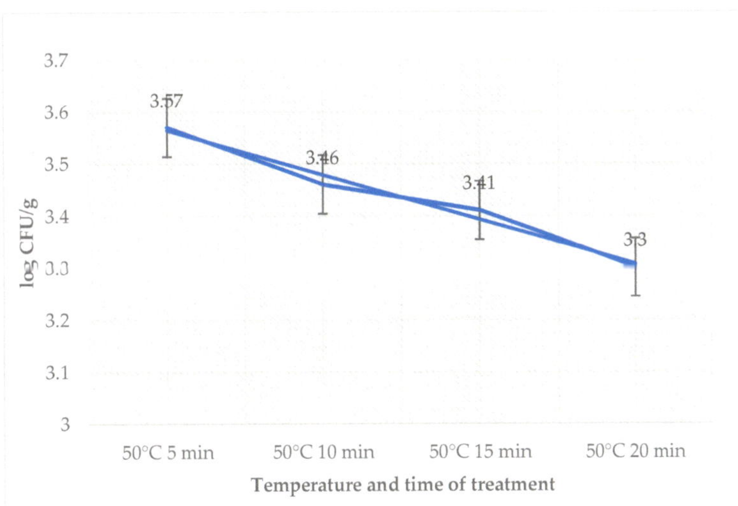

Figure 2. Number of *L. monocytogenes* (log CFU/g) on day 3 in the group treated with sage EO.

The antimicrobial effect of the treatment of sage EO, temperature, and time on day 6 are shown in Table 3. The total count of bacteria in the control group ranged from 2.30 ± 0.08 log CFU/g (55 °C, 20 min) to 2.87 ± 0.09 log CFU/g (50 °C, 5 min) and in the group with treatment of sage EO and *L. monocytogenes* from 2.09 ± 0.03 log CFU/g (55 °C, 5 min) to 2.70 ± 0.06 log CFU/g (50 °C, 5 min). The number of coliform bacteria on day

6 was zero. The number of *L. monocytogenes* decreased in the treated group with the time used at a temperature of 50 °C (Figure 3).

Table 3. Total count of bacteria in the control group and group treated with sage EO and *L. monocytogenes* (log CFU/g) on day 6.

Treatment	Temperature (°C)	Time (min)	Average	SD	*p* Value
BM	50	5	2.87	0.09	4.144×10^{-2}
BMLMEO	50	5	2.70	0.06	
BM	50	10	2.76	0.08	3.139×10^{-2}
BMLMEO	50	10	2.59	0.06	
BM	50	15	2.67	0.09	7.019×10^{-2} *
BMLMEO	50	15	2.51	0.06	
BM	50	20	2.60	0.10	1.830×10^{-1} *
BMLMEO	50	20	2.44	0.08	
BM	55	5	2.43	0.06	3.411×10^{-1} *
BMLMEO	55	5	2.39	0.04	
BM	55	10	2.42	0.08	5.202×10^{-2} *
BMLMEO	55	10	2.24	0.08	
BM	55	15	2.37	0.05	6.755×10^{-3}
BMLMEO	55	15	2.19	0.03	
BM	55	20	2.30	0.08	1.424×10^{-2}
BMLMEO	55	20	2.09	0.03	

BM: fresh beef meat was vacuum-packaged in polyethylene bags, stored anaerobically at 6 °C, and treated at 50–65 °C for 5–25 min; BMLMEO: fresh beef meat treated with *L. monocytogenes* and 1% of sage EO was vacuum-packaged in polyethylene bags, stored anaerobically at 6 °C, and treated at 50–65 °C for 5–25 min. * The data are not statistically significant at the 95% significance level.

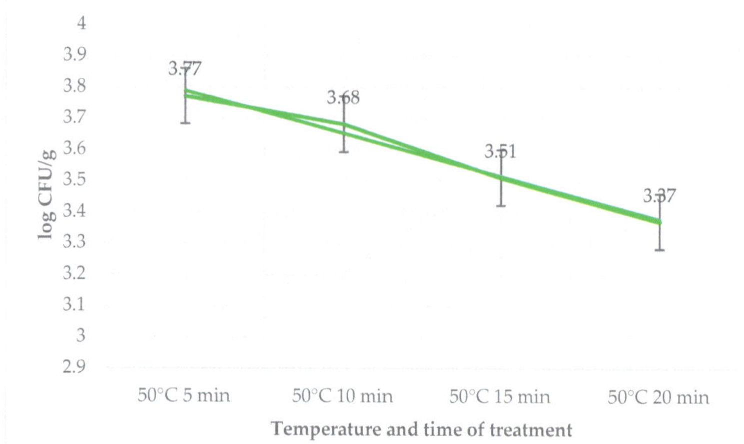

Figure 3. Number of *L. monocytogenes* (log CFU/g) on day 6 in the group treated with sage EO.

The total count of bacteria (Table 4) ranged in the control groups from 2.21 ± 0.12 log CFU/g (60 °C, 20 min) to 3.40 ± 0.04 log CFU/g (50 °C, 5 min) and in the treated groups from 1.17 ± 0.04 log CFU/g (60 °C, 20 min) to 3.22 ± 0.13 log CFU/g (50 °C, 5 min). The total count of bacteria for groups of sous-vide beef meat treated at a 65 °C temperature was zero. The total number of bacteria decreased on day 9 in counts from the time used for treatment. The number of coliform bacteria (Table 5) on day 9 ranged from 2.18 ± 0.12 log CFU/g (50 °C, 20 min) to 2.45 ± 0.04 log CFU/g (50 °C, 5 min) in the control groups and ranged from 1.28 ± 0.09 log CFU/g (50 °C, 20 min) to 1.85 ± 0.06 log CFU/g (50 °C, 5 min) in the treated groups. The number of *L. monocytogenes* decreased in the treated group by the time used at a temperature of 50 °C (Figure 4).

Table 4. Total count of bacteria in the control group and group treated with sage EO and *L. monocytogenes* (log CFU/g) on day 9.

Treatment	Temperature (°C)	Time (min)	Average	SD	p Value
BM	50	5	3.40	0.04	1.194×10^{-1} *
BMLMEO	50	5	3.22	0.13	
BM	50	10	3.31	0.08	2.117×10^{-3}
BMLMEO	50	10	2.89	0.07	
BM	50	15	3.26	0.08	3.525×10^{-4}
BMLMEO	50	15	2.70	0.03	
BM	50	20	3.08	0.08	7.764×10^{-4}
BMLMEO	50	20	2.58	0.05	
BM	55	5	3.05	0.06	6.324×10^{-4}
BMLMEO	55	5	2.49	0.08	
BM	55	10	2.87	0.11	8.11×10^{-4}
BMLMEO	55	10	2.20	0.07	
BM	55	15	2.80	0.06	5.917×10^{-2}
BMLMEO	55	15	2.14	0.04	
BM	55	20	2.75	0.06	4.779×10^{-2}
BMLMEO	55	20	1.95	0.06	
BM	60	5	2.38	0.07	3.222×10^{-4}
BMLMEO	60	5	1.66	0.08	
BM	60	10	2.307	0.02	1.546×10^{-2}
BMLMEO	60	10	1.59	0.05	
BM	60	15	2.23	0.06	2.175×10^{-2}
BMLMEO	60	15	1.37	0.05	
BM	60	20	2.21	0.12	1.350×10^{-4}
BMLMEO	60	20	1.17	0.04	

BM: fresh beef meat was vacuum-packaged in polyethylene bags, stored anaerobically at 6 °C, and treated at 50–65 °C for 5–25 min; BMLMEO: fresh beef meat treated with *L. monocytogenes* and 1% of sage EO was vacuum-packaged in polyethylene bags, stored anaerobically at 6 °C, and treated at 50–65 °C for 5–25 min. * The data are not statistically significant at the 95% significance level.

Table 5. Coliform bacteria in the control group and the group treated with sage EO and *L. monocytogenes* (log CFU/g) on day 9.

Treatment	Temperature (°C)	Time (min)	Average	SD	p Value
BM	50	5	2.45	0.05	2.065×10^{-4}
BMLMEO	50	5	1.85	0.06	
BM	50	10	2.35	0.04	6.221×10^{-2}
BMLMEO	50	10	1.71	0.06	
BM	50	15	2.27	0.04	1.257×10^{-2}
BMLMEO	50	15	1.52	0.04	
BM	50	20	2.18	0.04	5.653×10^{-2}
BMLMEO	50	20	1.28	0.09	

BM: fresh beef meat was vacuum-packaged in polyethylene bags, stored anaerobically at 6 °C, and treated at 50–65 °C for 5–25 min; BMLMEO: fresh beef meat treated with *L. monocytogenes* and 1% of sage EO was vacuum-packaged in polyethylene bags, stored anaerobically at 6 °C, and treated at 50–65 °C for 5–25 min.

The antimicrobial effects of sage EO, temperature, and time on day 12 are shown in Table 6. The total count of bacteria in the control group ranged from 1.90 ± 0.08 log CFU/g (65 °C, 10 min) to 3.90 ± 0.10 log CFU/g (50 °C, 5 min) and in the group with treatment of sage EO and *L. monocytogenes* from 2.05 ± 0.03 log CFU/g (65 °C, 10 min) to 3.64 ± 0.08 log CFU/g (50 °C, 5 min). The number of coliform bacteria (Table 7) on day 12 in the control group ranged from 2.20 ± 0.11 log CFU/g (55 °C, 5 min) to 3.26 ± 0.05 log CFU/g (50 °C, 5 min) and in the group with treatment of sage EO and *L. monocytogenes* from 2.41 ± 0.05 log CFU/g (55 °C, 5 min) to 2.82 ± 0.04 log CFU/g (50 °C, 5 min). The number of *L. monocytogenes* in the treated groups ranged from 4.95 ± 0.16 log CFU/g (50 °C, 20 min) to 5.41 ± 0.04 log CFU/g (50 °C, 5 min) (Figure 5).

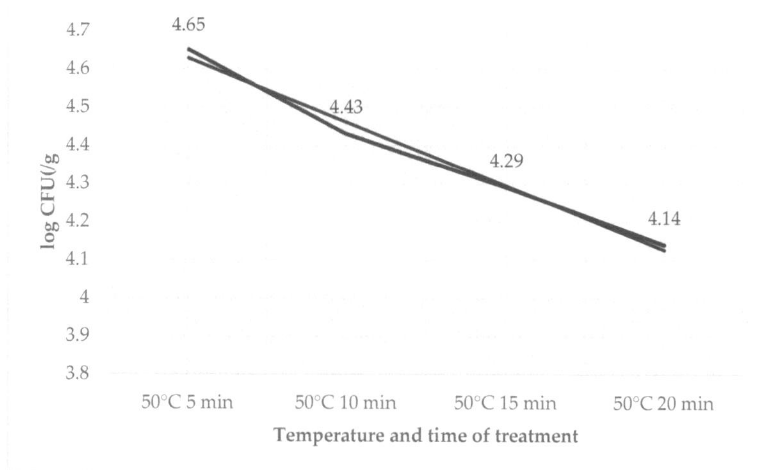

Figure 4. Number of *L. monocytogenes* (log CFU/g) on day 9 in the group treated with sage EO.

Table 6. Total count of bacteria in the control group and group treated with sage EO and *L. monocytogenes* (log CFU/g) on day 12.

Treatment	Temperature (°C)	Time (min)	Average	SD	*p* Value
BM	50	5	3.90	0.10	2.685×10^{-2}
BMLMEO	50	5	3.64	0.08	
BM	50	10	3.76	0.08	1.797×10^{-1} *
BMLMEO	50	10	3.58	0.17	
BM	50	15	3.64	0.07	4.083×10^{-2}
BMLMEO	50	15	3.40	0.12	
BM	50	20	3.56	0.08	4.058×10^{-2}
BMLMEO	50	20	3.35	0.09	
BM	55	5	3.47	0.07	6.631×10^{-3}
BMLMEO	55	5	3.16	0.07	
BM	55	10	3.24	0.07	1.572×10^{-2}
BMLMEO	55	10	3.04	0.05	
BM	55	15	3.14	0.06	1.110×10^{-1} *
BMLMEO	55	15	2.87	0.09	
BM	55	20	3.03	0.13	7.664×10^{-2} *
BMLMEO	55	20	2.64	0.11	
BM	60	5	2.71	0.07	1.133×10^{-1} *
BMLMEO	60	5	2.43	0.09	
BM	60	10	2.64	0.08	2.063×10^{-1} *
BMLMEO	60	10	2.40	0.10	
BM	60	15	2.45	0.08	4.512×10^{-1} *
BMLMEO	60	15	2.35	0.06	
BM	60	20	2.34	0.06	6.905×10^{-2} *
BMLMEO	60	20	2.17	0.03	
BM	65	5	2.20	0.10	1.325×10^{-1} *
BMLMEO	65	5	2.09	0.03	
BM	65	10	1.90	0.10	3.407×10^{-1} *
BMLMEO	65	10	2.05	0.08	

BM: fresh beef meat was vacuum-packaged in polyethylene bags, stored anaerobically at 6 °C, and treated at 50–65 °C for 5–25 min; BMLMEO: fresh beef meat treated with *L. monocytogenes* and 1% of sage EO was vacuum-packaged in polyethylene bags, stored anaerobically at 6 °C, and treated at 50–65 °C for 5–25 min. * The data are not statistically significant at the 95% significance level.

Table 7. Coliform bacteria in the control group and group treated with EO and *L. monocytogenes* (log CFU/g) on day 12.

Treatment	Temperature (°C)	Time (min)	Average	SD	*p* Value
BM	50	5	3.26	0.05	1.229×10^{-2}
BMLMEO	50	5	2.82	0.04	
BM	50	10	3.13	0.04	2.381×10^{-2}
BMLMEO	50	10	2.77	0.04	
BM	50	15	3.00	0.10	8.283×10^{-2} *
BMLMEO	50	15	2.66	0.05	
BM	50	20	2.81	0.07	5.750×10^{-2} *
BMLMEO	50	20	2.54	0.04	
BM	55	5	2.20	0.11	2.397×10^{-1} *
BMLMEO	55	5	2.41	0.05	

BM: fresh beef meat was vacuum-packaged in polyethylene bags, stored anaerobically at 6 °C, and treated at 50–65 °C for 5–25 min; BMLMEO: fresh beef meat treated with *L. monocytogenes* and 1% of sage EO was vacuum-packaged in polyethylene bags, stored anaerobically at 6 °C, and treated at 50–65 °C for 5–25 min. * The data are not statistically significant at the 95% significance level.

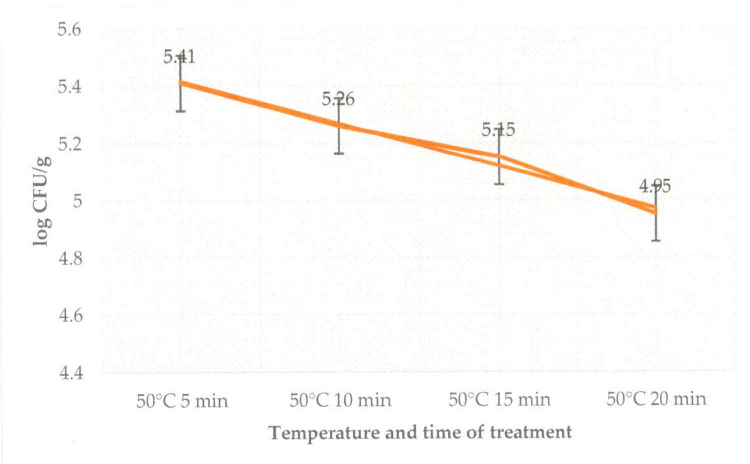

Figure 5. Number of *L. monocytogenes* (log CFU/g) on day 12 in the group treated with sage EO.

3.2. Isolated Species of Bacteria

A total of 381 isolates were identified from sous-vide beef meat from the control and treated groups of samples. A total of 10 families, 14 genera, and 25 species were isolated from the control group of samples (Figure 6). The most isolated species in this study were *Pseudomonas fragi* (21.53%), *Hafnia alvei* (10%), and *Pantotea agglomerans* (8.9%) followed by *Kocuria salcida* (7%). A total of 8 families, 13 genera, and 21 species were isolated from the treated group of sous-vide beef meat (Figure 7). The most isolated species was *L. monocytogenes* (28%), which was added to this group. The other most isolated species of bacteria from the treated group were *P. fragi* (10%), *Lysinibacillus xylanitaticus* (6%), *H. alvei* (5%), and *Pseudomonas graminis* (5%).

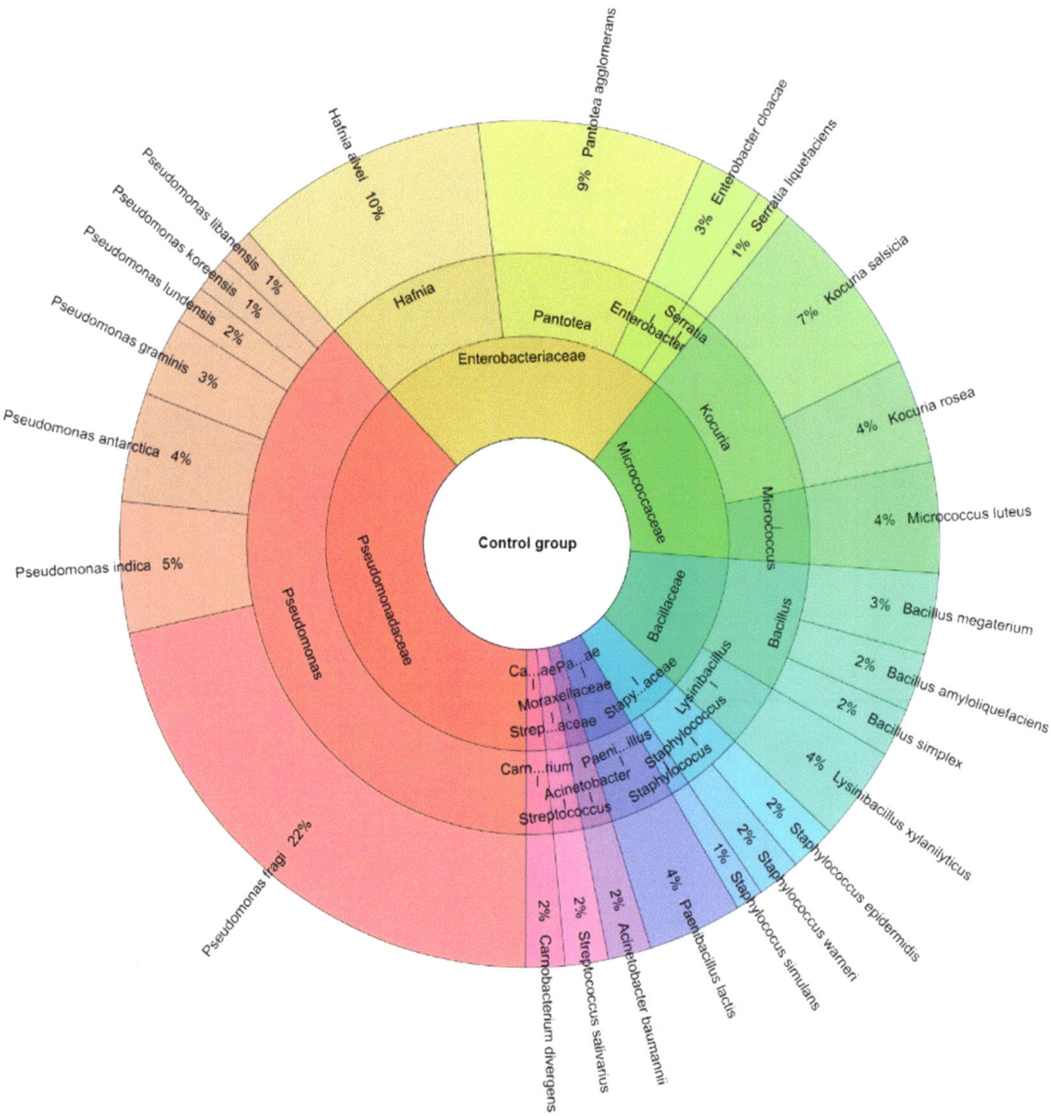

Figure 6. Krona diagram of isolated species of bacteria from the control group.

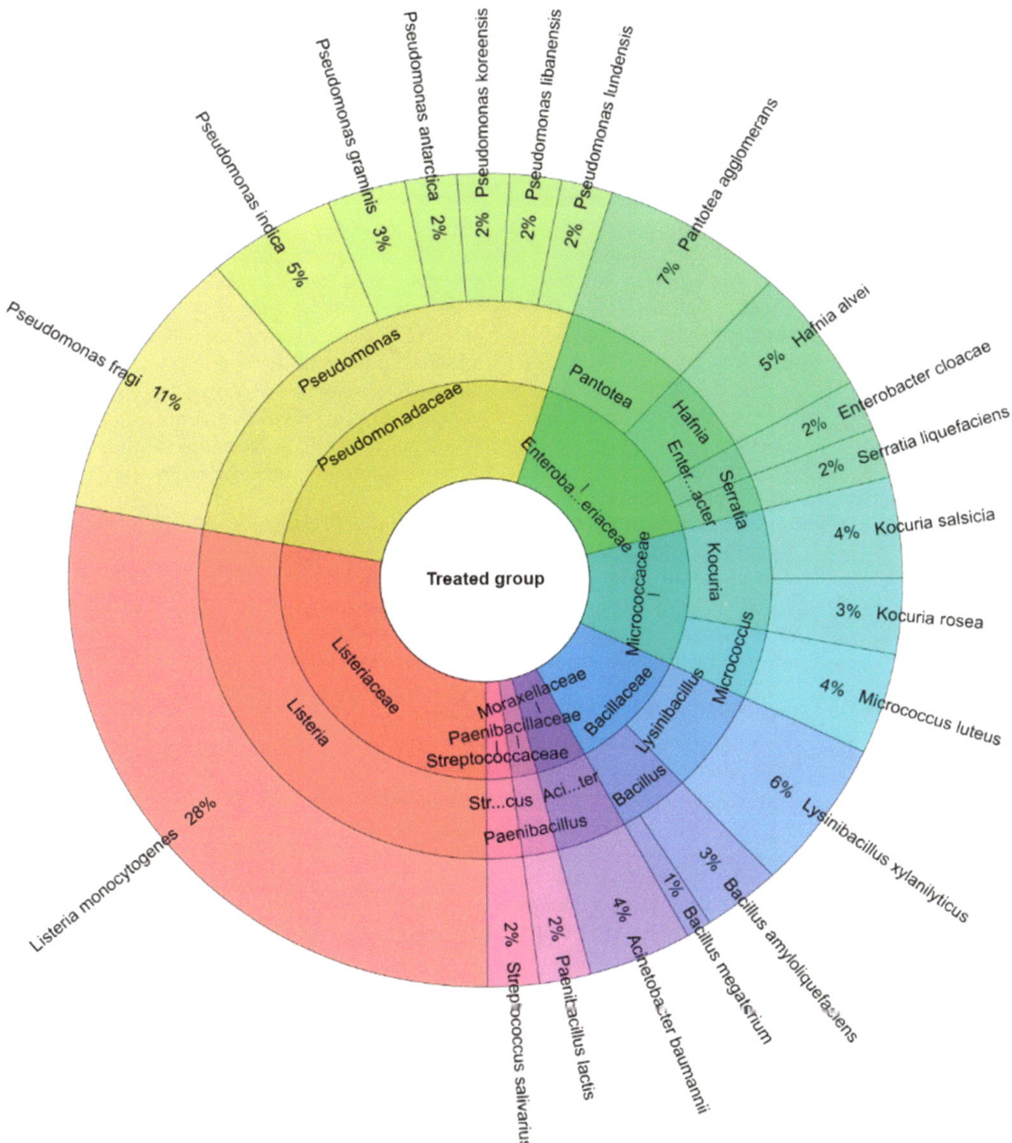

Figure 7. Krona diagram of isolated species of bacteria from the treated group.

4. Discussion

Antimicrobials can be added to food to ensure the microbial safety of ready-to-eat meals. The main role of antimicrobials is to prevent or eliminate pathogens and spoilage microorganisms. Due to concerns about synthetic preservatives, customers have recently preferred foods made with natural antimicrobials. Organic compounds or essential oils of plants, for example, have been used to prevent food from spoiling [18]. They have also been used to help ensure food safety, which was the focus of this research. Our results show the influence of the essential oil of *S. officinalis* at different temperatures and times. Higher temperatures and longer times have proven to be the most effective.

Sous-vide cooking involves the vacuum packing of raw or partially cooked food, which is then cooked at specific times and temperatures, chilled, and kept below 6 °C. This technique maintains the visual appeal and nutritional content of the food [19–22]. Vacuum sealing in low oxygen permeability pouches suppresses some of the odors associated with oxidation and prevents the evaporation of moisture and volatiles during cooking. Sarcoplasmic proteins are collected at temperatures up to 65 °C, increasing their softness, and nutritional losses decrease because nutrients are not sucked out by boiling water [19,21,23]. Sage is known for its anti-inflammatory effects [24] and antibacterial action against various bacteria, such as *Enterococcus faecalis* ATCC29212, *Klebsiella pneumoniae* ATCC700603, *Salmonella Paratyphi* A NCTC13, and *Staphylococcus aureus* ATCC29213) and spoilage microorganisms (*Proteus mirabilis, Photobacterium damselae, Vibrio vulnificus, Enterococcus faecalis, Pseudomonas luteola*, and *Serratia liquefaciens*) [25]. Several authors have documented this in various food matrices. The cause is reported to be 1,8-cineole, camphor, and thujone [6,26–29]. Sage EO in foods has an antilisterial effect; however, this effect is not well understood. According to a small number of writers, there is some evidence that sage exerts bactericidal and bacteriostatic effects on Gram-positive and Gram-negative bacteria [5]. Through a synergistic impact or simply through the additional effects that each component has on microorganisms, combining sage chemicals with heat treatment may reduce the resistance of *L. monocytogenes* [6].

The results of the study show the behavior of *Listeria monocytogenes* inoculation in beef with the addition of sage essential oil and vacuum packaging, over a period of 12 days and under different temperature settings. Our findings demonstrate the development of *L. monocytogenes*, coliform bacteria, and total bacteria from day 0 to day 12. In a separate study, long cooking times at low temperatures were used to mimic the conditions that arise in retail food service when processed foods are prepared sous-vide. According to Tangwatcharin et al. [30], different sous-vide temperatures for inoculated restructured goat steak had an impact on the D-values of *L. monocytogenes*. Its Z-value was 8.20 °C, and its D-value decreased as the temperature increased. Non-inoculated restructured goat sirloin was cooked using six D-values at 60, 65, and 70 °C to ensure the safety of the sous-vide product. The number of microorganisms in all samples was reduced, and pathogens were not found after cooking using various sous-vide techniques.

The optimal duration and temperature for cooking salmon sous-vide have been estimated to eliminate *L. monocytogenes*, and oregano oil and citric acid may aid by reducing the bacteria's ability to withstand heat. The study's findings are important for ensuring the safety of food and may help processing centers lower the risk of *L. monocytogenes* during thermal treatment. Our findings will also shed light on potential uses for heat treatment that could improve the results [31].

Low temperatures had a bacteriostatic impact on *L. monocytogenes*, according to Chan and Wiedmann [32]. This conclusion is consistent with this study's findings, which indicate that control samples stored at 2 °C significantly decreased from day 0 to day 28 by 1.23 log (on average). In our research, a consistent decrease in *L. monocytogenes* counts was observed during storage at 4 °C. An exponential increase was observed up to day 21.

The growth curves of *L. monocytogenes* inoculated in beef at storage temperatures ranging from 5 to 25 °C were observed in a study by Lee et al. [33]. At 5 °C, *L. monocytogenes* was discovered to thrive. Calculating the lag and stationary phase was impossible because of the very slow rate of development at this temperature. According to Farber and Peterkin [34], the *L. monocytogenes* lag phase in vacuum-packed roast beef held at 3 °C lasted 59 h, illustrating the effect of stress heat treatment on the organism and resulting in a log lag phase. The lack of growth of *L. monocytogenes* during storage prevented us from observing a lag period in our research at 2 °C. The multiplication of *L. monocytogenes* in a vacuum-packed mortadella kept at 4 °C and 8 °C was reported [35].

After heat treatment (grilling, microwave heating, or conventional cooking), Yilmaz et al. [36] noted a decrease in the number of mesophilic microorganisms. After 15 days of storage, the mesophilic bacterial counts in the study samples with additives were comparable to those found after thermal processing. The count of mesophilic bacteria did not alter significantly during the storage of vacuum-packed cooked turkey breast rolls, according to Smith and Alvarez [37]. In our study, different families, genera, and species were discovered in the sous-vide beef meat treatment group and the control test. The *P. fragi* species was the most isolated. The activity of various strains of this species under various packaging conditions may explain the seemingly contradictory findings. Consequently, more research is necessary to understand the prevalence and function of *Pseudomonas fragi* as meat spoilers [38]. The main spoiler of chilled beef kept in aerobic storage is thought to be *Pseudomonas fragi* [39]. Due to its fast development ability and the creation of strong odors associated with rotting, it quickly results in the meat being unacceptable for consumption [40]. It was also frequently discovered to be present in vacuum-packed beef [39,41–43], indicating the existence of several strains, likely with various genetic repertoires. The bacterium that was most frequently seen was *H. alvei*. The study by Sokołowicz et al. [44] discovered the same outcomes with isolated species.

5. Conclusions

The most effective combinations of sous-vide cooking time and temperature for beef meat to inactivate *L. monocytogenes* are determined by this study, and it is suggested that sage essential oil may be able to reduce the ability of *L. monocytogenes* to withstand heat in beef tenderloin treated with sous-vide technique. The results of this study are important for ensuring the safety of food and may help processing facilities lower the risk of *L. monocytogenes* during heat treatment. Our research will also shed light on potential uses that could improve the effects of thermal treatment when combined with proper times and sage essential oils.

Author Contributions: Conceptualization, R.G. and M.K.; methodology, R.G., N.Č. and M.K.; data curation, R.G., N.Č. and M.K.; writing—original draft preparation, R.G., N.Č., A.P. and M.K.; writing—review and editing, R.G., N.Č., A.P. and M.K.; supervision, M.K.; project administration, M.K. All authors have read and agreed to the published version of the manuscript.

Funding: This research was funded by the Internal Grant Agency of the Faculty of Technology, Tomas Bata University in Zlin, ref. No. IGA/FT/2022/003. This publication was supported by the Operational Program "Integrated Infrastructure" within the following project: Demand-driven research for sustainable and innovative food, Drive4SIFood 313011V336.

Data Availability Statement: The data presented in this study are available on request from the corresponding authors.

Conflicts of Interest: The authors declare no conflict of interest.

References

1. Zavadlav, S.; Blažić, M.; Van de Velde, F.; Vignatti, C.; Fenoglio, C.; Piagentini, A.M.; Pirovani, M.E.; Perotti, C.M.; Bursać Kovačević, D.; Putnik, P. Sous-Vide as a Technique for Preparing Healthy and High-Quality Vegetable and Seafood Products. *Foods* **2020**, *9*, 1537. [CrossRef] [PubMed]
2. Ruiz, J.; Calvarro, J.; Sánchez del Pulgar, J.; Roldán, M. Science and Technology for New Culinary Techniques. *J. Culin. Sci. Technol.* **2013**, *11*, 66–79. [CrossRef]
3. Stringer, S.C.; Metris, A. Spore Forming Pathogens and Sous Vide Food Safety. In *Sous Vide and Cook-Chill Processing for the Food Industry*, 2nd ed.; Springer: New York, NY, USA, 2018.
4. Hyytiä-Trees, E.; Skyttä, E.; Mokkila, M.; Kinnunen, A.; Lindström, M.; Lähteenmäki, L.; Ahvenainen, R.; Korkeala, H. Safety Evaluation of Sous Vide-Processed Products with Respect to Nonproteolytic *Clostridium botulinum* by Use of Challenge Studies and Predictive Microbiological Models. *Appl. Environ. Microbiol.* **2000**, *66*, 223–229. [CrossRef] [PubMed]
5. Ghorbani, A.; Esmaeilizadeh, M. Pharmacological Properties of *Salvia officinalis* and Its Components. *J. Tradit. Complement. Med.* **2017**, *7*, 433–440. [CrossRef] [PubMed]
6. Gurtler, J.B.; Fan, X.; Jin, T.; Niemira, B.A. Influence of Antimicrobial Agents on the Thermal Sensitivity of Foodborne Pathogens: A Review. *J. Food Prot.* **2019**, *82*, 628–644. [CrossRef]
7. Cui, H.; Zhang, X.; Zhou, H.; Zhao, C.; Lin, L. Antimicrobial Activity and Mechanisms of *Salvia sclarea* Essential Oil. *Bot. Stud.* **2015**, *56*, 16. [CrossRef]

8. Tuszyński, T.; Makarewicz, M. Effect of Herbal Extracts on the Growth of Selected Yeast (*Saccharomyces cerevisiae*) Strains. *Żywność* **2000**, *1*, 37–44.
9. Korczak, J.; Flaczyk, E.; Pazoła, Z. Effects of Spices on Stability of Minced Meat Products Kept in Cold Storage. *Fleischwirtschaft* **1988**, *1*, 64–66.
10. Madsen, H.L.; Andersen, L.; Christiansen, L.; Brockhoff, P.; Bertelsen, G. Antioxidative Activity of Summer Savory (*Satureja hortensis* L.) and Rosemary (*Rosmarinus officinalis* L.) in Minced, Cooked Pork Meat. *Z. Lebensm.-Unters. Forsch.* **1996**, *203*, 333–338. [CrossRef]
11. Kačániová, M.; Galovičová, L.; Valková, V.; Ďuranová, H.; Borotová, P.; Štefániková, J.; Vukovic, N.L.; Vukic, M.; Kunová, S.; Felsöciová, S.; et al. Chemical Composition and Biological Activity of *Salvia officinalis* Essential Oil. *Acta Hortic. Regiotect.* **2021**, *24*, 81–88. [CrossRef]
12. Zhang, L.; Lin, Y.H.; Leng, X.J.; Huang, M.; Zhou, G.H. Effect of Sage (*Salvia officinalis*) on the Oxidative Stability of Chinese-Style Sausage during Refrigerated Storage. *Meat Sci.* **2013**, *95*, 145–150. [CrossRef] [PubMed]
13. Mizi, L.; Cofrades, S.; Bou, R.; Pintado, T.; López-Caballero, M.E.; Zaidi, F.; Jiménez-Colmenero, F. Antimicrobial and Antioxidant Effects of Combined High Pressure Processing and Sage in Beef Burgers during Prolonged Chilled Storage. *Innov. Food Sci. Emerg. Technol.* **2019**, *51*, 32–40. [CrossRef]
14. Karpińska-Tymoszczyk, M. Effects of Sage Extract (*Salvia officinalis* L.) And A Mixture of Sage Extract And Sodium Isoascorbate On The Quality And Shelf Life Of Vacuum-Packed Turkey Meatballs. *J. Muscle Foods* **2007**, *18*, 420–434. [CrossRef]
15. Moura-Alves, M.; Gouveia, A.R.; de Almeida, J.M.M.M.; Monteiro-Silva, F.; Silva, J.A.; Saraiva, C. Behavior of Listeria Monocytogenes in Beef Sous Vide Cooking with *Salvia officinalis* L. Essential Oil, during Storage at Different Temperatures. *LWT* **2020**, *132*, 109896. [CrossRef]
16. Hernández-Macedo, M.L.; Barancelli, G.V.; Contreras-Castillo, C.J. Microbial Deterioration of Vacuum-Packaged Chilled Beef Cuts and Techniques for Microbiota Detection and Characterization: A Review. *Braz. J. Microbiol.* **2011**, *42*, 1–11. [CrossRef] [PubMed]
17. Kunová, S.; Sendra, E.; Haščík, P.; Vukovic, N.L.; Vukic, M.; Kačániová, M. Influence of Essential Oils on the Microbiological Quality of Fish Meat during Storage. *Animals* **2021**, *11*, 3145. [CrossRef]
18. Davidson, P.M.; Cekmer, H.B.; Monu, E.A.; Techathuvanan, C. The Use of Natural Antimicrobials in Food. In *Handbook of Natural Antimicrobials for Food Safety and Quality*; Elsevier: Amsterdam, The Netherlands, 2015; pp. 1–27. ISBN 978-1-78242-034-7.
19. Bongiorno, T.; Tulli, F.; Comi, G.; Sensidoni, A.; Andyanto, D.; Iacumin, L. Sous Vide Cook-Chill Mussel (*Mytilus galloprovincialis*): Evaluation of Chemical, Microbiological and Sensory Quality during Chilled Storage (3 °C). *LWT* **2018**, *91*, 117–124. [CrossRef]
20. Nyati, H. An Evaluation of the Effect of Storage and Processing Temperatures on the Microbiological Status of Sous Vide Extended Shelf-Life Products. *Food Control* **2000**, *11*, 471–476. [CrossRef]
21. Schellekens, M. New Research Issues in Sous-Vide Cooking. *Trends Food Sci. Technol.* **1996**, *7*, 256–262. [CrossRef]
22. Tornberg, E. Effects of Heat on Meat Proteins—Implications on Structure and Quality of Meat Products. *Meat Sci.* **2005**, *70*, 493–508. [CrossRef]
23. Eidi, M.; Eidi, A.; Bahar, M. Effects of *Salvia officinalis* L. (Sage) Leaves on Memory Retention and Its Interaction with the Cholinergic System in Rats. *Nutrition* **2006**, *22*, 321–326. [CrossRef] [PubMed]
24. Ballester-Costa, C.; Sendra, E.; Fernández-López, J.; Pérez-Álvarez, J.A.; Viuda-Martos, M. Chemical Composition and in Vitro Antibacterial Properties of Essential Oils of Four Thymus Species from Organic Growth. *Ind. Crops Prod.* **2013**, *50*, 304–311. [CrossRef]
25. Yazgan, H. Investigation of Antimicrobial Properties of Sage Essential Oil and Its Nanoemulsion as Antimicrobial Agent. *LWT* **2020**, *130*, 109669. [CrossRef]
26. Longaray Delamare, A.P.; Moschen-Pistorello, I.T.; Artico, L.; Atti-Serafini, L.; Echeverrigaray, S. Antibacterial Activity of the Essential Oils of *Salvia officinalis* L. and *Salvia triloba* L. Cultivated in South Brazil. *Food Chem.* **2007**, *100*, 603–608. [CrossRef]
27. Nezhadali, A.; Nabavi, M.; Rajabian, M.; Akbarpour, M.; Pourali, P.; Amini, F. Chemical Variation of Leaf Essential Oil at Different Stages of Plant Growth and in Vitro Antibacterial Activity of *Thymus vulgaris* Lamiaceae, from Iran. *Beni-Suef Univ. J. Basic Appl. Sci.* **2014**, *3*, 87–92. [CrossRef]
28. Singh, G.; Kapoor, I.P.S.; Singh, P. Effect of Volatile Oil And Oleoresin Of Anise On The Shelf Life Of Yogurt*: Anise Oil And Oleoresin: Effect On Yogurt. *J. Food Process. Preserv.* **2011**, *35*, 778–783. [CrossRef]
29. Selim, S. Antimicrobial Activity of Essential Oils against Vancomycin-Resistant Enterococci (VRE) and *Escherichia coli* O157:H7 in Feta Soft Cheese and Minced Beef Meat. *Braz. J. Microbiol.* **2011**, *42*, 187–196. [CrossRef]
30. Tangwatcharin, P.; Sorapukdee, S.; Kongsrirat, K. Sous-Vided Restructured Goat Steaks: Process Optimized by Thermal Inactivation of *Listeria monocytogenes* and Their Quality Characteristics. *Food Sci. Anim. Resour.* **2019**, *39*, 863–876. [CrossRef]
31. Dogruyol, H.; Mol, S.; Cosansu, S. Increased Thermal Sensitivity of Listeria Monocytogenes in Sous-Vide Salmon by Oregano Essential Oil and Citric Acid. *Food Microbiol.* **2020**, *90*, 103496. [CrossRef]
32. Chan, Y.C.; Wiedmann, M. Physiology and Genetics of *Listeria monocytogenes* Survival and Growth at Cold Temperatures. *Crit. Rev. Food Sci. Nutr.* **2008**, *49*, 237–253. [CrossRef]
33. Lee, S.-Y.; Kwon, K.-H.; Chai, C.; Oh, S.-W. Growth Behavior Comparison of *Listeria monocytogenes* between Type Strains and Beef Isolates in Raw Beef. *Food Sci. Biotechnol.* **2017**, *27*, 599–605. [CrossRef]

34. Farber, M.; Peterkin, P. Incidence and Behaviour of *Listeria monocytogenes* in Meat Products. In *Listeria, Listeriosis and Food Safety*; Marcel Dekker Inc.: New York, NY, USA, 1999; pp. 505–564, ISBN 978-1-4200-0087-0.
35. McCarthy, S.A. Pathogenicity of Nonstressed, Heat-Stressed, and Resuscitated *Listeria monocytogenes* 1A1 Cells. *Appl. Environ. Microbiol.* **1991**, *57*, 2389–2391. [CrossRef]
36. Yılmaz, İ.; Arıcı, M.; Gümüş, T. Changes of Microbiological Quality in Meatballs after Heat Treatment. *Eur. Food Res. Technol.* **2005**, *221*, 281–283. [CrossRef]
37. Smith, D.M.; Alvarez, V.B. Stability of Vacuum Cook-in-Bag Turkey Breast Rolls during Refrigerated Storage. *J. Food Sci.* **1988**, *53*, 46–48. [CrossRef]
38. De Filippis, F.; La Storia, A.; Villani, F.; Ercolini, D. Strain-Level Diversity Analysis of Pseudomonas Fragi after In Situ Pangenome Reconstruction Shows Distinctive Spoilage-Associated Metabolic Traits Clearly Selected by Different Storage Conditions. *Appl. Environ. Microbiol.* **2019**, *85*, e02212-18. [CrossRef]
39. Pennacchia, C.; Ercolini, D.; Villani, F. Spoilage-Related Microbiota Associated with Chilled Beef Stored in Air or Vacuum Pack. *Food Microbiol.* **2011**, *28*, 84–93. [CrossRef] [PubMed]
40. Casaburi, A.; Piombino, P.; Nychas, G.-J.; Villani, F.; Ercolini, D. Bacterial Populations and the Volatilome Associated to Meat Spoilage. *Food Microbiol.* **2015**, *45*, 83–102. [CrossRef] [PubMed]
41. Kiermeier, A.; Tamplin, M.; May, D.; Holds, G.; Williams, M.; Dann, A. Microbial Growth, Communities and Sensory Characteristics of Vacuum and Modified Atmosphere Packaged Lamb Shoulders. *Food Microbiol.* **2013**, *36*, 305–315. [CrossRef]
42. Botta, C.; Ferrocino, I.; Cavallero, M.C.; Riva, S.; Giordano, M.; Cocolin, L. Potentially Active Spoilage Bacteria Community during the Storage of Vacuum Packaged Beefsteaks Treated with Aqueous Ozone and Electrolyzed Water. *Int. J. Food Microbiol.* **2018**, *266*, 337–345. [CrossRef]
43. Jääskeläinen, E.; Johansson, P.; Kostiainen, O.; Nieminen, T.; Schmidt, G.; Somervuo, P.; Mohsina, M.; Vanninen, P.; Auvinen, P.; Björkroth, J. Significance of Heme-Based Respiration in Meat Spoilage Caused by *Leuconostoc Gasicomitatum*. *Appl. Environ. Microbiol.* **2013**, *79*, 1078–1085. [CrossRef]
44. Sokołowicz, Z.; Augustyńska-Prejsnar, A.; Krawczyk, J.; Kačániová, M.; Kluz, M.; Hanus, P.; Topczewska, J. Technological and Sensory Quality and Microbiological Safety of RIR Chicken Breast Meat Marinated with Fermented Milk Products. *Animals* **2021**, *11*, 3282. [CrossRef] [PubMed]

foods

MDPI

Communication

Prevalence of Foodborne Bacterial Pathogens and Antibiotic Resistance Genes in Sweets from Local Markets in Iran

Babak Pakbin [1,2], Zahra Amani [3], Zahra Rahimi [4], Somayeh Najafi [5], Behnaz Familsatarian [6], Alireza Khakpoor [7], Wolfram Manuel Brück [2,*] and Thomas B. Brück [1,*]

1 Werner Siemens Chair of Synthetic Biotechnology, Department of Chemistry, Technical University of Munich (TUM), Lichtenberg Str. 4, 85748 Garching bei München, Germany; b.pakbin@tum.de
2 Institute for Life Technologies, University of Applied Sciences Western Switzerland Valais-Wallis, 1950 Sion 2, Switzerland
3 Department of Food Hygiene and Quality of Control, Faculty of Veterinary Medicine, University of Tehran, Tehran 1417614411, Iran; z.amani@ut.ac.ir
4 Department of Food Safety and Health, School of Public Health, Qazvin University of Medical Sciences, Qazvin 34197-59811, Iran; food_supervisor@qums.ac.ir
5 Nutrition and Food Sciences Research Center, Faculty of Pharmacy and Pharmaceutical Sciences, Islamic Azad University, Tehran Medical University (IAUTMU), Tehran 19395-1495, Iran; somayeh_6586@yahoo.com
6 Medical Microbiology Research Center, Qazvin University of Medical Sciences, Qazvin 34197-59811, Iran; behnaz.sattarian@gmail.com
7 Department of Management, Ferdowsi University of Mashhad, Mashhad 9177948974, Iran; a.khakpour56@gmail.com
* Correspondence: wolfram.bruck@hevs.ch (W.M.B.); brueck@tum.de (T.B.B.)

Abstract: Background: This study aimed to investigate the prevalences of some important antibiotic-resistance genes (ARGs) and foodborne bacterial pathogens in sweet samples collected from local markets in Iran. Methods: Forty sweet samples were collected. Foodborne pathogens and ARGs were detected in the sweet samples by conventional and multiplex PCR assays using species-specific primers. Results: *Staphylococcus aureus*, *Cronobacter sakazakii*, *Shigella* spp., *Campylobacter jejuni*, and *Campylobacter coli* were detected and identified in 47.5%, 20%, 45%, 5%, and 30% of the sweet samples, respectively. We found *S. aureus* and *Shigella* spp. were the most prevalent bacterial pathogens. *S. aureus* was found to be the most frequent pathogenic bacteria profiled in these samples. We also found a significant correlation between the presence of *C. coli* and *Cr. sakazakii*. We detected the bla_{SHV} resistance gene in 97.5% of the sweet samples; however, bla_{TEM} was detected in only one sample (2.5%). Conclusions: Regarding these results, we suggest preventive strategies such as implementing automation of food processing; monitoring the personal hygiene and health of food handlers, and testing regularly for antibiotic resistance in raw materials and products.

Keywords: foodborne bacterial pathogens; antibiotic resistance genes; sweet products

Citation: Pakbin, B.; Amani, Z.; Rahimi, Z.; Najafi, S.; Familsatarian, B.; Khakpoor, A.; Brück, W.M.; Brück, T.B. Prevalence of Foodborne Bacterial Pathogens and Antibiotic Resistance Genes in Sweets from Local Markets in Iran. *Foods* **2023**, *12*, 3645. https://doi.org/10.3390/foods12193645

Academic Editor: Efstathios Giaouris

Received: 27 August 2023
Revised: 28 September 2023
Accepted: 30 September 2023
Published: 2 October 2023

1. Introduction

Foodborne diseases are defined as intestinal or extraintestinal disorders resulting from ingesting or consuming contaminated water, food, or food products [1]. Foodborne illnesses are toxic, infectious, or toxic infectious in nature, and are caused by foodborne pathogens, including different bacterial, fungal, parasitic, and viral species [2]. Consumption of contaminated or raw food products has been associated with several foodborne outbreaks worldwide. The World Health Organization recently reported an estimate that foodborne outbreaks and illnesses cause more than 600 million disease cases and 420,000 deaths annually around the world [3].

Regarding the global number of foodborne illness cases, the major foodborne hazards are *Campylobacter* spp., pathogenic *Escherichia coli* (*E. coli*), non-typhoidal *Salmonella* spp., and *Shigella* spp.; however, *S. typhi*, enteropathogenic *E. coli*, enterotoxigenic *E. coli*, *Vibrio cholerae*, and *Campylobacter* spp. are the major hazards considering the global number of deaths caused by bacterial foodborne pathogens. Also, other bacterial agents, such as *Listeria monocytogenes* (*L. monocytogenes*), *Clostridium botulinum*, *Brucella* spp., *Bacillus cereus* (*B. cereus*), *Staphylococcus aureus* (*S. aureus*), and *Clostridium perfringens* (*Cl. perfringens*), have been considered as other major causes of foodborne diseases worldwide [4].

On the other hand, resistance to different classes of antibiotics has been regarded as one of the most important challenges in public health and food safety. Foods and foodborne pathogens are the main routes of transmission of ARGs to the human gut microbiota, leading to intestinal drug-resistant infections [5,6]. Beta-lactam resistance genes, including *bla* genes, have been most frequently detected in food samples.

The safety level of foods also depends on the microbial quality of the raw materials used to produce these products. Different foodborne pathogens and antibiotic-resistance genes (ARGs) have been detected in the final products [6–8]. Food products such as sweets are primarily prepared and distributed by hand under relatively poor hygienic conditions, and they are susceptible to contamination with foodborne pathogens and ARGs from humans [5,9–11].

S. aureus strains and toxins are usually isolated from and detected in sweet products since the initial contamination sources of these bacteria are the mucous membranes and skin of humans [12].

S. aureus has mostly been known as one of the most important food safety challenges and public health concerns, causing intoxication, vomiting, and diarrhea in humans as a foodborne pathogen via the secretion of a heat-stable toxin, alpha-toxin [13]. Other known foodborne bacterial pathogens, such as *Campylobacter* spp., *Shigella* spp., and *Cronobacter sakazakii (Cr. sakazakii*), as a new emerging foodborne pathogen, have been isolated from low-moisture food products such as sweets [14,15]. These foodborne pathogens have been detected in sweet and confectionary products and are also regarded as the main concerns in food safety and public health since they contribute to acute intestinal and extraintestinal diseases in humans. The prevalences of these foodborne bacterial pathogens are significantly and directly linked to the safety levels and hygienic conditions during the production of sweet products [12–15].

Investigation of the prevalences of bacterial pathogens and ARGs in low-moisture food products, especially sweet samples, is strongly limited. On the other hand, there are also limited studies regarding the correlations among the presence of different types of foodborne pathogens in food samples [10]. Therefore, this study aimed to determine the prevalences of and the correlations among the presence of different new emerging foodborne bacterial pathogens, including *C. jejuni*, *C. coli*, *Shigella* spp., *Cr. sakazakii*, and *S. aureus*, and ARGs, in sweet samples.

2. Materials and Methods

2.1. Sample Collection and Preparation

Forty sweet samples (these traditional Iranian sweets are categorized as solid dry sweets with a moisture content range of 20–26% w/w), including 4 different subtypes and 3 various brands, were purchased and collected from local markets in Qazvin and Tehran cities, Iran (traditional sweets and confectionaries are mostly produced and consumed in these cities in the center of Iran), between January and April 2022. Samples were immediately transported aseptically in cool boxes containing ice packs to the central microbiology laboratory at the university.

Each sample was homogenized by using a stomacher BagMixer Lab-blender (Interscience, Saint-Nom-la-Bretèche, France) at 30 °C for 10 min. All samples were diluted 1:10 with phosphate-buffered saline and homogenized again with a stomacher for 5 min at 30 °C [16]. All samples were finally subjected to total DNA extraction and further PCR assays.

2.2. Total DNA Extraction

The total DNA in the prepared and diluted sweet samples was extracted by using a SinaClon commercial tissue total DNA extraction kit (SinaClon., Tehran, Iran) according to the manufacturer's instructions. The instructions of the kit have been optimized for total DNA extraction from food samples. The quantity and quality of the extracted total DNA were assessed by using a NanoDrop spectrophotometer model 1000 (ThermoFisher, Waltham, MA, USA). The optical density ratios of 260/280 of all DNA samples were observed in the range of 1.5–1.8, indicating the high purity of the extracted DNA. The concentrations of all extracted total DNA were adjusted to 50 ng. μL^{-1} by dilution with the addition of sterilized nuclease-free water. The DNA samples were kept at −20 °C for further molecular analysis.

2.3. PCR Assays for the Detection and Identification of Different Foodborne Bacterial Pathogens

In the present study, we detected and identified five different foodborne bacterial pathogens, including *C. jejuni*, *C. coli*, *Shigella* spp., *Cr. Sakazakii*, and *S. aureus* by conventional PCR methods (cPCR) using species-specific primers, as shown in Table 1, to directly detect the specific genes in the DNA samples. Identifying *S. aureus* in sweet samples was performed by detecting the *spa* gene in DNA templates according to the method and specific primers previously described and used by Larsen et al. (2008). The *spa* gene encodes Staphylococcal protein A in *S. aureus* strains. Detection of this genetic element proves species-specifically the presence of *S. aureus* in clinical, food, and environmental samples [17].

Table 1. Species-specific primers used to detect foodborne pathogens and beta-lactam resistance genes in sweet samples.

Gene	Primer	Sequence (5′ → 3′)	Reference
spa	*spa1*	TAAAGACGATCCTTCGGTGAGC	[17]
	spa2	CAGCAGTAGTGCCGTTTGCTT	
ompA	*ompAF*	GGATTTAACCGTGAACTTTTCC	[6]
	ompAR	CGCCAGCGATGTTAGAAGA	
ipaH	*ipaHF*	GTTCCTTGACCGCCTTTCCGATACCGTC	[18]
	ipaHR	GCCGGTCAGCCACCCTCTGAGAGTAC	
hipO	*hipOF*	AATGCACAAATTTGCCTTATAAAAGC	[19]
	hipOR	TNCCATTAAAATTCTGACTTGCTAAATA	
cadF	*cadFF*	GAGAAATTTTATTTTTATGGTTTAGCTGGT	[19]
	cadFR	ACCTGCTCCATAATGGCCAA	
bla$_{TEM}$	blaTEMF	CATTTCCGTGTCGCCCTTATTC	[18]
	blaTEMR	CGTTCATCCATAGTTGCCTGAC	
bla$_{SHV}$	blaSHVF	AGCCGCTTGAGCAAATTAAAC	[18]
	blaSHVR	ATCCCGCAGATAAATCACCAC	
bla$_{OXA}$	blaOXAF	GGCACCAGATTCAACTTTCAAG	[18]
	blaOXAR	GACCCCAAGTTTCCTGTAAGTG	
bla$_{CTX-M-1}$	CTXM1F	TTAGGAARTGTGCCGCTGYA	[18]
	CTXM1R	CGATATCGTTGGTGGTRCCAT	
bla$_{CTX-M-2}$	CTXM2F	CGTTAACGGCACGATGAC	[18]
	CTXM2R	CGATATCGTTGGTGGTRCCAT	
bla$_{CTX-M-8}$	CTXM8F	AACRCRCAGACGCTCTAC	[18]
	CTXM8R	TCGAGCCGGAASGTGTYAT	
bla$_{CTX-M-9}$	CTXM9F	TCAAGCCTGCCGATCTGGT	[18]
	CTXM9R	TGATTCTCGCCGCTGAAG	

Cr. sakazakii was detected in the samples according to the cPCR previously used to detect the *ompA* gene (outer membrane protein virulence factor encoding gene) in powdered infant formula samples by Pakbin et al. (2022) [6].

To identify and detect *Shigella* spp. in each sample, the invasion plasmid antigen encoding gene (*ipaH*) was detected by a PCR assay as previously described by Pakbin et al. (2021) [18].

C. coli and *C. jejuni* strains were identified in sweet samples by detecting the cadF and hipO species-specific genes, as described and employed before by Nafarrate et al. (2021) [19].

PCR assays were carried out with 10 μL of PCR 2X master mix (Ampliqon, Odense, Denmark), 2 μL of DNA template (50 ng. μL^{-1}), 1 μL of each specific primer (10 mM. μL^{-1}), and nuclease-free sterilized water up to the final reaction volume of 20 μL. PCR mixes were subjected to initial denaturation at 95 °C for 5 min, 40 cycles of amplification (denaturation at 95 °C for 1 min, the primer-specific annealing temperature for 30 s, and extension at 72 °C for 1 min), and a final elongation step at 72 °C for 5 min [6,18,19]. The PCR products were characterized by using electrophoresis at 100 v for 1 h on a 1.2% *w/v* agarose gel containing DNA safe-stain (Invitrogen, Carlsbad CA, USA). The PCR products were electrophoresed on agarose gels were evaluated and recorded by using a UV transillumination and gel documentation system (NovinPars Co., Tehran, Iran). Positive and negative samples were used as quality controls for detecting each foodborne pathogen in this study. DNA templates extracted from reference strains, including *S. aureus* (ATCC 25923), *C. jejuni* (ATCC 33291), *C. coli* (ATCC 43478), and *Sh. sonnei* (ATCC 29031), were used as the positive controls, and sterilized distilled water was used as the negative control in this study.

2.4. PCR Assays for the Detection of ARGs

Beta-lactamase resistance genes, including bla_{TEM}, bla_{SHV}, bla_{OXA}, $bla_{CTX\text{-}M\text{-}1}$, $bla_{CTX\text{-}M\text{-}2}$, $bla_{CTX\text{-}M\text{-}8}$, and $bla_{CTX\text{-}M\text{-}9}$, were detected in sweet samples by using cPCR with thermal cycling programs and specific primers, as shown in Table 1, as previously described by Pakbin et al. (2021) [18]. Reference strains *Klebsiella pneumonia* (ATCC 700603), *Escherichia coli* (ATCC 25922), and *Staphylococcus aureus* (ATCC 25923) harboring all of these ARGs were included and used as quality controls in this study.

2.5. Statistical Analysis

Correlations among the different foodborne bacterial pathogen prevalences in the sweet samples were evaluated by using Pearson's Chi-square test, with a significant difference defined as $p < 0.05$ and one degree of freedom. Fisher's exact test was used to evaluate significant differences ($p < 0.05$) between the prevalences of different bacterial pathogens, the prevalences of different ARGs, and different bacterial pathogen profiles. All statistical analyses were performed by using SPSS software version 22.0.1 (Chicago, IL, USA). All evaluations and measurements were carried out in triplicate.

3. Results

In the present study, we detected some important foodborne pathogens in sweet samples and evaluated the bacterial profiles and correlations between the presence of these pathogens in the collected samples. This study detected *S. aureus* (n = 19), *Cr. sakazakii* (n = 8), *Shigella* spp. (n = 18), *C. jejuni* (n = 2), and *C. coli* (n = 12) in 47.5%, 20.0%, 45.0%, 5.0%, and 30.0% of the sweet samples (N = 40), respectively. Among these bacterial pathogens, *S. aureus* and *Shigella* spp. significantly ($p < 0.05$) showed the highest prevalences compared to the other pathogens. Also, the results in this study illustrated that *C. coli* was significantly ($p < 0.05$) more often detected than *C. jejuni*.

The profiles of the foodborne bacterial pathogens detected in sweet samples in this study are shown in Table 2. A total of 34 (85%) out of 40 sweet samples were contaminated with at least one of the bacterial pathogens. Single bacterial species profiles were detected in 18 samples (45%). Only two foodborne bacterial pathogens were identified in 10 samples (25%). Multiple bacterial profiles, including three or more foodborne pathogens, were detected in 6 (15%) out of the 40 sweet samples. An *S. aureus* single bacterial profile was the significantly ($p < 0.05$) most prevalent profile in this study. A double bacterial profile, including *Shigella* spp. and *C. coli*, was significantly ($p < 0.05$) the most common non-single pathogenic bacterial profile detected in these samples.

Table 2. Profiles of different foodborne bacterial pathogens detected in sweet samples.

	Foodborne Bacterial Pathogen Profile [a]	Positive Samples (%) (n)
Single	SA	25 (10)
	SH	15 (6)
	CR	2.5 (1)
	CC	2.5 (1)
Double	SH + CC	10 (4)
	SA + SH	5 (2)
	SA + CC	2.5 (1)
	SA + CJ	2.5 (1)
	CR + CC	2.5 (1)
	SH + CR	2.5 (1)
Multiple	SA + CR + SH	2.5 (1)
	CR + SH + CC	2.5 (1)
	SA + SH + CC	2.5 (1)
	SA + CR + SH + CC	5 (2)
	SA + CR + CJ + CC	2.5 (1)

[a] S. aureus, SA; Shigella spp., SH; Cr. sakazakii, CR; C. jejuni, CJ; C. coli, CC.

We measured the possible correlations between the presence of different pathogens with Pearson's Chi-square test in this study. Table 3 presents the correlations among the presence of different bacterial pathogens in the sweet samples. Pearson's correlation analysis showed that the highest significant ($p < 0.05$) positive correlation (0.335) was between the presence of Cr. sakazakii and C. coli in the sweet samples. Significant correlations were not observed between the presence of other bacterial pathogens in this study.

Table 3. Correlations among the presence of different foodborne bacterial pathogens in the sweet samples.

	C. coli	*C. jejuni*	*Cr. sakazakii*	*Shigella spp.*	*S. aureus*
C. coli	1.000	0.100	0.355 *	0.285	−0.076
C. jejuni		1.000	0.172	−0.208	0.241
cr. sakazakii			1.000	0.176	0.025
Shigella spp.				1.000	−0.257
S. aureus					1.000

* $p < 0.05$.

This study detected beta-lactam resistance genes in total DNA extracted from the sweet samples by using cPCR. The bla_{SHV} gene was detected in 39 out of 40 sweet samples (97.5%); however, the bla_{TEM} gene was only detected in one sample (2.5%). bla_{OXA}, $bla_{CTX\text{-}M\text{-}1}$, $bla_{CTX\text{-}M\text{-}2}$, $bla_{CTX\text{-}M\text{-}8}$, and $bla_{CTX\text{-}M\text{-}9}$ resistance genes were detected in 25%, 37.5%, 37.5%, and 20% of the sweet samples, respectively.

4. Discussion

Foodborne bacterial pathogens are the main causes of intestinal (different types of diarrheal diseases and gastrointestinal disorders) and some extraintestinal illnesses around the world [20]. Regarding the fact that foods and foodborne pathogens are the main routes for the transmission of intestinal pathogens and ARGs to humans and their gut microbiota, the evaluation of the prevalence and presence of different bacterial foodborne pathogens and ARGs in these products is critically needed to implement regularly [21,22]. Considering these facts, we were motivated to investigate the prevalence and incidence of the most important and prevalent foodborne bacterial pathogens and ARGs in sweet products.

The relatively small sample size (N = 40) could be considered this study's main limitation. We observed the highest *S. aureus* and *Shigella* spp. prevalences in these samples. We detected at least one of these pathogens in 85% of the samples. *S. aureus* was also found to be the most prevalent profile detected in this study. *S. aureus* is an opportunistic pathogen and commensal colonizing the human mucous membranes and skin [23]. Since sweet products are usually prepared and produced traditionally under poor hygienic conditions, contamination of these products with *S. aureus* strains occurs causing gastrointestinal disorders caused by staphylococcal toxins [24]. Houng et al. (2010) detected *S. aureus* in 45 out of 212 ready-to-eat samples (21.2%) collected in Vietnam [25]. Kim et al. (2011) also found that 5.98% of traditional foods produced in Korea were contaminated with *S. aureus* [26]. Mahfoozi et al. (2019) investigated the prevalence of *S. aureus* in some food samples produced in Iran, including meat, dairy, and sweet products, and they detected *S. aureus* in 29.1% of the sweet samples [27]. Hassani et al. (2022) also recently reported contamination with *S. aureus* in 2.67% of the pastry sweets produced in Iran [28]. Using automatic instruments during sweet production and preparation and disinfection of workers' hands seem to be helpful strategies to reduce the prevalence of *S. aureus* in sweet products [13,23].

Shigella is a highly infectious foodborne pathogen belonging to the Enterobacteriaceae family. The main reservoir of this pathogen is the human intestinal system, and it causes mild to severe diarrhea in humans as a facultative intracellular pathogen. *Shigella* species are transmitted to foods via fecally contaminated water or human feces under deplorable hygienic and sanitation conditions [29]. Regarding the low-hygiene preparation conditions (by workers' hands and usually without any automation) of sweet products, a high prevalence of *Shigella* is predictable [18]. Nisa et al. (2021) also found significant levels of *Shigella* contamination in retail raw food samples in Pakistan. They reported that transmission via raw foods, hospital waste, and unhygienic food handling are the main risk factors for *Shigella* food contamination. This is the first study to report significantly higher levels of *Shigella* contamination in sweet samples; however, several other studies previously reported a high prevalence rate of *S. aureus* contamination in sweet and confectionery samples, as we also observed in the present study. The high prevalence rate of *Shigella* spp. indicates that the health and hygienic practices of personnel and food handlers during the manufacturing process of traditional sweet products should be considered more than other potential risk factors affecting the microbial quality of the products [24–31]. The same strategies can be used to decrease contamination with *Shigella* spp. and *S. aureus* in traditional foods [31]. The *ipaH* gene is also present in entero-invasive *E. coli*, indicating that these strains of *E. coli* might also be detected in this study in addition to *Shigella* spp. [32].

Cr. sakazakii is a Gram-negative foodborne pathogen belonging to the Enterobacteriaceae family. It causes severe intestinal and extraintestinal diseases (neonatal meningitis), and is associated with high mortality in infants and neonates [33]. This new emerging foodborne pathogen is relatively resistant to dry environmental conditions and is commonly isolated from dried and powdered foods, such as infant formula, sugar, wheat flour, and dry powdered ingredients [6,11]. In this study, we detected a significant level of contamination with *Cr. sakazakii* in sweet samples. Transmission of this pathogen to foods commonly occurs via contaminated raw food materials and insufficient thermal processing during food preparation [33]. Kim et al. (2011) reported a significant level of contamination (70%) with *Cr. sakazakii* in food samples (Sunshik) produced in Korea [34]. Hassani et al. (2022) also detected *Cr. sakazakii* in 7.14% of pastry sweet samples. Regular microbiological monitoring of raw food materials, especially in terms of detecting *Cr. sakazakii* by rapid methods, is the main strategy for protection against contamination with this foodborne pathogen [28].

Campylobacter is another foodborne pathogen associated with enterocolitis in humans. According to the global burden of diarrheal diseases, gastrointestinal disorders caused by this pathogen are very prevalent [35]. Poultry is recognized as the main reservoir of *Campylobacter* species, and poultry products are also known as the main source of contamination with this pathogen in other foods [36]. *C. jejuni* and *C. coli* are the prevalent species of this pathogen, frequently isolated from food products formulated with eggs. Egg whites and yolks are mainly used in the formulation of sweets; therefore, the presence of *Campylobacter* species in sweets is highly probable [37]. In this study, we detected both species of *Campylobacter* in sweet samples; however, the prevalence of *C. coli* was significantly more than that of *C. jejuni*. We also found a high correlation between the presence of *C. coli* and *Cr. sakazakii* in sweet samples. Consequently, the lack of thermal processing can also be considered a risk factor [38]. Systemic monitoring of raw materials, especially the egg whites and yolks used in the formulation of sweet products, in terms of the presence of these pathogens by using rapid assays, can also decrease the risk of contamination with *Campylobacter* species [37,39].

To produce sweet products with a high level of microbial quality and safety; automation of food processing, improving the personal hygiene of food handlers, and microbiological monitoring of raw food materials can be considered effective and practical strategies [40]. Notably, automation of sweet production may not be possible due to the fact that they are produced in small workshops using traditional methods; therefore, it should be suitable to suggest industrializing the production processes of confectionaries and sweets to improve the safety levels of these products. On the other hand, using uncontaminated raw materials such as flour, egg whites, spice powders, and dried fruits and nuts could significantly affect the hygienic production conditions of sweet products. Also, the correlations among the presence of different foodborne pathogens indicate that the same hygienic strategies can be considered to provide for the safety of these products.

Other foodborne pathogens, such as *Cl. perfringens, B. cereus*, and *L. monocytogenes*, and also some non-bacterial (fungal and viral) foodborne pathogens, have also been detected and identified in low-moisture sweet and confectionary products at relatively lower prevalence levels. However, bacterial pathogens are still considered the leading human health-threatening concern in these products [4,15,37,41].

In this study, we detected beta-lactam resistance genes, including bla_{TEM}, bla_{SHV}, bla_{OXA}, $bla_{CTX\text{-}M\text{-}1}$, $bla_{CTX\text{-}M\text{-}2}$, $bla_{CTX\text{-}M\text{-}8}$, and $bla_{CTX\text{-}M\text{-}9}$ in sweet samples. Beta-lactamases are the main enzymes responsible for the resistance of Gram-negative bacteria, such as *Cronobacter*, *Campylobacter*, and *Shigella* spp., against the beta-lactam antibiotics [42]. Pakbin et al. (2021) also detected the bla_{SHV} gene as the most prevalent ARG in low-moisture food samples. The bla_{SHV} gene, which was the dominant beta-lactam resistance gene in the sweet samples in this study, encodes resistance against amoxicillin and amoxicillin-clavulanic antibiotics [18]. There are limited studies concerning the presence of different ARGs in sweet samples; however, several studies reported and investigated the presence of ARGs in food samples such as processed dairy products. Wu et al. (2020) investigated the prevalences of ARGs in high-moisture sweet samples, and they found the largest prevalence of the *tetW, tetT,* and *tetA46* genes in these samples are associated with resistance to antibiotics belonging to the class of tetracyclines. They also detected the *erm41* and *tlrC* genes associated with resistance against the antibiotic class of macrolides [42]. Regarding the small sample size as the main limitation of this study, comprehensive epidemiological studies are suggested to be implemented in the future to investigate the prevalences of different foodborne bacterial, fungal, and viral pathogens in sweet products.

5. Conclusions

We found that sweet samples were significantly more contaminated with *S. aureus* and *Shigella* spp. *S. aureus* was the most frequent pathogenic bacteria in these profiles. We also found a significant correlation between the presence of *C. coli* and *Cr. sakazakii* in sweet samples. The bla_{SHV} gene was detected most frequently in sweet samples; adversely, the bla_{TEM} gene was only detected in one sample. Considering these results, we suggest automation of food processing, improving the personal hygiene of workers, evaluation of microbial contamination of raw materials, and regular monitoring of ARGs in raw materials and the final products to produce safe sweet products with higher levels of microbial quality.

This is the first study to report a high prevalence rate of *Shigella* spp. in traditional sweet products; consequently, preventive strategies regarding the health and hygiene of the manufacturing personnel are suggested to be more considered than other strategies during the production of traditional sweets. Comprehensive epidemiological studies concerning the prevalences of different foodborne bacterial, fungal, and viral pathogens in various sweet types are also highly recommended to be implemented in the future.

Author Contributions: Conceptualization, B.P., W.M.B. and T.B.B.; methodology, B.P., Z.A., Z.R., S.N. and B.F.; software, B.P.; validation, Z.A., and B.P.; formal analysis, B.P., W.M.B. and T.B.B.; investigation, A.K.; resources, A.K.; data curation, B.P.; writing—original draft preparation, B.P.; writing—review and editing, W.M.B. and T.B.B.; visualization, B.P.; supervision, A.K.; project administration, B.P.; funding acquisition, A.K. All authors have read and agreed to the published version of the manuscript.

Funding: This research received no external funding.

Data Availability Statement: We confirm that all data and findings of this study are available within the article.

Acknowledgments: We appreciate our colleagues and all technicians at the Medical Microbiology Research Center, Qazvin University of Medical Sciences, who assisted us in this research project.

Conflicts of Interest: The authors declare no conflict of interest.

References

1. Ashrafi Tamai, I.; Mohammadzadeh, A.; Zahraei Salehi, T.; Mahmoodi, P.; Pakbin, B. Investigation of antimicrobial susceptibility and virulence factor genes in Trueperella pyogenes isolated from clinical mastitis cases of dairy cows. *Food Sci. Nutr.* **2021**, *9*, 4529–4538. [CrossRef] [PubMed]
2. Faour-Klingbeil, D.; Todd, E. Prevention and control of foodborne diseases in Middle-East North African countries: Review of national control systems. *Int. J. Env. Res. Public Health.* **2020**, *17*, 70–73. [CrossRef] [PubMed]
3. Pires, S.M.; Desta, B.N.; Mughini-Gras, L.; Mmbaga, B.T.; Fayemi, O.E.; Salvador, E.M.; Gobena, T.; Majowicz, S.E.; Hald, T.; Hoejskov, P.S.; et al. Burden of foodborne diseases: Think global, act local. *Curr. Opin. Food Sci.* **2021**, *39*, 152–159. [CrossRef] [PubMed]
4. Bintsis, T. Foodborne pathogens. *AIMS Microbiol.* **2017**, *3*, 529. [CrossRef] [PubMed]
5. Pakbin, B.; Didban, A.; Brück, W.M.; Alizadeh, M. Phylogenetic analysis and antibiotic resistance of Shigella sonnei isolates. *FEMS Microbiol. Lett.* **2022**, *369*, fnac042. [CrossRef]
6. Pakbin, B.; Brück, W.M.; Allahyari, S.; Rossen, J.W.; Mahmoudi, R. Antibiotic Resistance and Molecular Characterization of Cronobacter sakazakii Strains Isolated from Powdered Infant Formula Milk. *Foods* **2022**, *11*, 1093. [CrossRef]
7. Foddai, A.C.; Grant, I.R. Methods for detection of viable foodborne pathogens: Current state-of-art and future prospects. *Appl. Microbiol. Biotechnol.* **2020**, *104*, 4281–4288. [CrossRef]
8. Bordeleau, S.; Asselin, H.; Mazerolle, M.J.; Imbeau, L. "Is it still safe to eat traditional food?" Addressing traditional food safety concerns in aboriginal communities. *Sci. Total Environ.* **2016**, *565*, 529–538. [CrossRef]
9. Menz, G.; Vriesekoop, F.; Zarei, M.; Zhu, B.; Aldred, P. The growth and survival of food-borne pathogens in sweet and fermenting brewers' wort. *Int. J. Food Microbiol.* **2010**, *140*, 19–25. [CrossRef]
10. Mashak, Z.; Sodagari, H.; Moradi, B. Microbiological and chemical quality of sohan: An Iranian traditional confectionary product. *J. Food Qual. Hazards Control* **2014**, *1*, 56–60.
11. Pakbin, B.; Mahmoudi, R.; Mousavi, S.; Allahyari, S.; Amani, Z.; Peymani, A.; Qajarbeygi, P.; Hoseinabadi, Z. Genotypic and antimicrobial resistance characterizations of Cronobacter sakazakii isolated from powdered milk infant formula: A comparison between domestic and imported products. *Food Sci. Nutr.* **2020**, *8*, 6708–6717. [CrossRef] [PubMed]
12. Chen, Q.; Xie, S. Genotypes, enterotoxin gene profiles, and antimicrobial resistance of Staphylococcus aureus associated with foodborne outbreaks in Hangzhou, China. *Toxins* **2019**, *11*, 307. [CrossRef] [PubMed]
13. Argudín, M.Á.; Mendoza, M.C.; Rodicio, M.R. Food poisoning and Staphylococcus aureus enterotoxins. *Toxins* **2010**, *2*, 1751–1773. [CrossRef] [PubMed]
14. Beuchat, L.R.; Komitopoulou, E.; Beckers, H.; Betts, R.P.; Bourdichon, F.; Fanning, S.; Joosten, H.M.; Ter Kuile, B.H. Low-water activity foods: Increased concern as vehicles of foodborne pathogens. *J. Food Prot.* **2013**, *76*, 150–172. [CrossRef]
15. Sánchez-Maldonado, A.F.; Lee, A.; Farber, J.M. Methods for the control of foodborne pathogens in low-moisture foods. *Annu. Rev. Food Sci. Technol.* **2018**, *9*, 177–208. [CrossRef]

16. Arabestani, M.R.; Kamarehei, F.; Dini, M.; Jalilian, F.A.; Moradi, A.; Shokoohizadeh, L. Characterization of Staphylococcus aureus isolates from pastry samples by rep-PCR and phage typing. *Iran. J. Microbiol.* **2022**, *14*, 76. [CrossRef]
17. Larsen, A.; Stegger, M.; Sørum, M. spa typing directly from a mecA, spa and pvl multiplex PCR assay—A cost-effective improvement for methicillin-resistant Staphylococcus aureus surveillance. *Clin. Microbiol. Infect.* **2008**, *14*, 611–614. [CrossRef] [PubMed]
18. Legese, H.; Kahsay, T.; Gebrewahd, A.; Berhe, B.; Fseha, B.; Tadesse, S.; Gebremariam, G.; Negash, H.; Mardu, F.; Tesfay, K. Prevalence, antimicrobial susceptibility pattern, and associated factors of Salmonella and Shigella among food handlers in Adigrat University student's cafeteria, northern Ethiopia, 2018. *Trop. Dis. Travel Med. Vaccines* **2020**, *6*, 19. [CrossRef]
19. Nafarrate, I.; Lasagabaster, A.; Sevillano, E.; Mateo, E. Prevalence, molecular typing and antimicrobial susceptibility of Campylobacter spp. isolates in northern Spain. *J. Appl. Microbiol.* **2021**, *130*, 1368–1379. [CrossRef] [PubMed]
20. Pakbin, B.; Brück, W.M.; Rossen, J.W. Virulence factors of enteric pathogenic Escherichia coli: A review. *Int. J. Mol. Sci.* **2021**, *22*, 9922. [CrossRef]
21. Bennani, H.; Mateus, A.; Mays, N.; Eastmure, E.; Stärk, K.D.; Häsler, B. Overview of evidence of antimicrobial use and antimicrobial resistance in the food chain. *Antibiotics* **2020**, *9*, 49. [CrossRef] [PubMed]
22. Caniça, M.; Manageiro, V.; Abriouel, H.; Moran-Gilad, J.; Franz, C.M. Antibiotic resistance in foodborne bacteria. *Trends Food Sci. Technol.* **2019**, *84*, 41–44. [CrossRef]
23. Fetsch, A.; Johler, S. Staphylococcus aureus as a foodborne pathogen. *Curr. Clin. Microbiol. Rep.* **2018**, *5*, 88–96. [CrossRef]
24. Rahimi, E. Enterotoxigenicity of Staphylococcus aureus isolated from traditional and commercial dairy products marketed in Iran. *Braz. J. Microbiol.* **2013**, *44*, 393–399. [CrossRef] [PubMed]
25. Huong, B.T.M.; Mahmud, Z.H.; Neogi, S.B.; Kassu, A.; Van Nhien, N.; Mohammad, A.; Yamato, M.; Ota, F.; Lam, N.T.; Dao, H.T.A. Toxigenicity and genetic diversity of Staphylococcus aureus isolated from Vietnamese ready-to-eat foods. *Food Control* **2010**, *21*, 166–171. [CrossRef]
26. Kim, N.; Yun, A.R.; Rhee, M. Prevalence and classification of toxigenic Staphylococcus aureus isolated from refrigerated ready-to-eat foods (sushi, kimbab and California rolls) in Korea. *J. Appl. Microbiol.* **2011**, *111*, 1456–1464. [CrossRef] [PubMed]
27. Mahfoozi, A.; Shirzad-Aski, H.; Kaboosi, H.; Ghaemi, E. Identification of the classical enterotoxin genes of Staphylococcus aureus in various foods by multiplex PCR assay. *Iran. J. Vet. Res.* **2019**, *20*, 209. [PubMed]
28. Hassani, S.; Pakbin, B.; Allahyari, S.; Mahmoudi, R.; Mousavi, S.; Ghajarbeygi, P. Prevalence and antimicrobial susceptibility of foodborne bacterial pathogens isolated from baghlava an iranian exporting pastry sweet. *Carpathian J. Food Sci. Technol.* **2022**, *14*, 57–69.
29. Pakbin, B.; Amani, Z.; Allahyari, S.; Mousavi, S.; Mahmoudi, R.; Brück, W.M.; Peymani, A. Genetic diversity and antibiotic resistance of Shigella spp. isolates from food products. *Food Sci. Nutr.* **2021**, *9*, 6362–6371. [CrossRef]
30. Nisa, I.; Qasim, M.; Driessen, A.; Nijland, J.; Ali, A.; Mirza, M.R.; Khan, M.A.; Khan, T.A.; Jalal, A.; Rahman, H. Prevalence and associated risk factors of Shigella flexneri isolated from drinking water and retail raw foods in Peshawar, Pakistan. *J. Food Sci.* **2021**, *86*, 2579–2589. [CrossRef]
31. Dranenko, N.O.; Tutukina, M.N.; Gelfand, M.S.; Kondrashov, F.A.; Bochkareva, O.O. Chromosome-encoded IpaH ubiquitin ligases indicate non-human enteroinvasive Escherichia. *Sci. Rep.* **2022**, *12*, 6868. [CrossRef]
32. Ling, N.; Jiang, X.; Forsythe, S.; Zhang, D.; Shen, Y.; Ding, Y.; Wang, J.; Zhang, J.; Wu, Q.; Ye, Y. Food safety risks and contributing factors of Cronobacter spp. *Engineering* **2022**, *12*, 128–138. [CrossRef]
33. Kim, J.-B.; Park, Y.-B.; Kang, S.-H.; Lee, M.-J.; Kim, K.-C.; Jeong, H.-R.; Kim, D.-H.; Yoon, M.-H.; Lee, J.-B.; Oh, D.-H. Prevalence, genetic diversity, and antibiotic susceptibility of Cronobacter spp.(Enterobacter sakazakii) isolated from Sunshik, its ingredients and soils. *Food Sci. Biotechnol.* **2011**, *20*, 941–948. [CrossRef]
34. Endtz, H.P. Campylobacter infections. In *Hunter's Tropical Medicine and Emerging Infectious Diseases*; Elsevier: Amsterdam, The Netherlands, 2020; pp. 507–511.
35. Asuming-Bediako, N.; Parry-Hanson Kunadu, A.; Abraham, S.; Habib, I. Campylobacter at the human–food interface: The african perspective. *Pathogens* **2019**, *8*, 87. [CrossRef] [PubMed]
36. Gharbi, M.; Béjaoui, A.; Ben Hamda, C.; Alaya, N.; Hamrouni, S.; Bessoussa, G.; Ghram, A.; Maaroufi, A. Campylobacter spp. in Eggs and Laying Hens in the North-East of Tunisia: High Prevalence and Multidrug-Resistance Phenotypes. *Vet. Sci.* **2022**, *9*, 108. [CrossRef] [PubMed]
37. Aaliya, B.; Sunooj, K.V.; Navaf, M.; Akhila, P.P.; Sudheesh, C.; Mir, S.A.; Sabu, S.; Sasidharan, A.; Hlaing, M.T.; George, J. Recent trends in bacterial decontamination of food products by hurdle technology: A synergistic approach using thermal and non-thermal processing techniques. *Food Res. Int.* **2021**, *147*, 110514. [CrossRef] [PubMed]
38. Shane, S. Campylobacter infection of commercial poultry. *Rev. Sci. Tech.-Off. Int. Epizoot.* **2000**, *19*, 376–385. [CrossRef] [PubMed]
39. Malavi, D.N.; Abong, G.O.; Muzhingi, T. Effect of food safety training on behavior change of food handlers: A case of orange-fleshed sweetpotato purée processing in Kenya. *Food Control* **2021**, *119*, 107500. [CrossRef]
40. Lee, G.-I.; Lee, H.-M.; Lee, C.-H. Food safety issues in industrialization of traditional Korean foods. *Food Control* **2012**, *24*, 1–5. [CrossRef]

41. Papp-Wallace, K.M. The latest advances in β-lactam/β-lactamase inhibitor combinations for the treatment of Gram-negative bacterial infections. *Expert Opin. Pharmacother.* **2019**, *20*, 2169–2184. [CrossRef]
42. Wu, Z.; Luo, Y.; Bao, J.; Luo, Y.; Yu, Z. Additives affect the distribution of metabolic profile, microbial communities and antibiotic resistance genes in high-moisture sweet corn kernel silage. *Bioresour. Technol.* **2020**, *315*, 123821. [CrossRef] [PubMed]

foods

MDPI

Article

Mitigation of *Salmonella* in Ground Pork Products through Gland Removal in Pork Trimmings

Reagan L. Jiménez, Mindy M. Brashears, Rossy Bueno López, David A. Vargas and Marcos X. Sanchez-Plata *

International Center for Food Industry Excellence, Department of Animal and Food Sciences, Texas Tech University, Lubbock, TX 79409, USA; mindy.brashears@ttu.edu (M.M.B.)
* Correspondence: marcos.x.sanchez@ttu.edu; Tel.: +1-8068346503

Abstract: Bio-mapping studies conducted in pork harvest and fabrication facilities have indicated that *Salmonella* is prevalent and mitigations are needed to reduce the pathogen in trim and ground products. *Salmonella* can be isolated from the lymph nodes and can cause contamination in comminuted pork products. The objective of this study was to determine if physically removing topical and internal lymph nodes in pork products prior to grinding would result in the mitigation of *Salmonella* and a reduction in indicators in the final ground/comminuted products. In total, three treatment groups were assigned in a commercial pork processing facility as follows: (1) untreated control, (2) topical (surface) glands removed before grinding, and (3) topical, jowl, and internal lymph nodes and glands removed before grinding. Indicator microorganisms were determined using the BioMérieux TEMPO® system and the quantification of *Salmonella* was performed using the BAX® System Real-Time *Salmonella* SalQuant® methodology. The removal of lymph nodes located on the topical and internal surfaces and in the jowl significantly ($p < 0.05$) reduced the presence of *Salmonella* and also reduced the presence of indicator organisms according to this study. Briefly, 2.5-Log CFU/sample of *Salmonella* was initially observed in the trim samples, and the ground samples contained 3.8-Log CFU/sample of *Salmonella*. The total numbers were reduced to less than 1-Log CFU/sample in both trim and ground products. This study indicates a need for lymph node mitigation strategies beginning prior to harvest, in order to prevent contamination in further-processed pork products.

Keywords: lymph node; risk assessment; quantification; *Salmonella*

Citation: Jiménez, R.L.; Brashears, M.M.; Bueno López, R.; Vargas, D.A.; Sanchez-Plata, M.X. Mitigation of *Salmonella* in Ground Pork Products through Gland Removal in Pork Trimmings. *Foods* **2023**, *12*, 3802. https://doi.org/10.3390/foods12203802

Academic Editor: Nicoletta Cristiana Quaglia

Received: 21 September 2023
Revised: 29 September 2023
Accepted: 13 October 2023
Published: 17 October 2023

1. Introduction

Pork is now number two in meat consumption globally. As of 2020, 106.3 million tons of pork are consumed annually around the world [1]. The National Pork Producers Council has reported that in the United States (U.S.), more than 2.2 million metric tons of pork and pork-related products are exported annually [2]. The value of these exports is approximately USD 7.7 billion [3]. In the United States, there were 129.9 million pigs slaughtered in the U.S. in 2019, all of which entered the food supply chain. The United States Department of Agriculture (USDA) reports that pork is consumed as fresh cuts of meat such as chops, ribs, roasts, or hams and the remaining is consumed in the form of processed pork such as sausages, hot dogs, and bacon [4].

Salmonella is often reported as the leading cause of foodborne illness in U.S. populations [5]. Every year in the U.S., non-typhoidal *Salmonella* is responsible for approximately 1,027,561 cases, 19,336 hospitalizations, and 378 deaths [6], resulting in USD 3.7 billion worth of costs to the United States economy [7]. The Interagency Food Safety Analytics Collaboration (IFSAC) reported these attributions in October 2021 for 2019 in collaboration with data from the Centers for Disease Control and Prevention (CDC), the U.S. Food and Drug Administration (FDA), and the United States Department of Agriculture's Food Safety and Inspection Service (USDA-FSIS). They reported that 75.9% of *Salmonella* cases were attributed to the following seven major food categories: chicken, fruits, pork, seeded vegetables, other produce, turkey, and eggs. In total, 12.8% of these cases were attributed to pork [8]. These data indicate that pork is the second highest contributor to salmonellosis cases from FSIS-regulated products, and the third highest contributor from all food products.

Previous studies have biomapped microorganisms in pork processing facilities to determine the prevalence of *Salmonella*, generic *E. coli*, and other indicator organisms at various stages of the pork processing chain. These studies reported that pathogen and indicator organism prevalence were reduced throughout the processing line, but increased again in trim and further processing stages, creating a U-shaped curve of the biomapped organism [4]. The facility in this study was operating as a HACCP-Based Inspection Models Project (HIMP) facility prior to the New Swine Inspection System (NSIS) for over 20 years. *Salmonella* prevalence, as determined via the previously developed baseline biomapping study, is 19% within the entire facility. Out of the 650 samples collected during the baseline biomapping study, 125 samples were positive at 24 h under prevalence testing. However, only 47 out of the 125 were positive at the 6 h SalQuant® time point, meaning that only these select samples were quantifiable [4]. Two samples were selected in this facility for sample collection: boneless picnic trim and final product brick sausage. The previous study concluded that the selected testing locations for further processed products had a 24% prevalence of *Salmonella* positives. This increase in bacterial prevalence in further processed products when compared to the facility average indicates a need for intervention strategies at the later pork processing stages. To mitigate *Salmonella* in trimmings and ground pork products, novel intervention strategies must be studied for efficacy and cost effectivity for industry applications because to date, there are few interventions available.

Lymph nodes are embedded within various muscle tissue groups, and these glands are used to filter lymphatic fluid during the lifespan of the animal. According to recent research, lymph nodes in both pork and beef have been shown to carry a high load of *Salmonella* after harvest [9]. Currently, pork trim and ground products are processed with the following lymph nodes and glands included in the muscle tissue.

- Topical
    - This group includes superficial popliteal lymph nodes located in the back legs, as well as superficial inguinal lymph nodes located in the fat on the medioventral surface of the hind leg. This group may also include various accessory glands and notably includes subiliac nodes.
- Jowl
    - The pork jowl contains 2–3 salivary glands, such as the paratoid and submaxillary glands, and several lymph nodes down the jowl, cheeks, and neck.
- Internal
    - The se deep tissue lymph nodes are located in the fat in the crease between the semitendinosus (eye of round) and gastrocnemius (knuckle). This group also includes the gluteal and ischiadic lymph nodes located on the sarcotuberal ligament. This group also contains various glands from the loin region such as the vesicular and bulbourethral glands.

The removal of glands and lymph nodes from boneless picnic hams prior to grinding for sausage production is a possible means to reduce *Salmonella* prevalence in further processed pork products, but little data exist on node/gland removal as a mitigation strategy. The objective of this study was to determine if physically removing topical and internal lymph nodes in pork products prior to grinding would result in the mitigation of *Salmonella* and a reduction in indicators in the final ground/comminuted products, thus identifying the targeted glands and lymph nodes as sources of *Salmonella* contamination in further processed pork products. In the future, the authors of this study will aim to further investigate mitigation strategies to reduce *Salmonella* prevalence in these structures.

2. Materials and Methods

2.1. Sample Collection and Treatment

Pork trim and ground sausage samples were collected online from a large-scale USDA-inspected hog processing facility. The facility where samples were collected is a large hog processing facility that is currently processing approximately 10,400 per head per day and 1250 per head per hour on average, located in the United States. This facility is USDA-inspected and is currently operating under the New Swine Inspection System (NSIS), as proposed in 2019. To operate under the NSIS, the processing plant was required to implement specific worker safety measures, including an agreement with a workers' union to represent their employees [10]. These safety measures allow the processing plant to work at higher line speeds under supervision and regular testing requirements to ensure safety and quality standards while increasing productivity. According to a constituent update published by the USDA FSIS in 2021, facilities that are not operating under the time-limited trial of the NSIS can only process up to 1106 per head per hour. Picnic trim comprises lean muscle and fat trimmings that come from the picnic shoulder of the carcass. The shoulder is a muscle of the hog that typically is tougher due to the amount of work put on the muscle. This trim, along with formulated fat and spices, is most often used as the meat component for ground sausage products within this facility. Ground sausage samples are taken after final packing in vacuum brick packages with easy-open seals. This sausage is formulated in accordance with the company's recipe independent of the treatment group. In total, 15 samples were taken from each sampling location for each treatment group (n = 90) per repetition. The entire study was replicated 5 times over a period of 4 months to account for natural variation and seasonality (N = 450). The individual sampling dates (23 March 2022, 29 March 2022, 4 April 2022, 20 May 2022, 1 June 2022, and 16 June 2022) were used to account for seasonality and natural variation within the product. FSIS directive number 65-20 was utilized to collect trim and ground samples. Protocols for whole pork cuts (intact and non-intact) and comminuted pork aseptic grab sample not in final packaging were followed as they were for the nationwide sampling program. As written in the protocol, a 375 g sample of fresh, not frozen, raw pork was collected and placed into a single sterile Whirl-Pak bag (Millipore Sigma, Burlington, MA, USA). A 1 lb portion of ground pork sausage was collected in its final packaging from the facility. To conduct raw material sample collection, 15 samples of each of the three treatment groups were taken after gland removal on each day of sampling with five replications.

Samples were shipped overnight to the food microbiology laboratory at Texas Tech located in the International Center for Food Industry Excellence (ICFIE) after immediate chilling. Samples were processed and evaluated for Enterobacteriaceae (EB), Aerobic Count Bacteria (AC), and *Salmonella* concentrations. Additionally, retainer samples were kept at the plant for further evaluation. Chilled pork carcasses were fabricated through a standard process. As the shoulder passed through the production line, the foot, jowl, and neck bone were removed. The butt was then separated from the picnic to generate a bone in, skin on picnic. The picnic was then transferred to the boning department through conveyors and equipment. The treatment of sample groups 2 and 3 was performed on belts, conveyors, and other equipment that only received gland-free products during processing to avoid cross-contamination. Between treatments, the belts and equipment were thoroughly sanitized by trained personnel in order to reduce any potential cross-contamination. All picnic trim samples were collected from a singular vat per treatment group. The treatment groups are defined as follows:

- Treatment 1—standard trim on boneless picnic hams—control
    - This treatment included standard trim without removing additional glands/ defects. Skin, bone, meat, trim and inedible tissue were removed. The skin, meat, bone, trim, and inedible tissue were collected and weighed for the vat to obtain yield information. The vat of picnics was identified and they were sent for sausage production.

- Treatment 2—retail trim on boneless picnic hams
    - Here, additional trimming was required; blood clots, and all surface/exposed glands were removed regardless of color. Skin, bone, meat, trim, and inedible tissue were removed. The skin, meat, bone, trim, and inedible tissue were collected and weighed for the vat to obtain yield information. The vat of picnics was identified and they were sent for sausage production.
- Treatment 3—export trim on boneless picnic hams
    - In this treatment, the bones in picnics were removed for standard trimming with the addition of removing exposed glands and surface blood clots regardless of size and color. Glands associated with the jowl and glands inside the boneless picnic were removed, as were skin, bone, meat, trim and inedible tissues. The skin, meat, bone, trim, and inedible tissues were collected and weighed to obtain the vat for yield information. The vat of picnics was identified and they were sent for sausage production.

These treatment groups were consistent for both the trim and group samples.

2.2. Processing Methodology

Upon arrival at Texas Tech University, the samples were evaluated for any leaking, damage, or potential temperature abuse. A 50 g aliquot of the sampled pork cut was weighed into a filtered Whirl-Pak bag (55 oz). A 200 mL portion of 45 °C (pre-warmed) of BAX® MP media (Hygiena™, Camarillo, CA, USA) was added to the sample bag. A stomacher (Model 400 Circulator, Seward, West Sussex, UK) was used to homogenize the trim samples at 230 rpm for 30 s. The processing of the ground pork sausage samples followed a similar protocol. First, 50 g of the product was weighed into a Whirl-Pak bag (55 oz, filtered) and a pre-warmed 200 mL aliquot (45 °C) of BAX® MP media was added. Ground pork samples were then homogenized in a stomacher for 1 min at 230 rpm. A 30 mL aliquot of the homogenate from the primary Whirl-Pak bag was aseptically transferred into another filtered Whirl-Pak bag (24 oz) using a disposable serological pipette (Fisher Scientific, Foods 2022, 11, 2580 5 of 20 Waltham, MA, USA). To the aliquot in this bag, a 30 mL portion of BAX® MP Media containing 1 mL of Quant solution (Hygiena™, Camarillo, CA, USA) was added to the 30 mL pure homogenized sample. An additional 10 mL aliquot of each sample type (both ground and trim) was transferred using a serological pipette into sterile tubes to enumerate indicator microorganisms, which was conducted before the samples were incubated for *Salmonella* enumeration and prevalence. The utilized processing methodology was adapted from a previous study conducted to biomap *Salmonella* and indicator organisms at each step of the pork processing line [4]. This study was replicated 3 times over a period of four months to account for the natural seasonality and variability of the pathogens (N = 450).

2.3. Microbial Analyses

Indicator bacteria were enumerated using the TEMPO® system (BioMérieux, Paris, France). The method of the Association of Official Agricultural Chemists (AOAC) 121204 was used to enumerate AC. Briefly, the method calls for the incubation of TEMPO cards for 22–28 h at 35 ± 1 °C. For EB, enumeration cards were incubated for 22 h at 35 °C. For the TEMPO enumeration of indicator organisms, the original sample was diluted to a 1/20 dilution in all sample and indicator types. To prepare this dilution, 3 mL of water and 1 mL of the sample rinsate was added to a dehydrated media vial. This dilution was then filled into the correlating TEMPO card for each indicator, EB or AC, and incubated according to the directions for each organism. Once incubated appropriately, the cards were read using TEMPO Reader. Results were converted into Log10 values for interpretation and evaluation.

In order to quantify Salmonella levels in the collected pork samples, trim samples were placed into a 42 °C incubator for 6 h, and ground pork sausage samples were incubated for a

7 h period for quantification. Following the AOAC 081201 protocol, after incubation for 7 h, *Salmonella* was enumerated using the BAX® System SalQuant® (Hygiena, Camarillo, CA, USA).The AOAC Level 2 validation of BAX® System Real-Time Polymerase Chain Reaction (RT-PCR) Assay for *Salmonella* and BAX® system SalQuant® (Certification No. 081201) followed. An aliquot of each sample was taken for the enumeration protocol, and then the original sample bags (containing the homogenate) were put back into the incubator at 42 °C for 18–24 h to detect any *Salmonella* that might have been present but below the limits of quantification..

The sample preparation protocol for the BAX® System Real-Time PCR Assay for *Salmonella* has 3 stages for the workflow: preparation, lysis, and PCR. The first stage, sample preparation, consisted of preparing the lysis reagent in accordance with the provided protocol and thermal blocks that were pre-heated to 37 °C and 95 °C. The lysis step was completed by transferring 5 µL of the sample to cluster tubes, and then a heating step at 37 °C for 20 min was carried out. Additionally, a subsequent heating step was conducted at 95 °C for 10 min. Upon the completion of the steps, samples were cooled for 5 min. The PCR stage of this protocol involved hydrating PCR tables with 30 µL of the lysate and running the BAX® Q7 thermocycler with the appropriate assay parameters.

2.4. Data Analysis

To evaluate the microbiological results, all data were analyzed using R (Version 4.1.2) statistical software. Each treatment was compared to the control. Counts of indicator organisms were converted into LogCFU/g and *Salmonella* counts were reported as LogCFU/sample. A one-way ANOVA (analysis of variance) was performed on the data, which compared the pathogen counts from each of the treatment groups, followed by pairwise multiple comparison *t*-tests, and adjusted via the Bonferroni method. *p*-values of 0.05 or less were used to determine significant differences.

Data were arranged into boxplots, with a horizontal line within the box to represent the median of the data. The lower (0.25) and upper (0.75) quartiles are represented by the top and bottom lines of the box. The upper and lower lines represent 1.5 times the interquartile range. The dots present on the plots represent the actual collected data points. For each matrix, boxes indicated with different letters are reported as significantly different between treatments according to *t*-test analysis at p-value < 0.05.

3. Results

The LogCFU/g counts of AC indicate significant differences between the control samples, the retail trim, and the export trim. Enterobacteriaceae results show a statistically significant difference among the control trim and both treatment groups, but there was not a significant difference between EB counts obtained from retail trim and export trim. *Salmonella* counts were recorded and presented in Log_{10}CFU/sample using a 50 g sample basis. Trim samples had overall higher counts of both indicator organisms and *Salmonella* for all treatment groups when compared to ground samples from the same treatment groups. Indicator organisms, especially EB, show a wide range of variance for each set of samples, which indicates an overall need for better process control methods within this facility in order to reduce the variation.

3.1. Detection and Quantification of Salmonella

In total, 72/450 samples tested positive for Salmonella (16%). Table 1 shows *Salmonella* prevalence from each treatment group. In total, 38 samples (52.7%) were suitable for enumeration with the majority being detected from treatment group 1, the control group. The breakdown of collected positives from each treatment group and matrix can be found in Table 1.

Table 1. Percentage of samples testing positive for *Salmonella* upon SalQuant® analysis using the BAX® system using commercially obtained ground pork and pork trimmings with and without node removal.

Product	Control (%)	Retail Trim (%)	Export Trim(%)
Ground Pork with Seasonings	30.5 n = 22	19.4 n = 14	5.6 n = 4
Boneless Picnic Meat with Different Trim Levels	29.2 n = 21	11.1 n = 8	4.2 n = 3
Total	59.7% n = 43	30.5% n = 22	9.8% n = 7

Overall, 72 of the 450 samples tested positive for *Salmonella* in the prevalence assay after 24 h of enrichment (n = 72). Of these, 43 samples were part of the control group, 22 of the positive samples were from the retail trim (topical gland removal only) group, and 7 positives were a part of the export trim category, which had the topical, jowl, and internal glands removed. Overall, more *Salmonella* positives were detected in ground samples as opposed to trim samples, and this was displayed across each treatment group.

Salmonella counts were very low in the majority of the samples as analyzed on a per gram basis. Therefore, for visualization purposes, all data were transformed into LogCFU/sample, which is equivalent to LogCFU/50 g, to facilitate data interpretation. The limit of quantification (LOQ) for SalQuant® on pork trim and ground pork was 0.1 CFU/mL and 0.1 CFU/g or 0.70 LogCFU/sample. When samples were negative for quantification, they were reported as 50% of the LOQ (0.35 LogCFU/sample). Samples that were not quantifiable or detectable were reported as 0 LogCFU/sample but these are not reflected on Figure 1 as including these values would have altered the mean values of the positive samples.

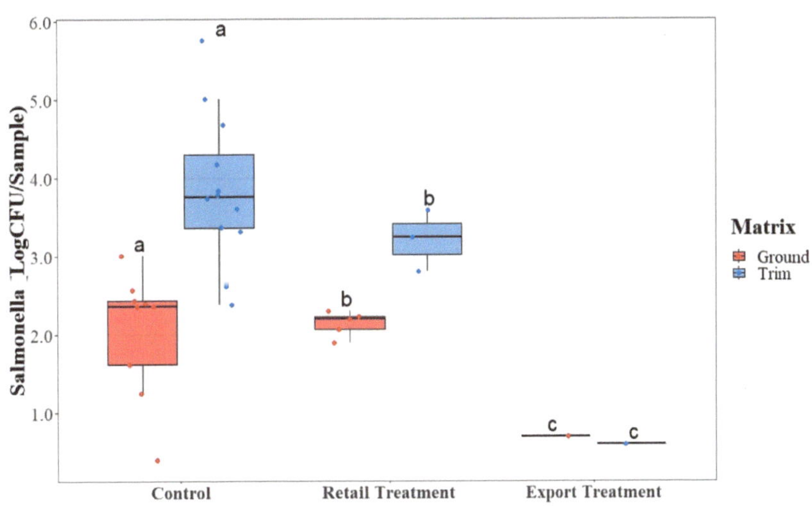

Figure 1. *Salmonella* quantification as determined via SalQuant® on pork trimmings and ground product subjected to gland and node removal. Horizontal lines are the median; upper and lower quartiles are represented by the top and bottom lines of the box. Each dot represents a data point. a, b, c: for each matrix, boxes with different letters are significantly different among treatments according to a *t*-test analysis at *p*-value < 0.05.

Of the 450 collected samples across five replications, 38 samples were positive at the SalQuant® time point for quantification (n = 38). The recorded quantitative values are displayed in Figure 1. The control group averaged at 2.5 Log CFU/sample and

3.8 Log CFU/sample of *Salmonella* in ground and trim samples, respectively. The export trim group held the lowest average of *Salmonella* counts for both matrices at less than 1 Log CFU/sample. There were statistical differences among each of the three treatment groups for both detection and quantification methodologies. Of the 31 quantifiable samples, 3 were from the export trim, 12 were from retail trim, and 16 were from the control trim. The mean of each sample point was used to determine significant differences between the sample groups. There was a significant difference ($p < 0.05$) between the control samples and each of the treatment groups. However, there was not a significant difference between the retail trim and export trim treatment groups.

3.2. Enumeration of Enterobacteriaceae and Aerobic Count Bacteria

AC counts are described in Figure 2. Total ACs were statistically compared across mean values for each treatment group and matrix.

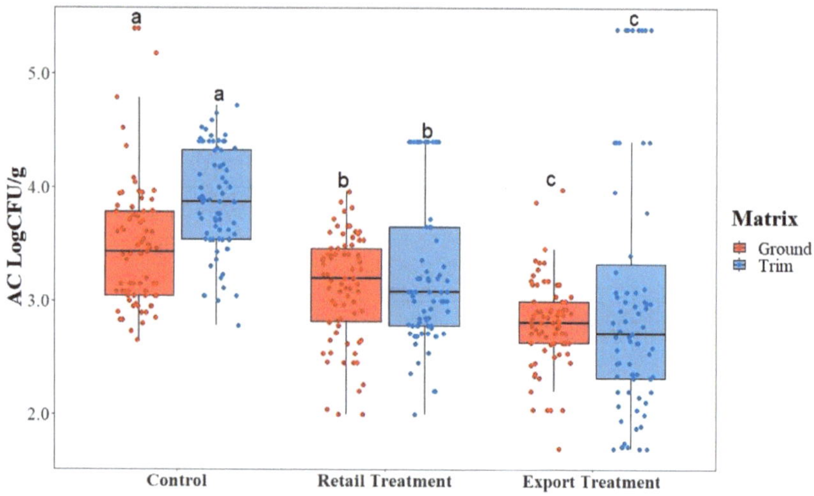

Figure 2. Aerobic plate count (AC) of ground pork and pork trimmings collected over a 4-month period (n = 450) with and without gland removal. The mean of each sample point was utilized to determine significant differences among treatment groups. a, b, c: for each matrix, boxes with different letters are significantly different among treatments ac-cording to a *t*-test analysis at *p*-value < 0.05.

As shown in Figure 1, there was a statistical difference ($p < 0.05$) among the treatment groups for both matrices with the control having the highest, retail having a smaller difference than that of the control but a higher difference than that of the export, and export having the lowest. Export trim, which was composed of boneless picnic trim with the topical, jowl, and internal glands removed, had the lowest average AC counts for aerobic plate counts for both the ground and trim matrices. The lowering of AC counts indicates a reduction in overall microbial activity within the samples collected from each treatment group as the glands were removed.

EB counts, as detailed in Figure 3, were compared in terms of mean value for each treatment group and matrix.

Enterobacteriaceae counts were determined using the TEMPO system and converted into Log_{10}CFU/g values. While the range for each treatment group remains wide across treatments, the median value for each group decreases as the lymph nodes and glands are removed. The means of each sample point were used to determine significant differences among treatments. There was a statistical difference ($p < 0.05$) among EB counts collected from the control group, and counts collected from each of the treatment groups. Unlike the case for AC counts, there was not a statistical difference between treatment retail trim and export trim. a reduction in EB microorganisms is beneficial to the product as it indicates a lower amount of potential pathogenic presence.

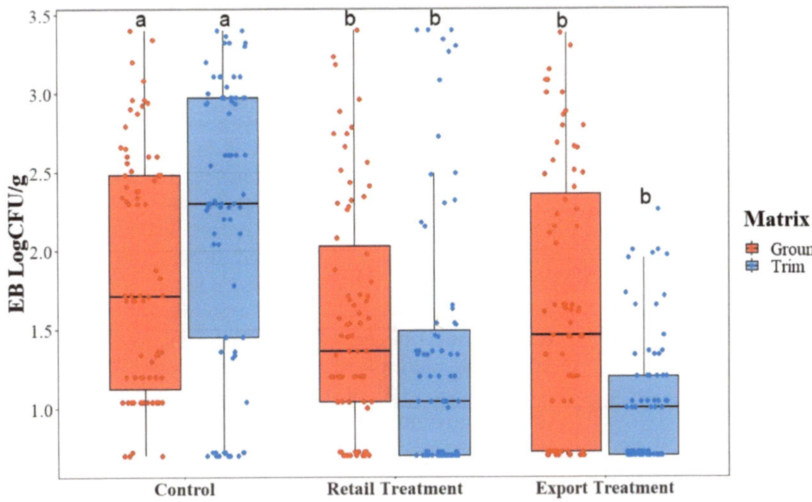

Figure 3. Enterobacteriaceae (EB) results for pork trim and ground pork collected from a commercial pork facility with and without lymph node removal. a, b, c: for each matrix, boxes with different letters are significantly different among treatments ac-cording to a *t*-test analysis at *p*-value < 0.05.

The pattern of the detected and quantified *Salmonella* correlates closely with the pattern of the EB and AC indicator organisms measured within this study. While these values correlate in pattern, there is not an exact ratio between the relationships. Since the organisms follow similar patterns, indicator organisms can be observed to suggest the presence of *Salmonella* in pork products; however, the observation of indicator organisms cannot be utilized in place of *Salmonella* testing in this particular operation. The similar pattern followed by *Salmonella* and the indicator organisms additionally suggests that the removal of lymph nodes affected the organisms directionally.

4. Discussion

The quantification of *Salmonella* in pork samples from commercial industry establishments may be limited because of pathogen recovery, due to pathogen stress caused by the processing environment, and the application of antimicrobial interventions. The quantification techniques utilized within this project have been validated to recover pathogens from positive samples as a result of a recovery stage. The use of short enrichment steps strengthens the quantification data via the recovery of injured cells [11,12]. Additionally, *Salmonella* quantification may offer an opportunity to make risk-based and data-driven decisions based on the prevalence and overall concentration at specific processing seps in the process, rather than the presence or absence of the pathogen [13]. The quantification of pathogens can benefit the pork processing industry as indicated by the results of this study, which provides evidence for novel uses of emerging pathogen detection technologies. The utilization of a rapid PCR-based enumeration methods for *Salmonella*, in conjunction with the enumeration of indicators, provides the pork industry with a tool to make data-driven decisions to reduce pathogenic prevalence in trim and further processed pork products and to mitigate the risk of public health of foodborne illness.

Furthermore, the results of this study indicate that the removal of topical lymph nodes and glands from boneless picnic trim was an effective method for reducing *Salmonella* in boneless picnic pork trim and ground sausage products in this operation. Furthermore, the results of this study indicate that the removal of topical, jowl, and internal glands and lymph nodes further reduced the prevalence of *Salmonella* and other indicator organisms in boneless picnic trim and ground sausage when compared to that with the removal of topical glands and lymph nodes alone. It is important to note that in this study, strict sanitary measures were used in node and gland removal and the amount of time taken to remove the nodes and glands was significant; thus, the same procedure may be difficult to implement in commercial operations. There could be a risk of cross contamination if the nodes are not carefully removed and proper sanitation protocols are not implemented. If this method is chosen to mitigate *Salmonella*, effective and efficient methods for gland and lymph node removal should be determined and developed before the implementation of these strategies within the industry. It is also critical to understand which nodes contributed to the most reduction. Nodes were not isolated and it could be a single node or combination that resulted in the reductions. Finally, the serotype and the pathogenicity of the *Salmonella* was not determined. In order to make an impact on public heath, serotypes of the highest concern that are related to human illnesses should be considered. Additional mitigation strategies should also be observed, such as pre-harvest strategies to prevent node contamination, or chemical/physical applications of interventions for reducing pathogenic prevalence within further processed pork products. This study clearly establishes that lymph nodes contribute to *Salmonella* presence in ground products. Additionally, the results of this study indicate that the lymph nodes and glands identified within this study and removed from the treatment groups were probable sources of *Salmonella* contamination in further processed pork products. The authors of this study further recommend the identification of mitigation strategies to reduce *Salmonella* prevalence within the lymph nodes as the removal of these structures in a large-scale pork processing facility is not currently feasible due to the high labor demand of the removal process.

Pork facilities also face proposed performance standards from the FSIS that must be met. While not yet implemented, the current proposed performance standards determine that the "pass or fail" status of a processing plant are based on the total number of *Salmonella* positives taken from samples over a 52-week rolling window. However, the positives are based on the detection methodology that determines if any amount of *Salmonella* is present within the tested sample. Pork processing plants are currently being tested for the presence or absence of *Salmonella* alone and not for quantification, determination, serotype, or pathogenicity. It is currently estimated that the infectious dose of *Salmonella* is relatively high when compared to that of other pathogens, being estimated to be between 10^5 and 10^6 cells [14]. The quantifiable *Salmonella* positives detected within this study showed levels far below 10^6 CFU. This suggests that pathogenic loads of the product might be below or far below the number of cells that would cause human infection in a healthy adult upon the consumption of a fully cooked product. Therefore, this data could be used to inform future decisions regarding performance standards and their dependency on quantification-based methodologies for *Salmonella* testing rather than presence–absence alone in order to make more informed decisions about the safety of a product. Implementing quantification methodologies within the industry for pathogenic testing may provide further insights and data to make risk-based public health decisions.

In conclusion, the results of this study indicate that the lymph nodes and glands targeted and removed in the treatment groups of this study could be identified as probable sources of *Salmonella* in further processed pork products. This information can be used as foundational support for further research to be conducted on the implementation of mitigation strategies to reduce pathogenic prevalence within these structures.

Author Contributions: Conceptualization, M.X.S.-P. and M.M.B.; methodology, R.L.J., R.B.L., D.A.V. and M.X.S.-P.; validation, R.L.J. and R.B.L.; formal analysis, R.L.J. and D.A.V.; investigation, R.L.J., R.B.L. and D.A.V.; resources, M.X.S.-P.; data curation, R.L.J. and D.A.V.; writing—original draft preparation, R.L.J.; writing—review and editing, R.L.J., R.B.L., D.A.V., M.M.B. and M.X.S.-P.; supervision, M.X.S.-P.; project administration, R.L.J. and M.X.S.-P.; funding acquisition, M.M.B. and M.X.S.-P. All authors have read and agreed to the published version of the manuscript.

Funding: The current study was funded by the International Center for Food Industry Excellence (ICFIE) at Texas Tech University.

Data Availability Statement: Data are available upon request from the corresponding author. The data are not publicly available due to privacy from the pork processing partner which allowed the project to be conducted within their facility.

Acknowledgments: We would like to acknowledge the help of all ICFIE Food Microbiology personnel that helped with sample processing and helped in any other way to make this project a reality.

Conflicts of Interest: The authors declare no conflict of interest.

References

1. Shahbendeh, M. Meat Consumption Worldwide from 1990 to 2021, by Meat Type. *Retrieved Novemb.* **2021**, *28*, 2021.
2. National Pork Producers Council. New Report Highlights U.S. Pork Industry Contributions to American Jobs and the Economy. 2022. Available online: https://nppc.org/press-releases/2021_economic_impact_report/ (accessed on 3 May 2023).
3. United States Department of Agriculture. *Food Safety and Inspection Service Sampling Requirements to Demonstrate Process Control in Slaughter Operations*; United States Department of Agriculture: Washington, DC, USA, 2020.
4. Bueno López, R.; Vargas, D.A.; Jimenez, R.L.; Casas, D.E.; Miller, M.F.; Brashears, M.M.; Sanchez-Plata, M.X. Quantitative Bio-Mapping of *Salmonella* and Indicator Organisms at Different Stages in a Commercial Pork Processing Facility. *Foods* **2022**, *11*, 2580. [CrossRef] [PubMed]
5. Center for Food Safety and Applied Nutrition. Release of 2019 Annual Report on the Sources of Foodborne Illness by the Interagency Food Safety Analytics Collaboration. 2021. Available online: https://www.fda.gov/food/cfsan-constituent-updates/release-2019-annual-report-sources-foodborne-illness-interagency-food-safety-analytics-collaboration (accessed on 26 August 2022).
6. Scallan, E.; Hoekstra, R.M.; Angulo, F.J.; Tauxe, R.V.; Widdowson, M.A.; Roy, S.L.; Jones, J.L.; Griffin, P.M. Foodborne Illness Acquired in the United States—Major Pathogens. *Emerg. Infect. Dis.* **2011**, *17*, 7–15. [CrossRef] [PubMed]
7. Hoffmann, S.; Maculloch, B.; Batz, M. *Economic Burden of Major Foodborne Illnesses Acquired in the United States*; United States Department of Agriculture: Washington, DC, USA, 2015.
8. Interagency Food Safety Analytics Collaboration. *Foodborne Illness Source Attribution Estimates for 2019 for Salmonella, Escherichia Coli O157, Listeria Monocytogenes, and Campylobacter Using Multi-Year Outbreak Surveillance Data, United States*; U.S. Department of Health and Human Services: Washington, DC, USA; CDC: Atlanta, GA, USA; FDA: Silver Spring, MD, USA; USDA: Washington, DC, USA, 2018.
9. Webb, H.E.; Brichta-Harhay, D.M.; Brashears, M.M.; Nightingale, K.K.; Arthur, T.M.; Bosilevac, J.M.; Kalchayanand, N.; Schmidt, J.W.; Wang, R.; Granier, S.A.; et al. Salmonella in Peripheral Lymph Nodes of Healthy Cattle at Slaughter. *Front. Microbiol.* **2017**, *8*, 2214. [CrossRef] [PubMed]
10. United States Department of Agriculture. *Food Safety and Inspection Service Modernization of Swine Slaughter Inspection*; United States Department of Agriculture: Washington, DC, USA, 2019.
11. Applegate, S.F.; Englishbey, A.K.; Stephens, T.P.; Sanchez-Plata, M.X. Development and Verification of a Poultry Management Tool to Quantify *Salmonella* from Live to Final Product Utilizing RT-PCR. *Foods* **2023**, *12*, 419. [CrossRef] [PubMed]
12. Vargas, D.A.; Betancourt-Barszcz, G.K.; Blandon, S.E.; Applegate, S.F.; Brashears, M.M.; Miller, M.F.; Gragg, S.E.; Sanchez-Plata, M.X. Rapid Quantitative Method Development for Beef and Pork Lymph Nodes Using BAX® System Real Time Salmonella Assay. *Foods* **2023**, *12*, 822. [CrossRef]
13. De Villena, J.F.; Vargas, D.A.; Bueno López, R.; Chávez-Velado, D.R.; Casas, D.E.; Jiménez, R.L.; Sanchez-Plata, M.X. Bio-Mapping Indicators and Pathogen Loads in a Commercial Broiler Processing Facility Operating with High and Low Antimicrobial Intervention Levels. *Foods* **2022**, *11*, 775. [CrossRef]
14. Kurtz, J.R.; Goggins, J.A.; McLachlan, J.B. *Salmonella* Infection: Interplay Between the Bacteria and Host Immune System. *Immunol. Lett.* **2017**, *190*, 42–50. [CrossRef] [PubMed]

foods

MDPI

Article

Unveiling Fresh-Cut Lettuce Processing in Argentine Industries: Evaluating *Salmonella* Levels Using Predictive Microbiology Models

Sofia Griselda Cuggino [1], Arícia Possas [2,*], Guiomar Denisse Posada-Izquierdo [2], Martin Gustavo Theumer [3,4,†] and Fernando Pérez-Rodríguez [2,†]

1 Departamento de Fundamentación Biológica, Facultad de Ciencias Agropecuarias, Universidad Nacional de Córdoba, Córdoba X5000, Argentina; sofiacuggino@gmail.com
2 Department of Food Science and Technology, UIC Zoonosis y Enfermedades Emergentes ENZOEM, CeiA3, Universidad de Córdoba, 14014 Córdoba, Spain; bt2poizg@uco.es (G.D.P.-I.); b42perof@uco.es (F.P.-R.)
3 Departamento de Bioquímica Clínica, Facultad de Ciencias Químicas, Universidad Nacional de Córdoba, Córdoba X5000, Argentina; mgtheumer@unc.edu.ar
4 Consejo Nacional de Investigaciones Científicas y Técnicas (CONICET), Centro de Investigaciones en Bioquímica Clínica e Inmunología (CIBICI), Córdoba X5000, Argentina
* Correspondence: g12mepoa@uco.es
† These authors contributed equally to this work.

Abstract: A survey was performed to gather information on the processing steps, conditions, and practices employed by industries processing ready-to-eat (RTE) leafy vegetables in Argentina. A total of seven industries participated in the survey. A cluster analysis of the data obtained was performed to identify homogeneous groups among the participating industries. The data collected were used as inputs of two predictive microbiology models to estimate *Salmonella* concentrations after chlorine washing, during storage and distribution of final products, and to rank the different practices according to the final estimated *Salmonella* levels. Six different clusters were identified by evaluating the parameters, methods, and controls applied in each processing step, evidencing a great variability among industries. The disinfectant agent applied by all participating industries was sodium hypochlorite, though concentrations and application times differed among industries from 50 to 200 ppm for 30 to 110 s. Simulations using predictive models indicated that the reductions in *Salmonella* in RTE leafy vegetables would vary in the range of 1.70–2.95 log CFU/g during chlorine-washing depending on chlorine concentrations applied, washing times, and vegetable cutting size, which varied from 9 to 16 cm^2 among industries. Moreover, *Salmonella* would be able to grow in RTE leafy vegetables during storage and distribution, achieving levels of up to 2 log CFU/g, considering the storage and transportation temperatures and times reported by the industries, which vary from 4 to 14 °C and from 18 to 30 h. These results could be used to prioritize risk-based sampling programs by Food Official Control or determine more adequate process parameters to mitigate *Salmonella* in RTE leafy vegetables. Additionally, the information gathered in this study is useful for microbiological risk assessments.

Keywords: predictive microbiology; disinfection; food safety; cross-contamination; foodborne pathogens

Citation: Cuggino, S.G.; Possas, A.; Posada-Izquierdo, G.D.; Theumer, M.G.; Pérez-Rodríguez, F. Unveiling Fresh-Cut Lettuce Processing in Argentine Industries: Evaluating *Salmonella* Levels Using Predictive Microbiology Models. *Foods* **2023**, *12*, 3999. https://doi.org/10.3390/foods12213999

Academic Editor: Alberto Cepeda Sáez

Received: 10 September 2023
Revised: 22 October 2023
Accepted: 26 October 2023
Published: 1 November 2023

1. Introduction

The World Health Organization recommends the consumption of at least 400 g of fresh produce per day, while the Dietary Guidelines for the Argentine Population recommend the daily consumption of five servings of fruits and vegetables in a variety of types and colors [1,2]. However, it has been reported that only 6% of the Argentine population follows this recommendation [3]. Many factors contribute to consumers not incorporating fruits and vegetables into their diets, including socioeconomic status, availability of and accessibility to quality products, educational level, and, in some cases, the time required to prepare these foods [4,5]. The horticultural sector in Argentina is committed and challenged to optimize its production to promote an increase in fresh produce consumption, not only in terms of quantity and diversity but also in terms of quality [5–8].

While most vegetables are marketed as commodities or in bulk, another part is used for canning and dehydration. However, the production of fresh and ready-to-eat (RTE) products is still an emerging segment in Argentina, with little development in the agro-industrial sector [9,10]. These products require industrial processing that involves steps including selection and classification, cutting, washing-disinfection, centrifugations, and packaging. They are marketed as products for direct consumption in homes, restaurants, hotels, food chains, and catering services, among others [11]. In countries such as Spain, the production of fresh and RTE vegetables has been widely consolidated, offering the consumer healthy, fresh, and easy-to-consume or prepared foods, with high organoleptic quality [9,12–14].

Fresh-cut vegetables are consumed raw, without being subjected to any additional disinfection steps, which can completely eliminate any remaining pathogen contamination. Therefore, these products have been frequently implicated in foodborne outbreaks primarily related to contamination with pathogens such as *Escherichia coli* O157:H7 and *Salmonella* spp. [15–18]. According to Callejón et al. [15], *Salmonella* is a key pathogen responsible for foodborne illness outbreaks linked to fresh produce. Studies in South America have also reported the presence of *Salmonella* spp. in minimally processed vegetables [19–22].

The production chain of RTE vegetables is complex and encompasses several critical steps in which food safety may be compromised from a microbiological point of view [23,24]. Hence, a systematic approach, including all aspects of the production chain from "farm to table" is required, starting with the quality of the raw material, the effectiveness of washing, disinfection, and cross-contamination control through production, packaging, transport, and storage [25–29]. Additionally, several researchers have indicated that washing and sanitizing treatments constitute one of the most critical steps impacting the product quality, safety, and shelf-life [30–33]. Also, disinfection efficacy is affected by many factors such as the disinfectant agent, dosage, residual concentration, contact time, cut size, temperature, pH, and wash water/surface ratio where the sanitizer is applied [34–37].

The application of predictive microbiology models allows for estimation of the responses of microorganisms under specific environmental conditions along the food production chain, storage, and distribution [38–40]. Therefore, appropriate predictive models can be used to anticipate the growth, survival, and inactivation of microorganisms during the production–distribution chains of RTE vegetables, in order to evaluate the effect of different factors associated with each processing step (e.g., temperatures, times). In this sense, predictions can be useful for testing the efficacy of control measures or modifications of process parameters to prevent the exposure of consumers to microbial hazards.

A more specific knowledge of the practices and conditions employed by industries processing RTE vegetables can help to understand, prevent, and reduce contamination with pathogens throughout the production chain and ensure their microbiological safety. Therefore, the aims of this study were: (i) to collect information on the practices, parameters, and methods applied by RTE vegetable processing industries from different provinces of Argentina; and (ii) to use predictive models to estimate the impact of the washing and disinfection process with chlorine on the concentration of *Salmonella* in RTE vegetables, and evaluate whether storage and distribution practices applied in the industries affect pathogen survival or growth in the final product.

2. Materials and Methods

2.1. Participating Industries and Survey

Ten major RTE vegetables industries (processing plants, companies) located in different regions of Argentina were contacted for this study. First, contact was made through the e-mails published on their websites, and subsequent communications were established with the quality managers of each company. A questionnaire regarding the processing steps and parameters of RTE leafy vegetables production was elaborated, considering the most important factors affecting microbial fate in these types of products (e.g., temperature, time, sanitizer concentration ranges). The final version of the survey consisted of 28 questions with multiple answers, which were subdivided into six sections: (1) Description of industry and the processing steps carried out; (2 and 3) Data on the washing step and disinfection of RTE-leafy greens; (4) Details on the packaging conditions; (5) Information on post processing in-plant storage; and (6) Data on transport and distribution of the RTE leafy-green vegetables. The questionnaire was emailed to the quality managers of ten Argentine industries via GoogleForms® (Google LLC, Mountain View, CA, USA) in March 2021.

2.2. Data Analysis

The R software (version 4.0.0) was used for graphical representation of the data derived from the questionnaire responses [41]. A cluster analysis was carried out using the InfoStat software, version 2017 (Grupo InfoStat, Córdoba, Argentina) [42], to identify the similar patterns among the surveyed industries considering the processing steps and conditions adopted by them.

2.3. Simulation Scenarios Using Predictive Models in the Processing and Conditioning Stages

Predictive models were used to quantitatively evaluate the impact of the processing steps, conditions, and practices adopted by the surveyed industries on levels of *Salmonella* in RTE leafy green vegetables. Thus, two models were selected for predicting the effect of distinct processing steps.

First, the polynomial model by Cuggino et al. [36] was used to evaluate the disinfection efficacy of chlorine washing of fresh-cut lettuce, assuming products were contaminated with *Salmonella*. This model was developed based on the Response Surface Methodology and describes the *Salmonella* Thompson inactivation during chlorine washing as a function of four independent variables: free chlorine concentration (FCC, 0–200 ppm), contact time (30–110 s), cutting size (9–21 cm^2), and benzyl isothiocyanate concentration (BITC, 0–80 ppm). It is important to highlight that the data to develop this model were obtained using model water at 4 °C and a water:lettuce ratio of 8.5 L/kg; therefore, these parameters were considered when evaluating the scenarios.

Second, to evaluate the growth potential of *Salmonella* in the post-processing steps of fresh-cut lettuce, the secondary model by Cuggino et al. [43] was coupled with the primary model by Baranyi and Roberts [44]. Briefly, to develop this secondary model, *Salmonella* Thompson growth data over time were obtained in artificially contaminated fresh-cut lettuce at different storage temperatures (9–18 °C). The *Salmonella* growth parameters (i.e., growth rates, maximum population density) were estimated by fitting the Baranyi and Roberts model to the growth data. Then, the secondary model by Ratkowsky et al. [45] was used to relate *Salmonella* growth rates to the environmental temperature.

The abovementioned predictive models were implemented in the software MicroHibro (www.microhibro.com) for simulations. Once implemented into the software database, models become available in a user-friendly interface, in which users must define the conditions of predictions (e.g., initial level of contamination, chlorine concentration, environmental temperature) and the software returns *Salmonella* concentrations and estimated kinetic parameters. For simulations using the predictive models, the initial level of *Salmonella* contamination lettuce was set to be 3 log CFU/g, that is, the highest level that was reported in this type of product [36,46]. The parameters used as inputs for the predictive models to evaluate the different scenarios represented by each industry were the following: the average concentration of the disinfectant (ppm) applied during washing, the average disinfection time, and the cut size of the pieces of leafy vegetables (cm^2), which varied among companies. The BITC was set to zero since none of the industries apply it in the disinfection process. The highest storage and distribution temperatures (°C) reported by the industries were used in the simulations. Therefore, worst-case scenarios in the disinfection and storage stages were used for the simulations using the predictive models.

3. Results and Discussion

3.1. Description of the Participating Industries and the Processing Stages of Ready-to-Eat Leafy Vegetables

Seven out of the ten contacted industries answered the survey. They were identified with the letters A to G. Three of them are located in the province of Córdoba; two are located in Buenos Aires; one, in Mendoza; and one in Santa Fe. All of them are small enterprises with a maximum of 25 employees.

Two main stages were differentiated in all the companies: processing and conditioning. The processing stage included common processing steps, such as selection and classification of raw material, cutting process, washing-disinfection and centrifugation/dewatering; all of them being necessary for the transformation of vegetables into RTE vegetables [47]. Moreover, the conditioning stage included packaging, storage, and transportation of final products. The process stages informed by industries are those essential for obtaining vegetables ready for consumption, and have been previously reported [47–49].

A flow diagram showing the general process for RTE leafy vegetable production in Argentina was developed, based on the information provided by the seven participating industries (Figure 1). Despite the similarities in the processing and conditioning stages among industries, it should be noted that in some of them such as C, D, E, and F, the processing steps are totally automated, while in others, like A, B and G, some stages are manual.

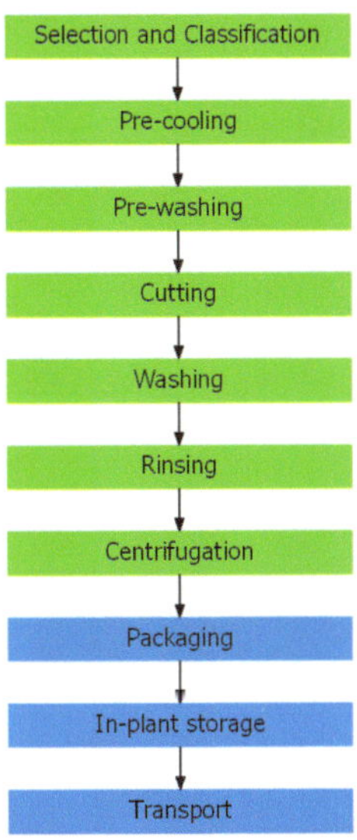

Figure 1. General process flow diagram of the elaboration of RTE leafy vegetables in Argentina. Processing (green boxes) and conditioning (blue boxes) stages.

3.2. Assessment of the Processing and Conditioning Stages of RTE Leafy Vegetables in Argentine Industries

To identify similarities or differences between the seven processing industries which participated in the survey, a cluster analysis was carried out using information about the processing and conditioning steps of RTE leafy vegetables available in the questionnaire responses. The results of the cluster analysis with regard to the steps carried out in the seven companies showed three major clusters, namely groups 1–3 (Figure 2). Group 1 (line blue), comprising industry B, showed a stark difference compared to the other groups, since it does not perform pre-washing, pre-cooling, or rinsing stages during the processing of leafy greens. Group 2 (line green), comprising industries D, E, F, and G, reported all the same processing stages shown in Figure 1. Group 3 (line red), comprising companies A and C, differ from company B in that they perform a rinsing stage, and from Group 2 in that they do not perform the pre-washing and pre-cooling stages.

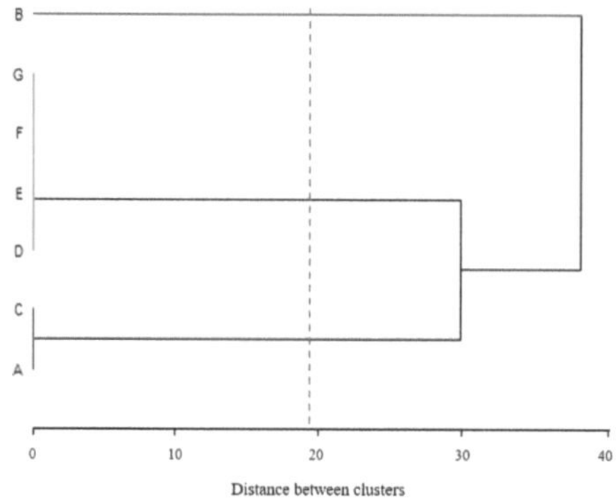

Figure 2. Cluster analysis of the processing and conditioning steps of ready-to-eat leafy vegetables performed by the seven participating industries (A–G).

Furthermore, a cluster analysis was conducted, encompassing all parameters (e.g., temperature, times), methods, and controls employed at each stage. This yielded six distinct clusters (six different color lines), indicating significant variability among the practices and methods utilized by the various industries which participated in the survey (Figure 3).

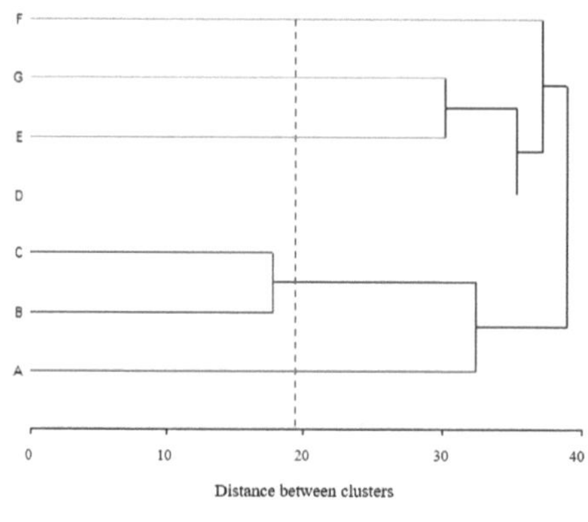

Figure 3. Cluster analysis of the parameters and methods used for the processing stages of ready-to-eat leafy vegetables in seven Argentine industries (A–G).

The similarities and differences found in the processing and conditioning stages are detailed below.

3.2.1. Processing Stage: Selection and Classification, Cutting, Disinfection, and Centrifugation Process

When analyzing in detail the common steps performed by the seven industries, similarities and differences were observed in some specific parameters and in the processing operations. For example, all participating companies select and classify their leafy greens, removing outer damaged leaves before processing. Although all seven industries perform the cutting operation, in companies A, B, C, D, E, and F lettuce is cut prior to disinfection, while in company G, lettuce is first disinfected and then cut. Concerning the vegetable cut sizes, they varied from 9 to 18 cm^2, as shown in Table 1.

Table 1. Parameters of the processing and conditioning stages of RTE leafy vegetables of seven Argentine industries (A–G).

Parameter	A	B	C	D	E	F	G
Cut size (cm^2)	16	12	12	9	16	18	16
pH Control/pH	NA	NA	NA	6 to 8	NA	6 to 8	6 to 8
Packaging	Tray with film without vacuum	Tray with film without vacuum	Plastic bags MAP	PET containers with lid	Sealed plastic bags	Sealed plastic bags	Plastic bags MAP
Shelf life (h)	144	96	216	144	96	144	192

NA: not applied. MAP: modified atmosphere packaging.

Another common step among all surveyed industries is disinfection; however, the parameters utilized vary among the different companies. For example, the temperature of the water used for the disinfection process was greater than or equal to 8 °C, and, in some cases, wash water at room temperature was used. As indicated by Gil et al. [30], leafy greens should cool quickly (less than 90 min) after harvest. This would slow down microbial growth in the case of contamination, and can be partially achieved by using wash water at low temperatures [32,50].

Also, all industries reported using sodium hypochlorite as a disinfectant for leafy vegetables during the washing operation, at concentrations ranging from 50 to 200 ppm. These results are in line with the fact that sodium hypochlorite remains the most widely utilized disinfectant in the fresh produce industry [32,51–54], due to its comparatively affordable price, ease of application, and extensive range of antimicrobial effectiveness [55,56].

On the other hand, disinfection times also vary among companies, spanning from 30 to 110 s, as shown in Figure 4. This difference could be attributed to various factors unique to each industry's processing needs and practices. Interestingly, despite this wide range of disinfection times, our analysis uncovered an unexpected outcome: no significant correlation existed between the concentration of chlorine and the duration of washing (Figure 4).

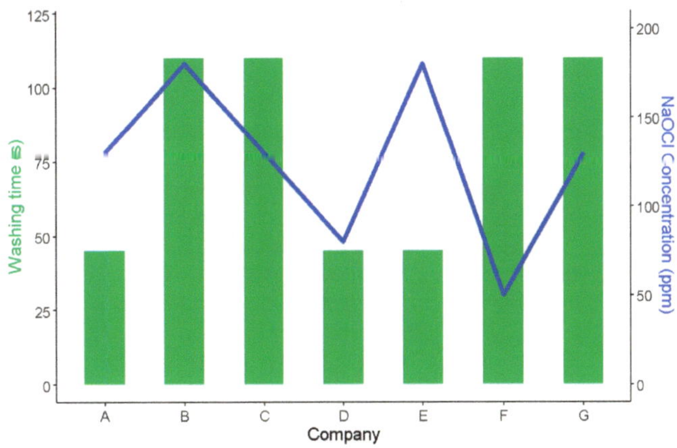

Figure 4. Sodium hypochlorite concentration and washing time used in seven ready-to-eat vegetable industries in Argentina (A–G). The bars represent the washing times, and the line represents the NaOCl concentration.

Addressing these differences in disinfection times and the disinfection concentrations used, optimization of the disinfection process takes on a crucial role. Ensuring the minimization of cross-contamination risks in the wash water while simultaneously curbing the formation of disinfection by-products (DBPs) is a pivotal consideration [57,58]. Consequently, maintaining a stringent and vigilant approach to monitoring the concentration of the disinfectant employed becomes imperative [31]. Also, it is noteworthy that the introduction of higher concentrations of chlorine, while potentially enhancing disinfection efficacy, also leads to heightened levels of DBPs in the wash water [31], requiring a thoughtful balance in their control as a key parameter within the disinfection process.

According to the information provided by the quality managers of the industries, it is revealed that in some companies the pH of the wash water is not monitored (Table 1). It should be noted that there is much research showing that wash water pH is an essential parameter to control since the effective action of chlorine is highly dependent on pH, and its highest efficiency is at pH 7.5 [55,56]. It should also be highlighted that only company F adds citric acid as a pH regulator to adjust the value to 6.0–6.5.

Concluding the analysis of survey results regarding the processing stages, following disinfection, all companies reported conducting a centrifugation process to eliminate the remaining wash water. By implementing this step, companies ensure the removal of excess moisture, which not only enhances the product's shelf life but also contributes to preserving the desired sensory attributes of fresh produce [27,49].

3.2.2. Conditioning Stage: Packaging, In-Plant Storage, and Transport

Packaging under hygienic conditions, carried out immediately after dewatering, plays an essential role in the microbiological protection of fresh-cut products [59,60]. The selection of the material, the conditions generated by the packaging, and the weight/volume relationship of the packaging are very important. For this reason, survey questions pertaining to packaging material and packaging conditions were included. Surprisingly, the findings demonstrate variations in the type of packaging material and conditions utilized, as specified by each industry (Table 1).

Industries A and B use plastic trays with film, without vacuum, while only companies C and G reported the use of plastic bags with modified atmosphere packaging (MAP), which is one of the most recommended methods for this type of products [34,56,61]. Modified atmosphere packaging has been introduced as an enhancement technology to extend shelf life of RTE vegetables, but it cannot always be implemented in small and medium-sized companies [62], which prioritize packaging solutions that require simple and economic technologies and materials.

Regarding in-plant storage, the finished products are stored for 12 h in all cases, except for industry A, where they are stored from 12 to 24 h. After storage, as reported by industry D, the products are transported for 12 h to the point of sale. In contrast, the other industries reported conducting transportation within 6 h. It is important to highlight that the RTE vegetables produced in industries A and B are transported at temperatures exceeding 8 °C. As described above, the temperature which RTE vegetables are exposed to has a direct impact on the sensory and microbiological quality of the products.

Regarding the declared shelf life of the products in the surveys, the finished products can be consumed while ensuring their quality and maintaining an appealing appearance for consumers in terms of color, hydration, crisp texture, and absence of discoloration, within a timeframe from 96 to 216 h (4 to 9 days) (Table 1). The reported findings do not show correlation between the type of packaging atmosphere used, storage and distribution times, and the declared shelf life (Table 1, Figure 5).

The exposure temperatures of leafy vegetables during processing and conditioning are depicted in Figure 6. As observed in the graph, there is significant variability in the temperature values reported by the industries.

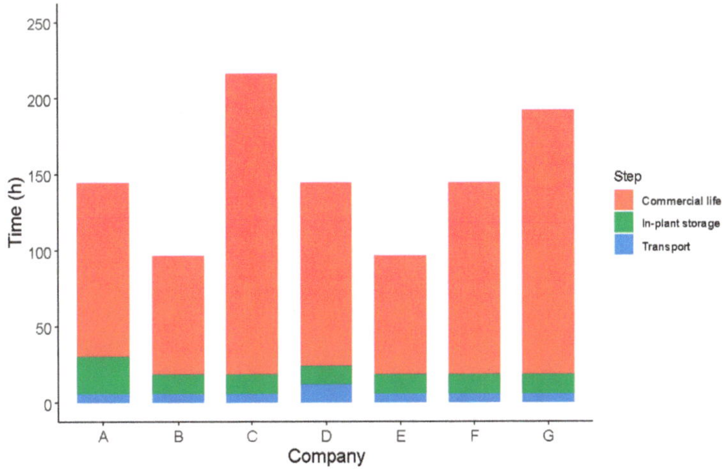

Figure 5. Relationship between packaging, storage and distribution time, and shelf life established for RTE leafy vegetables for industries A–G.

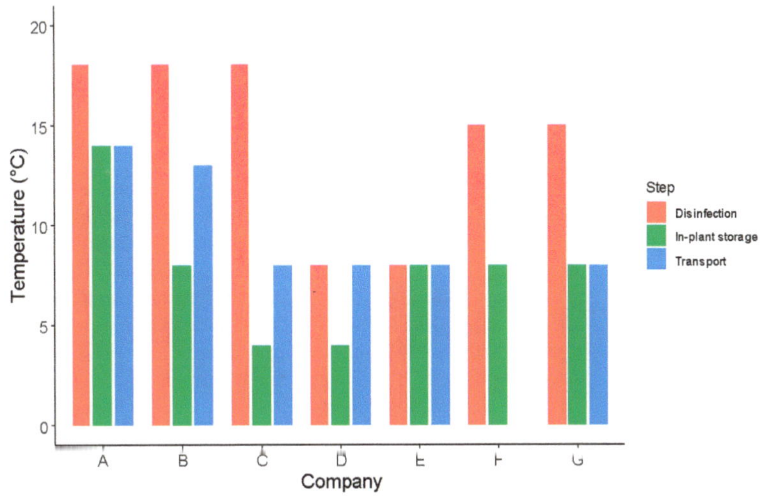

Figure 6. Exposure temperatures of leafy vegetables during processing (disinfection) and conditioning (in-plant storage and distribution) in industries A–G.

Regarding the disinfection process temperature, none of the industries conduct this stage at temperatures below 8 °C. Concerning the storage stage of the final product in the industry, only industries C and D adhere to storage at temperatures below 4 °C, a temperature that inhibits microbial growth and reduces the respiration rate [13,63,64]. Particularly, company A indicated that products are disinfected, stored, and transported at temperatures exceeding 8 °C, posing a potential risk to the final product's quality. This creates a favorable environment for microbial growth [34,61,62,65], which, in turn, impacts the organoleptic quality and the safety of the vegetables [34,48,61,64,66–68].

In addition, many studies have demonstrated that refrigeration is the most convenient and effective method to preserve organoleptic properties and extend the shelf life of fresh products [62,69–71]. However, to maintain a constant low temperature throughout the processing and distribution of RTE vegetables is a challenge and many small industries cannot economically afford to [48].

3.3. Evaluation of *Salmonella* Levels in Fresh-Cut Lettuce Using Predictive Models

3.3.1. Disinfection Efficacy Using Chlorine Washing

The model by Cuggino et al. [36] was applied to obtain estimations of *Salmonella* reductions in fresh-cut lettuce during chlorine washing. Seven different scenarios were considered for model simulations, each one representing the processing conditions of every participating industry (A–G). Table 2 shows the results obtained from the simulations of the predictive model.

Table 2. Reductions in *Salmonella* according to disinfection parameters set by the industries.

Industry	Chlorine Concentration (ppm)	Disinfection Time (s)	Cut Size (cm^2)	Reductions in *Salmonella* (log CFU/g)
A	130	45	16	2.12
B	180	110	12	2.65
C	130	110	12	2.73
D	80	45	9	1.70
E	180	45	16	1.87
F	50	110	18	2.29
G	130	110	16	2.95

As shown in Table 2, *Salmonella* reductions would range from 1.70 to 2.95 log CFU/g, depending on the conditions of the chlorine washing step. In other previous investigations, the effectiveness of chlorine in combination with disinfection time on the concentration of *Salmonella* resulted in similar reductions in fresh-cut produce [37,72]. Additionally, in accordance with the results of model simulations, in the study by Possas et al. [37] which modelled the *Salmonella* inactivation in fresh-cut lettuce, a maximum of 2.6 log-decrease in *Salmonella* levels were computed during chlorine washings at 50 to 150 mg/L for 0 to 2.5-min.

Specifically, the combination of parameters applied in industry G leads to a higher *Salmonella* reduction (i.e., 2.95 log CFU/g). On the contrary, the parameters set by company D show the lowest pathogen reduction (i.e., 1.7 log CFU/g). When comparing in detail the parameters and stages declared by industries A and G, it was observed that an increase in the disinfection time resulted in a greater reduction in *Salmonella*. Moreover, when comparing companies C and G, which apply the same chlorine concentration and washing times, it can be noted that the cut size can exert a slight effect in *Salmonella* levels, with a higher reduction when the size is bigger. This is interesting and evidences the need to consider vegetable surface area, in addition to other well-known intrinsic and extrinsic factors when estimating microbial growth, survival, or inactivation in fresh-cut lettuce.

The provisions of Article 925 of the Food Code [73] in Argentina, and the Commission Regulation N° 2073/2005 [74] in Europe lay down that the absence (non-detection) of *Salmonella* must be ensured in 25 g of RTE vegetables. Thus, in a worst-case scenario of *Salmonella* contamination at 3 log CFU/g in unprocessed lettuce, none of the combinations of the parameters set by the seven industries would ensure the absence of the pathogen in 25 g. Although in company G the estimated *Salmonella* concentration after washing would be close to 0 (i.e., 0.05 log CFU/g), considering the uncertainty in the model predictions, absence of the pathogen cannot be assured.

All steps of RTE vegetable processing play an important role in the quality of the final product, but the washing-disinfection step is key to achieving a reduction in microbial loads, while removing dirtiness and cell exudates [75]. However, if the chlorine concentrations are not well controlled, cross-contamination events during washing may occur and play a paramount role in the microbiological safety of the final products [25].

3.3.2. *Salmonella* Growth Potential during In-Plant Storage and Distribution

The effects of in-plant storage and transport temperatures and times on *Salmonella* growth in fresh-cut lettuce following the chlorine washing step were evaluated by applying the model by Cuggino et al. [43]. The temperatures and storage and distribution times reported by the seven participating industries were used for model simulations (Table 3, Figure 6). This model has been developed with *Salmonella* growth data obtained from fresh-cut lettuce previously subjected to chlorine washing. The results of the model simulations are shown in Table 3.

Table 3. Estimation of *Salmonella* growth in fresh-cut lettuce under different conditions of in-plant storage and transport.

Industries	Concentration of *Salmonella* after Disinfection [1] (CFU/g)	Storage Temperature (°C)	Storage Time (h)	Concentration of *Salmonella* after Storage (CFU/g)	Transport Temperature (°C)	Transportation and Distribution Time (h)	Concentration of *Salmonella* after Transport (CFU/g)
A	0.88	14	24	1.74	14	6	1.95
B	0.35	8	12	0.50	13	6	0.69
C	0.27	4	12	0.27	8	6	0.35
D	1.30	4	12	1.30	8	12	1.45
E	1.13	8	12	1.28	8	6	1.36
F	0.71	8	12	0.86	NR	6	-
G	0.05	8	12	0.20	8	6	0.28

[1] Simulations conducted considering an initial *Salmonella* contamination of 3 CFU/g. NR: not reported.

Since the growth model was developed in MAP fresh-cut lettuce, it was assumed for the simulations that all industries trade their products in this format. In addition, as the model was made for temperatures greater than or equal to 9 °C; a temperature of 9 °C was considered in model simulations for the companies that indicated carrying out storage and distribution at 8 °C, as this could be a possible scenario in the summertime.

Industries C and D indicated that the storage temperature of the products in the plant is 4 °C, in accordance with the recommended storage temperature for these products [76–78]. Therefore, a growth of *Salmonella* would not be expected during storage in companies C and D. However, since the disinfection treatment was not sufficient to fully eliminate *Salmonella* from their products, pathogen growth is expected during product distribution (Table 3). Considering the storage conditions (14 °C for 24 h), the highest increase in *Salmonella* levels would occur in products from company A (i.e., 0.86 CFU/g), compared to the other industries evaluated. On the other hand, the storage conditions of companies B, E, and F would also favor the growth of *Salmonella*.

After in-plant storage, RTE vegetables are distributed to retailers. Table 3 shows that, in all the scenarios evaluated, there would be an increase in the *Salmonella* concentration during the distribution step. In this sense, industry G should ensure no risk of cross-contamination with *Salmonella* during storage, since, after this step, the products are distributed at a higher temperature than recommended.

The application of the predictive microbiology models allowed for estimation of the responses of *Salmonella* in all the scenarios considered. It also highlights the importance of effectively combining an efficient disinfection step with the optimal control of storage and transport temperature of RTE vegetables to ensure the final quality and safety of the product. Overall, the relevance of applying predictive models implemented into user-friendly software lies in their ability to transfer knowledge to all food players involved in food safety, enabling a wide range of industries to make informed decisions, optimize processes, and enhance product quality and safety. More detail on the features and functionalities of MicroHibro can be found in Cubero-González et al. [79]. In addition to allowing a quantitative assessment of microbial responses in foods by means of microbiological models, MicroHibro includes a risk assessment module, a module for the design of sampling plans, and also a module which allows for food shelf-life assessment and estimation by means of quality models. Finally, an overall description of predictive microbiology models and risk assessments developed in recent years for fresh-produce can be found in Possas et al. [25].

4. Conclusions

The data collected in this study offer valuable insights into the diverse processes, methodologies, and practices employed throughout the production chain of RTE leafy vegetables in Argentina. Furthermore, in conjunction with previously published studies, this study plays a pivotal role in the identification and comprehensive evaluation of factors and process parameters that could significantly impact the safety of freshly processed products. The survey results highlight a significant disparity in practices and parameters employed by diverse industries involved in the production of RTE leafy vegetables in Argentina, confirming that knowledge and scientific development are often at odds with the industrial reality. The application of mathematical models enabled us to assess the impact of chlorine disinfection, as well as storage and distribution conditions, on the *Salmonella* levels in RTE lettuce, considering the conditions reported by fresh produce industries. Although chlorine disinfection is a crucial step for controlling pathogens in these products, model simulations confirmed that it is not enough to fully eliminate the pathogen since it can survive and then grow during subsequent storage and distribution stages. Despite the limited availability of public health data pertaining to diseases transmitted by RTE leafy vegetables, this research underscores the valuable role of surveys and predictive microbiology tools in guiding decision making and the evaluation of control measures within the fresh produce industry.

Author Contributions: Conceptualization, S.G.C., A.P., F.P.-R. and M.G.T.; methodology, S.G.C. and A.P.; software, S.G.C. and A.P.; validation, S.G.C. and A.P.; formal analysis, S.G.C. and A.P.; investigation, S.G.C. and A.P.; resources, S.G.C. and A.P.; data curation, S.G.C. and A.P.; writing—original draft preparation, S.G.C. and A.P.; writing—review and editing, S.G.C., A.P., G.D.P.-I., F.P.-R. and M.G.T.; visualization, G.D.P.-I., F.P.-R. and M.G.T.; supervision, F.P.-R. and M.G.T.; project administration, F.P.-R.; funding acquisition, F.P.-R. All authors have read and agreed to the published version of the manuscript.

Funding: This work has been developed as part of the project BioFreshCloud, with reference PRIMA-S2-2019-PCI2020-112015, which is part of the PRIMA programme supported by the European Union and funded by MCIN/AEI/10.13039/501100011033 and European Union "NextGenerationEU/PRTR". Sofia Cuggino is the holder of a doctoral scholarship of the Asociación Universitario Iberoamericana de Posgrado (AUIP). Martín Gustavo Theumer is a career investigator from the National Research Council from Argentina (CONICET).

Data Availability Statement: The data presented in this study are available on request from the corresponding author.

Acknowledgments: The authors would like to thank the industries surveyed for their selfless and committed participation.

Conflicts of Interest: The authors declare that they have no known competing financial interest or personal relationships that could have appeared to influence the work reported in this paper.

References

1. Ministerio de Salud de la Nación. *Guías Alimentarias para la Población Argentina*; Ministerio de Salud de la Nación: Buenos Aires, Argentina, 2020.
2. Food and Agriculture Organization. *Fruit and Vegetables—Your Dietary Essentials*; FAO: Rome, Italy, 2020; ISBN 978-92-5-133709-7.
3. Instituto Nacional de Estadística y Censos—INDEC; Secretaría de Gobierno de Salud de la Nación. *4° Encuesta Nacional de Factores de Riesgo*; Dirección Nacional de Promoción de la Salud y Control de Enfermedades Crónicas No Transmisibles: Buenos Aires, Argentina, 2019.
4. Adrogué, C.; Orlicki, M.E. Factores relacionados al consumo de frutas y verduras en base a la encuesta nacional de factores de riesgo en Argentina. *Rev. Pilquen* **2019**, *22*, 70–82.
5. Giacobone, G.; Castronuovo, L.; Tiscornia, V.; Allemandi, L. *Análisis de la Cadena de Suministro de Frutas y Verduras en Argentina*; Fundación InterAmericana del Corazón-Argentina: Buenos Aires, Argentina, 2018.

6. Castagnino, A.M.; Díaz, K.; Fernández Lozano, J.; Guisolis, A.; Liverotti, O.; Rosini, M.B.; Sasale, S. Panorama of the Argentine Horticultural Sector: Characterization and Priorities of National Horticulture. *Hortic. Argent.* **2020**, *39*, 76–102.
7. Castronuevo, L.; Tiscornia, M.V.; Gutkowski, P.; Allemadi, L. Perceived barriers and facilitators of fruit and vegetable consumption in Argentina: A Qualitative Study. *Rev. Argent. Salud Pública* **2019**, *10*, 14–21.
8. Garcia, G.; Vàzquez, L. *Guía de Prácticas Correctas de Higiene para Vegetales y Derivados, Frescos, Mondados, Troceados o Envasados;* Agencia de Salud Pública de Cataluña, Ed.; Agencia de Salud Pública de Catalunya: Cataluña, Spain, 2015.
9. Galizio, R.Y.; Diaz, K. Tecnología de los productos hortofrutícolas: Hortalizas procesadas (IV Gama). *Hortic. Argent.* **2020**, *39*, 1851–9342.
10. Rodríguez, S.D.C.; Gutiérrez, D.R.; Torales, A.C.; Questa, A.G. Vegetales IV Gama: Producción, comercialización y aspectos sanitarios en la región Noroeste Argentino y en la Argentina. In *Aportes de la Faya Para el Desarrollo Agropecuario y Agroindustrial del Noa;* Orientación Grafica: Buenos Aires, Argentina, 2017; pp. 137–147. ISBN 978-987-1676-68-2.
11. Rojas-Graü, M.A.; Garner, E.; Martin-Belloso, O. The fresh-cut fruit and vegetables industry current situation and market trends. In *Advances in Fresh-Cut Fruits and Vegetables Processing*, 1st ed.; Martín-Belloso, O., Soliva-Fortuny, R., Eds.; CRC Press: BocaRaton, FL, USA, 2011; pp. 1–12.
12. Ma, L.; Zhang, M.; Bhandari, B.; Gao, Z. Recent developments in novel shelf life extension technologies of fresh-cut fruits and vegetables. *Trends Food Sci. Technol.* **2017**, *64*, 23–38. [CrossRef]
13. Mir, S.A.; Shah, M.A.; Mir, M.M.; Dar, B.N.; Greiner, R.; Roohinejad, S. Microbiological contamination of ready-to-eat vegetable salads in developing countries and potential solutions in the supply chain to control microbial pathogens. *Food Control* **2018**, *85*, 235–244. [CrossRef]
14. Yousuf, B.; Deshi, V.; Ozturk, B.; Siddiqui, M.W. Fresh-Cut fruits and vegetables: Quality issues and safety concerns. In *Fresh-Cut Fruits and Vegetables;* Siddiqui, M.W., Ed.; Academic Press: Waltham, MA, USA, 2020; pp. 1–15. ISBN 978-0-12-816184-5. [CrossRef]
15. Callejón, R.M.; Rodríguez-Naranjo, M.I.; Ubeda, C.; Hornedo-Ortega, R.; Garcia-Parrilla, M.C.; Troncoso, A.M. Reported foodborne outbreaks due to fresh produce in the united states and european union: Trends and causes. *Foodborne Pathog. Dis.* **2015**, *12*, 32–38. [CrossRef]
16. EFSA. Scientific Opinion on Establishing Food-Based Dietary Guidelines. *EFSA J.* **2010**, *8*, 1460. [CrossRef]
17. González, J.; Cadona, J.S.; Sanz, M.; Bustamante, A.V.; Sanso, A.M. Molecular characterization of diarrheagenic *Escherichia coli* isolated from vegetables in Argentina. *Int. J. Food Microbiol.* **2017**, *261*, 57–61. [CrossRef]
18. Moretti, C.I.; Rodríguez, L.E. *Cumplimiento de las Normativas del Código Alimentario Argentino en Cuanto a Calidad Microbiológica, Física y Organoléptica en Bandejas Mixtas de Repollo Blanco y Morado, Zanahoria y Lechuga/Achicoria Expendidas en Comercios de la Ciudad de Córdoba;* Universidad Nacional de Córdoba, Facultad de Ciencias Médicas Escuela de Nutrición: Córdoba, Argentina, 2020.
19. Gentili, A.; Marzocca, M.; Oriani, S.; Baldini, M. Calidad bacteriologica de ensaladas de zanahoria rallada y eficacia de tratamientos previos a su consumo. *RESPYN Rev. Salud Pública Nutr.* **2017**, *16*, 9–15. [CrossRef]
20. Gómez Albanys, D.; Toledo Lisette, S.; Quintero Giovanna, B.; Donado Yadira, H.; Roo Yeiny, Á.; Leal Kutchynskaya, V. Calidad Microbiológica de Ensaladas Crudas que se Expenden en Puestos Ambulantes de Comida Rápida de la Ciudad de Maracaibo-Venezuela. *Kasmera* **2018**, *46*, 116–126.
21. Maistro, L.C.; Miya, N.T.N.; Sant'Ana, A.S.; Pereira, J.L. Microbiological quality and safety of minimally processed vegetables marketed in Campinas, SP—Brazil, as Assessed by traditional and alternative methods. *Food Control* **2012**, *28*, 258–264. [CrossRef]
22. de Oliveira, M.A.; Maciel de Souza, V.; Morato Bergamini, A.M.; De Martinis, E.C.P. Microbiological Quality of Ready-to-Eat Minimally Processed Vegetables Consumed in Brazil. *Food Control* **2011**, *22*, 1400–1403. [CrossRef]
23. Pang, H.; Lambertini, E.; Buchanan, R.L.; Schaffner, D.W.; Pradhan, A.K. Quantitative microbial risk assessment for *Escherichia coli* O157:H7 in Fresh-Cut Lettuce. *J. Food Prot.* **2017**, *80*, 302–311. [CrossRef] [PubMed]
24. Bozkurt, H.; Bell, T.; van Ogtrop, F.; Phan-Thien, K.-Y.; McConchie, R. Assessment of Microbial risk during Australian industrial practices for *Escherichia coli* O157:H7 in fresh cut-cos lettuce: A stochastic quantitative approach. *Food Microbiol.* **2021**, *95*, 103691. [CrossRef]
25. Possas, A.; Pérez-Rodríguez, F. New insights into cross-contamination of fresh-produce. *Curr. Opin. Food Sci.* **2023**, *49*, 100954. [CrossRef]
26. Beharielal, T.; Thamaga-Chitja, J.; Schmidt, S. Pre-and Post-Harvest practices of smallholder farmers in rural Kwazulu-natal, South Africa: Microbiological quality and potential market access implications. *Food Control* **2018**, *92*, 53–62. [CrossRef]
27. Castro-Ibáñez, I.; López-Gálvez, F.; Gil, M.I.; Allende, A. Identification of sampling points suitable for the detection of microbial contamination in fresh-cut processing lines. *Food Control* **2016**, *59*, 841–848. [CrossRef]
28. Goodburn, C.; Wallace, C.A. The Microbiological Efficacy of Decontamination methodologies for fresh produce: A review. *Food Control* **2013**, *32*, 418–427. [CrossRef]
29. Koseki, S. Risk Assessment of microbial and chemical contamination in fresh produce. In *Woodhead Publishing Series in Food Science, Technology and Nutrition, Global Safety of Fresh Produce;* Woodhead Publishing: Cambridge, UK, 2014; pp. 153–171. ISBN 9781782420187.
30. Gil, M.I.; Selma, M.V.; Suslow, T.; Jacxsens, L.; Uyttendaele, M.; Allende, A. Pre- and postharvest preventive measures and intervention strategies to control microbial food safety hazards of fresh leafy vegetables. *Crit. Rev. Food Sci. Nutr.* **2015**, *55*, 453–468. [CrossRef]

31. López-Gálvez, F.; Tudela, J.A.; Allende, A.; Gil, M.I. Microbial and chemical characterization of commercial washing lines of fresh produce highlights the need for process water control. *Innov. Food Sci. Emerg. Technol.* **2019**, *51*, 211–219. [CrossRef]
32. Meireles, A.; Giaouris, E.; Simões, M. Alternative disinfection methods to chlorine for use in the fresh-cut industry. *Food Res. Int.* **2016**, *82*, 71–85. [CrossRef]
33. Posada-Izquierdo, G.D.; Zurera, G.; Pérez-Rodríguez, F. *Quantitative Microbial Risk Assessment Methods for Food Safety in rte Fresh Vegetables*; Ravishankar Rai, J.V., Bai, A., Eds.; CRC Press Taylor & Francis Group: Boca Raton, FL, USA, 2014; ISBN 9781466593060.
34. Posada-Izquierdo, G.D.; Pérez-Rodríguez, F.; López-Gálvez, F.; Allende, A.; Selma, M.V.; Gil, M.I.; Zurera, G. Modelling growth of *Escherichia coli* O157:H7 in fresh-cut lettuce submitted to commercial process conditions: Chlorine washing and modified atmosphere packaging. *Food Microbiol.* **2013**, *33*, 131–138. [CrossRef]
35. Van Haute, S.; López-Gálvez, F.; Gómez-López, V.M.; Eriksson, M.; Devlieghere, F.; Allende, A.; Sampers, I. Methodology for modeling the disinfection efficiency of fresh-cut leafy vegetables wash water applied on peracetic acid combined with lactic acid. *Int. J. Food Microbiol.* **2015**, *208*, 102–113. [CrossRef] [PubMed]
36. Cuggino, S.G.; Bascón-Villegas, I.; Rincón, F.; Pérez, M.A.; Posada-Izquierdo, G.; Marugán, J.; Pablos Carro, C.; Pérez-Rodríguez, F. Modelling the combined effect of chlorine, benzyl isothiocyanate, exposure time and cut size on the reduction of *Salmonella* in fresh-cut lettuce during washing process. *Food Microbiol.* **2020**, *86*, 103346. [CrossRef]
37. Possas, A.; Posada-Izquierdo, G.D.; Tarlak, F.; Jiménez-Jiménez, F.; Pérez-Rodríguez, F. Inactivation of *Salmonella* Typhimurium in fresh-cut lettuce during chlorine washing: Assessing the impacts of free chlorine concentrations and exposure times. *LWT* **2023**, *184*, 115069. [CrossRef]
38. Allende, A.; Bover-Cid, S.; Fernández, P.S. Challenges and opportunities related to the use of innovative modelling approaches and tools for microbiological food safety management. *Curr. Opin. Food Sci.* **2022**, *45*, 100839. [CrossRef]
39. Pérez-Rodríguez, F.; Valero, A. Predictive microbiology in foods. In *Predictive Microbiology in Foods*; Springer: New York, NY, USA, 2013; pp. 1–10.
40. Ross, T.; McMeekin, T.A.; Baranyi, J. Predictive microbiology and food safety. In *Encyclopedia of Food Microbiology*; Academic Press: Cambridge, MA, USA, 2014; pp. 59–68. [CrossRef]
41. R Core Team. *A Language and Environment for Statistical Computing*; R Core Team: Vienna, Austria, 2020.
42. Di Rienzo, J.A.; Casanoves, F.; Balzarini, M.G.; Gonzalez, L.; Tablada, M.; Robledo, C.W. InfoStat Versión 2017. Grupo InfoStat, FCA, Universidad Nacional de Córdoba, Argentina. Available online: http://www.infostat.com.ar (accessed on 1 January 2021).
43. Cuggino, S.G.; Posada-Izquierdo, G.; Bascón Villegas, I.; Gustavo Theumer, M.; Pérez-Rodríguez, F. Effects of chlorine and peroxyacetic acid wash treatments on growth kinetics of *Salmonella* in fresh-cut lettuce. *Food Res. Int.* **2023**, *167*, 112451. [CrossRef]
44. Baranyi, J.; Roberts, T.A. A Dynamic approach to predicting bacterial growth in food. *Int. J. Food Microbiol.* **1994**, *23*, 277–294. [CrossRef]
45. Ross, T.; Dalgaard, P. Secondary Models. In *Modelling Microbial Responses in Food*; McKellar, R.C., Lu, X., Eds.; CRC Press: New York, NY, USA, 2004; pp. 63–150.
46. Sant'Ana, A.S.; Landgraf, M.; Destro, M.T.; Franco, B.D.G.M. Prevalence and counts of *Salmonella* spp. in minimally processed vegetables in São Paulo, Brazil. *Food Microbiol.* **2011**, *28*, 1235–1237. [CrossRef]
47. Jideani, A.I.O.; Anyasi, T.A.; Mchau, G.R.A.; Udoro, E.O.; Onipe, O.O. Processing and preservation of fresh-cut fruit and vegetable products. *Postharv. Handl.* **2017**, *47*, 69763. [CrossRef]
48. Castro-Ibáñez, I.; Gil, M.I.; Allende, A. Ready-to-Eat vegetables: Current problems and potential solutions to reduce microbial risk in the production chain. *LWT Food Sci. Technol.* **2017**, *85*, 284–292. [CrossRef]
49. Raffo, A.; Paoletti, F. Fresh-Cut Vegetables Processing: Environmental sustainability and food safety issues in a comprehensive perspective. *Front. Sustain. Food Syst.* **2022**, *5*, 681459. [CrossRef]
50. Maffei, D.F.; Alvarenga, V.O.; Sant'Ana, A.S.; Franco, B.D.G.M. Assessing the Effect of washing practices employed in brazilian processing plants on the quality of ready-to-eat vegetables. *LWT Food Sci. Technol.* **2016**, *69*, 474–481. [CrossRef]
51. De Corato, U. Improving the shelf-life and quality of fresh and minimally-processed fruits and vegetables for a modern food industry: A comprehensive critical review from the traditional technologies into the most promising advancements. *Crit. Rev. Food Sci. Nutr.* **2019**, *60*, 940–975. [CrossRef]
52. López-Gálvez, F.; Allende, A.; Selma, M.V.; Gil, M.I. Prevention of *Escherichia coli* cross-contamination by different commercial sanitizers during washing of fresh-cut lettuce. *Int. J. Food Microbiol.* **2009**, *133*, 167–171. [CrossRef]
53. Weng, S.; Luo, Y.; Li, J.; Zhou, B.; Jacangelo, J.G.; Schwab, K.J. Assessment and speciation of chlorine demand in fresh-cut produce wash water. *Food Control* **2016**, *60*, 543–551. [CrossRef]
54. Van Haute, S.; Sampers, I.; Holvoet, K.; Uyttendaele, M. Physicochemical quality and chemical safety of chlorine as a reconditioning agent and wash water disinfectant for fresh-cut lettuce washing. *Appl. Environ. Microbiol.* **2013**, *79*, 2850–2861. [CrossRef]
55. Chen, X.; Hung, Y.-C. Effects of organic load, sanitizer ph and initial chlorine concentration of chlorine-based sanitizers on chlorine demand of fresh produce wash waters. *Food Control* **2017**, *77*, 96–101. [CrossRef]
56. Ramos, B.; Miller, F.A.; Brandão, T.R.S.; Teixeira, P.; Silva, C.L.M. Fresh fruits and vegetables—An overview on applied methodologies to improve its quality and safety. *Innov. Food Sci. Emerg. Technol.* **2013**, *20*, 1–15. [CrossRef]

57. Gómez-López, V.M.; Lannoo, A.-S.; Gil, M.I.; Allende, A. Minimum free chlorine residual level required for the inactivation of *Escherichia coli* o157:h7 and trihalomethane generation during dynamic washing of fresh-cut spinach. *Food Control* **2014**, *42*, 132–138. [CrossRef]
58. Zhou, B.; Luo, Y.; Nou, X.; Millner, P. Development of an algorithm for feed-forward chlorine dosing of lettuce wash operations and correlation of chlorine profile with *Escherichia coli* O157:H7 inactivation. *J. Food Prot.* **2014**, *77*, 558–566. [CrossRef] [PubMed]
59. Turatti, A. Process design, facility and equipment requirements. In *Advances in Fresh Cut Fruits and Vegetables Processing*; Food Preservation Technology Series; Taylor & Francis Group: Abingdon-on-Thames, UK, 2011; pp. 339–361.
60. WHO/FAO (World Health Organization/Food and Agriculture Organization). Exposure assessment of microbiological hazards in food. In *Microbiological Risk Assessment Series 7*; WHO: Geneva, Switzerland; FAO: Rome, Italy, 2008.
61. Oliveira, M.; Usall, J.; Solsona, C.; Alegre, I.; Viñas, I.; Abadias, M. Effects of packaging type and storage temperature on the growth of foodborne pathogens on shredded 'romaine' lettuce. *Food Microbiol.* **2010**, *27*, 375–380. [CrossRef]
62. Arienzo, A.; Murgia, L.; Fraudentali, I.; Gallo, V.; Angelini, R.; Antonini, G. Microbiological quality of ready-to-eat leafy green salads during shelf-life and home-refrigeration. *Foods* **2020**, *9*, 1421. [CrossRef] [PubMed]
63. EFSA. European Food Safety Authority. Scientific opinion on the risk posed by pathogens in food of non-animal origin. Part 2 (*Salmonella* and Norovirus in tomatoes). *EFSA J.* **2014**, *12*, 3832. [CrossRef]
64. Sant'Ana, A.S.; Franco, B.D.G.M.; Schaffner, D.W. Modeling the growth rate and lag time of different strains of *Salmonella enterica* and *Listeria monocytogenes* in ready-to-eat lettuce. *Food Microbiol.* **2012**, *30*, 267–273. [CrossRef]
65. Kim, Y.J.; Park, J.Y.; Suh, S.H.; Kim, M.G.; Kwak, H.S.; Kim, S.H.; Heo, E.J. Development and validation of a predictive model for pathogenic *Escherichia coli* in fresh-cut produce. *Food Sci. Nutr.* **2021**, *9*, 6866–6872. [CrossRef]
66. de Frias, J.A.; Luo, Y.; Zhou, B.; Turner, E.R.; Millner, P.D.; Nou, X. Minimizing pathogen growth and quality deterioration of packaged leafy greens by maintaining optimum temperature in refrigerated display cases with doors. *Food Control* **2018**, *92*, 488–495. [CrossRef]
67. Luo, Y.; He, Q.; McEvoy, J.L. Effect of storage temperature and duration on the behavior of *Escherichia coli* O157:H7 on packaged fresh-cut salad containing romaine and iceberg lettuce. *J. Food Sci.* **2010**, *75*, M390–M397. [CrossRef]
68. Tsironi, T.; Dermesonlouoglou, E.; Giannoglou, M.; Gogou, E.; Katsaros, G.; Taoukis, P. Shelf-Life prediction models for ready-to-eat fresh cut salads: Testing in real cold chain. *Int. J. Food Microbiol.* **2017**, *240*, 131–140. [CrossRef]
69. Hoffmann, T.G.; Ronzoni, A.F.; da Silva, D.L.; Bertoli, S.L.; de Souza, C.K. Impact of household refrigeration parameters on postharvest quality of fresh food produce. *J. Food Eng.* **2021**, *306*, 110641. [CrossRef]
70. Maffei, D.F.; Sant'Ana, A.S.; Franco, B.D.G.M.; Schaffner, D.W. Quantitative assessment of the impact of cross-contamination during the washing step of ready-to-eat leafy greens on the risk of illness caused by *Salmonella*. *Food Res. Int.* **2017**, *92*, 106–112. [CrossRef]
71. Manzocco, L.; Alongi, M.; Lagazio, C.; Sillani, S.; Nicoli, M.C. Effect of temperature in domestic refrigerators on fresh-cut iceberg salad quality and waste. *Food Res. Int.* **2017**, *102*, 129–135. [CrossRef]
72. Huang, R.; Chen, H. Evaluation of inactivating *Salmonella* on iceberg lettuce shreds with washing process in combination with pulsed light, ultrasound and chlorine. *Int. J. Food Microbiol.* **2018**, *285*, 144–151. [CrossRef] [PubMed]
73. Código Alimentario Argentino CAA. Capítulo XI. Alimentos Vegetales. Artículo 925 Quarter. Available online: http://www.anmat.gov.ar/alimentos/codigoa/Capitulo_XI.pdf (accessed on 10 June 2022).
74. European Commission. Commission Regulation (EC) No 2073/2005 of 15 November 2005 on microbiological criteria for foodstuffs. *Off. J. Eur. Union* **2005**, *50*, 1–26.
75. D'Acunzo, F.; Del Cimmuto, A.; Marinelli, L.; Aurigemma, C.; De Giusti, M. Ready-to-eat vegetables production with low-level water chlorination. An evaluation of water quality and of its impact on end products. *Ann. Ist. Super. Sanita* **2012**, *48*, 151–160. [CrossRef]
76. FDA. *FDA Food Code*; FDA: Silver Spring, MD, USA, 2012.
77. FDA. *Guidance for Industry: Guide to Minimize Microbial Food Safety Hazards of Fresh-Cut Fruits and Vegetables*; FDA: Silver Spring, MD, USA, 2008.
78. Rediers, H.; Claes, M.; Peeters, L.; Willems, K.A. Evaluation of the cold chain of fresh-cut endive from farmer to plate. *Postharv. Biol. Technol.* **2009**, *51*, 257–262. [CrossRef]
79. González, S.C.; Possas, A.; Carrasco, E.; Valero, A.; Bolívar, A.; Posada-Izquierdo, G.D.; García-Gimeno, R.M.; Zurera, G.; Pérez-Rodríguez, F. "MicroHibro": A Software tool for predictive microbiology and microbial risk assessment in foods. *Int. J. Food Microbiol.* **2019**, *290*, 226–236. [CrossRef]

foods

MDPI

Article

Dynamic Thermal Treatments in Green Coconut Water Induce Dynamic Stress Adaptation of *Listeria innocua* That Increases Its Thermal Resistance

Gerardo A. González-Tejedor [1,2], Alberto Garre [3], Asunción Iguaz [3], Ricardo Wong-Zhang [2], Pablo S. Fernández [3,*] and Arícia Possas [4,*]

1 Sistema Nacional de Investigación (SNI), Senacyt, Ciudad de Panamá, Panama; gerardo.gonzalez@utp.ac.pa
2 Centro de Producción e Investigaciones Agroindustriales, Universidad Tecnológica de Panamá, Ciudad de Panamá, Panama; ricardo.wong2@utp.ac.pa
3 Departamento de Ingeniería Agronómica, Instituto de Biotecnología Vegetal, Universidad Politécnica de Cartagena, 30202 Cartagena, Spain; alberto.garre@upct.es (A.G.); asun.iguaz@upct.es (A.I.)
4 Department of Food Science and Technology, UIC Zoonosis y Enfermedades Emergentes ENZOEM, CeiA3, Universidad de Córdoba, 14014 Córdoba, Spain
* Correspondence: pablo.fernandez@upct.es (P.S.F.); g12mepoa@uco.es (A.P.)

Abstract: The global coconut water market is projected to grow in the upcoming years, attributed to its numerous health benefits. However, due to its susceptibility to microbial contamination and the limitations of non-thermal decontamination methods, thermal treatments remain the primary approach to ensure the shelf-life stability and the microbiological safety of the product. In this study, the thermal inactivation of *Listeria innocua*, a *Listeria monocytogenes* surrogate, was evaluated in coconut water and in tryptone soy broth (TSB) under both isothermal (50–60 °C) and dynamic conditions (from 30 to 60 °C, with temperature increases of 0.5, 1 and 5 °C/min). Mathematical models were used to analyse the inactivation data. The Geeraerd model effectively described the thermal inactivation of *L. innocua* in both TSB and coconut water under isothermal conditions, with close agreement between experimental data and model fits. Parameter estimates and analysis revealed that acidified TSB is a suitable surrogate medium for studying the thermal inactivation of *L. innocua* in coconut water, despite minor differences observed in the shoulder length of inactivation curves, likely attributed to the media composition. The models fitted to the data obtained at isothermal conditions fail to predict *L. innocua* responses under dynamic conditions. This is attributed to the stress acclimation phenomenon that takes place under dynamic conditions, where bacterial cells adapt to initial sub-lethal treatment stages, leading to increased thermal resistance. Fitting the Bigelow model directly to dynamic data with fixed z-values reveals a three-fold increase in D-values with lower heating rates, supporting the role of stress acclimation. The findings of this study aid in designing pasteurization treatments targeting *L. innocua* in coconut water and enable the establishment of safe, mild heat treatments for refrigerated, high-quality coconut water.

Keywords: inactivation; plant-based beverages; stress response; foodborne pathogens; modelling

Citation: González-Tejedor, G.A.; Garre, A.; Iguaz, A.; Wong-Zhang, R.; Fernández, P.S.; Possas, A. Dynamic Thermal Treatments in Green Coconut Water Induce Dynamic Stress Adaptation of *Listeria innocua* That Increases Its Thermal Resistance. *Foods* **2023**, *12*, 4015. https://doi.org/10.3390/foods12214015

Academic Editors: Wei Li and Danijela Bursać Kovačević

Received: 29 September 2023
Revised: 23 October 2023
Accepted: 27 October 2023
Published: 3 November 2023

1. Introduction

Coconut water is the sweet and transparent fluid that can be found inside young coconuts (*Cocos nucifera* L.) [1]. This plant-based product is mainly composed of water (around 95%), and also contains sugars such as sucrose, glucose or fructose and small amounts of proteins and lipids [2,3]. It also contains calcium, magnesium, potassium, sodium or phosphorus, among other inorganic ions, as well as vitamins such as vitamin C, thiamine, riboflavin, niacin, pantothenic acid, pyridoxine and folates [4].

The global coconut water market is expected to grow with a compound annual growth rate of 10.8% from 2023 to 2028 [5]. As a plant-based product, it has gained popularity in recent years, offering a myriad of health benefits to consumers. The nutrients present in coconut water include active components that could prevent oxidative stress [6,7]. Moreover, this refreshing drink possesses anti-inflammatory activity, cardio-protectant properties, and has shown to improve the lipidic profile and reduce high blood pressure [8–11]. It has also been used for oral rehydration, besides being proposed as an intravenous solution to hydrate patients in emergency situations [10,12].

Due to its nutritious and physicochemical characteristics (e.g., pH 5.6), once the water is extracted from coconuts, it can be easily deteriorated by the activity of microorganisms and enzymes [13]. Additionally, this beverage can become contaminated with bacterial pathogens, such as spore-forming bacteria, *Salmonella* spp. and *Listeria monocytogenes*, during harvest, since coconuts are usually placed on the soil, and during handling and extraction [14]. Once contaminated, these pathogens could persist over the coconut shelf-life. For instance, *L. monocytogenes* has shown great growth capacity in the product at temperatures ranging from 4 to 35 °C [14].

Different non-thermal decontamination techniques for coconut water have been investigated, including pulsed light [15], UV-C light [16], atmospheric cold plasma [17] and high hydrostatic pressure, among others [13]. These non-thermal treatments have shown different levels of effectiveness, but experience limitations for their commercial implementation, such as high costs and scaling-up [13]. Despite the negative effects of thermal treatments on coconut water quality, they remain as the most applied method to mitigate microbial contamination in this beverage, mainly for yielding shelf-stable products [2,15].

In most industrial heat treatment processes, foods are subjected to dynamic heating, meaning that temperatures are not immediately raised from room to pasteurization values [18]. Mathematical models have been applied to describe *L. monocytogenes* thermal inactivation kinetics in different food products under isothermal and dynamic conditions, reflecting real life scenarios. In addition, the empirical research has demonstrated that models relying on isothermal data frequently fail in accurately predicting microbial inactivation when dynamic conditions are involved [19,20].

In situations involving dynamic heating, microbial populations may adapt to stress conditions [18,21]. This adaptation involves bacterial cells activating a range of stress response mechanisms that enhance their chances of surviving. When the heating rate is slow, these mechanisms may kick in during the early stages of dynamic treatment, leading to an adaptation to stress and greater resistance to the lethal phase of treatment than what would be anticipated based solely on isothermal experiments [18].

Therefore, this study develops a dynamic model to predict the response of *L. monocytogenes* during the pasteurization of coconut milk, designed to be combined with subsequent refrigeration. The model is based on *Listeria innocua*, a common surrogate for *L. monocytogenes* [22,23]. Due to its ability to predict the number of bacterial reductions for different temperature profiles, accounting for dynamic stress acclimation, the model can be a valuable tool for the design of thermal pasteurization treatments for this product.

2. Materials and Methods

2.1. Substrate

Coconuts (*Cocos nucifera*) were obtained from a local producer from Panama. They were washed with chlorinated water, and coconut water was extracted mechanically with a tool that perforated the external parts and allowed for the extraction. It was collected in a sterile recipient and used directly for heat resistance experiments or stored in a refrigerator until use.

2.2. Microorganism and Preparation

Listeria innocua serovar 6a (CECT 910) was obtained as a lyophilized culture from the Spanish Type Culture Collection following the methodology previously described [24]. A stock containing approximately 10^9 CFU/mL was prepared. It was mixed with glycerol (30%) and kept frozen at −80 °C until use.

To perform the experiments, *L. innocua* was grown overnight from the frozen stock, adding 200 μL to 50 mL of Tryptone Soya Broth (TSB; Scharlab Chemie S.A., Barcelona, Spain). It was then incubated for 12 h to reach the stationary phase with a concentration of 10^8 CFU/mL, that was used as inoculum for heat resistance experiments. The cell concentration was confirmed by viable plate counts, performing decimal dilutions in peptone water, and growing the samples in Tryptone Soya Agar (TSA; Scharlab Chemie S.A., Barcelona, Spain) for 48 h at 37 °C. Plates containing between 30 and 300 CFU were then counted.

2.3. Microbial Inactivation Experiments

A 400 mL volume of TSB or coconut water was introduced in the vessel of a thermoresistometer Mastia [25], previously sterilized, and heated at the target temperature. Then, a 200 μL volume of the *L. innocua* suspension was injected with a Hamilton pipette into the vessel and samples were taken at the time intervals selected. The temperatures for isothermal inactivation experiments were 50, 52.5, 55, 57.5 and 60 °C. For non-isothermal conditions, ramps starting at 30 °C up to 57.5 or 60 °C were programmed, with temperature increases of 0.5, 1 and 5 °C/min. Samples were cooled immediately and viable plate counts were performed as explained above.

2.4. Microbial Inactivation under Isothermal Conditions

The data on microbial inactivation of *L. innocua* on both media were analysed using the Geeraerd model without a tail [26] due to the nonlinearity of the survivor curves. This model describes the relationship between the microbial concentration (N) and the treatment time (t), as shown in Equation (1), where SL is the shoulder length and k is the inactivation rate during the exponential phase.

$$N = N_0 \cdot e^{-k \cdot t} \cdot e^{k \cdot SL} / \left(1 + \left(e^{k \cdot SL} - 1\right) \cdot e^{-k \cdot t}\right) \tag{1}$$

To ease the interpretation, the models were fitted using the D-value (D), which represents the treatment time required to reduce the microbial concentration by 1 decimal logarithm, instead of the inactivation rate (k), using the identity $D = \ln(10)/k$.

We used a Bigelow-type secondary model to describe the relationship between the D-value and the treatment temperature, as shown in Equation (2). This model introduces a reference temperature (T_{ref}) that improves the parameter identifiability but without any biological meaning. This parameter was set at the mean of the temperature range (55 °C), as previously suggested [27]. Then, this model is defined by the D-value at the reference temperature (D_{ref}) and the z-value (z), which defines the temperature increase required to reduce the D-value by 90%.

$$logD = logD_{ref} - \frac{T - T_{ref}}{z} \tag{2}$$

The relationship between the shoulder length duration (SL) and the storage temperature was described using the model from Equation (3), where a and b are two unknown regression coefficients.

$$logSL = a - b\left(T - T_{ref}\right) \tag{3}$$

Primary inactivation models were fitted to the isothermal data using the functions included in the *bioinactivation* package for R [28], using its web interface [29] currently available at: https://foodlab-upct.shinyapps.io/bioinactivation4/ (accessed on 4 August 2023). Then, the relationship between the model parameters and the treatment temperature was evaluated visually to suggest secondary models. Finally, a global inactivation model was fitted using a one-step fitting algorithm to the complete isothermal dataset, as this approach often provides more accurate parameter estimates than the two-step method [30].

The global model was fitted using R version 4.2.3 [31] using the Levenberg–Marquard algorithm [32] implemented in the *minpack.lm* package version 1.2-3 [33].

2.5. Microbial Inactivation under Dynamic Heating Conditions

Microbial inactivation under dynamic conditions was studied using both the Bigelow and Geeraerd inactivation models. The Bigelow model (Equation (4)) is a generalization of the Bigelow model for isothermal conditions (Equation (1)), so it assumes first-order inactivation kinetics. The inactivation rate is quantified by the D-value (D), whose relationship with temperature is given by the same secondary model as for isothermal conditions (Equation (2)).

$$dN/dt = -\ln(10)/D(T)\cdot N \tag{4}$$

The Geeraerd inactivation model [24] is an extension of first-order kinetics based on similar arguments as the Baranyi growth model [34]. As shown in Equation (5), it introduces two correction factors in the differential equation: one to describe the shoulder (α) and one for the tail (β).

$$dN/dt = -\alpha\cdot k\cdot \beta\cdot N \tag{5}$$

The shoulder is supported on a theoretical substance (C) that must be inactivated before bacterial inactivation takes place. In this model, it is assumed that C follows first-order kinetics (Equation (6)) and affects the population inactivation rate as shown in Equation (7). Accordingly, the shoulder length is defined by the initial value of this theoretical substance (C_0) by the identity shown in Equation (8).

$$dC/dt = -k\cdot C \tag{6}$$

$$\alpha = 1/(1 + C) \tag{7}$$

$$SL = (\ln (C_0) + 1)/k \tag{8}$$

The tail is described in the Geeraerd model assuming that there is a horizontal asymptote on the bacterial concentration (N_{res}) and that the population is self-regulated by a logistic term (Equation (9)).

$$\beta = (1 - N_{res}/N) \tag{9}$$

The parameter values estimated from isothermal conditions (Section 2.4) were used for making predictions under dynamic conditions, comparing them against the experimental data. This was carried out using the web version of *bioinactivation* (available at https://foodlab-upct.shinyapps.io/bioinactivation4/, accessed on 4 August 2023), which solves the differential equation numerically. Due to the lack of a tail in the experimental data, parameter β was fixed to one. Regarding the shoulder length, it is not clear how isothermal results can be extrapolated to dynamic conditions, due to the latter being defined by C_0 (and different SL resulting in different C_0; Equation (8)). Consequently, predictions were calculated for the maximum and minimum values of C_0 corresponding to the shoulder lengths estimated from the isothermal data.

Moreover, to further analyse the differences between predictions based on isothermal conditions and dynamic observations, the dynamic Bigelow model was fitted to the dynamic observations using the functions included in *bioinactivation*. To simplify comparison against isothermal models, the models were fitted fixing the z-value to the one estimated from isothermal conditions.

3. Results and Discussion

3.1. Inactivation of *Listeria innocua* in TSB and Coconut Water under Isothermal Conditions

The Geeraerd model was successful at describing the thermal inactivation of *L. innocua* in both TSB and coconut water under isothermal conditions at all the temperatures tested.

Figure 1 illustrates both the experimental data and the fitted models, showing a good agreement between the data and model fits. These plots also show that the inactivation kinetics in TSB were similar to those in coconut water, evidencing that acidified TSB could be a good surrogate medium for the thermal inactivation of *L. innocua* in coconut water.

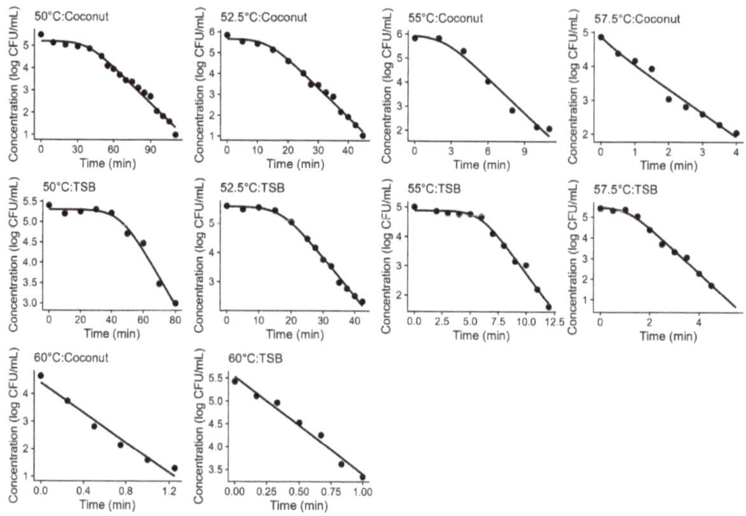

Figure 1. Fitting of the Geeraerd model (solid line) to the inactivation data of *Listeria innocua* (•) under isothermal conditions. The subplots show the results at different temperatures for acidified TSB and coconut water (conditions showed in titles).

The similarities between the media are further confirmed in Table 1, which provides the parameter estimates of the primary models fitted to each isothermal condition. These parameters are illustrated in Figure 2, showing that there is practically no difference between the *D*-values obtained in both media (Figure 2). Although there is a small difference in the shoulder length observed in coconut water and acidified TSB (with the laboratory media showing slightly longer shoulder lengths than the food product), these results also support TSB as a good surrogate for thermal inactivation of *L. innocua* in coconut water, as models based on TSB would provide conservative predictions ("fail-safe").

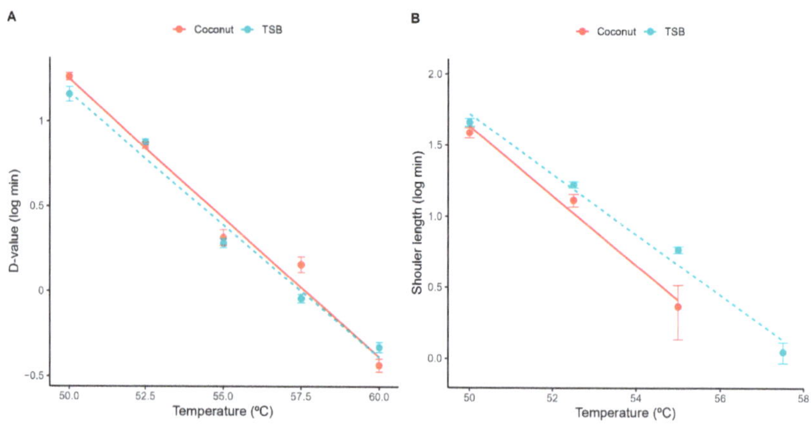

Figure 2. Parameter estimates of the Geeraerd model fitted to each individual isothermal inactivation experiment. *D*-value (**A**) and shoulder length (**B**).

Table 1. Parameter estimates (estimate ± standard error) for the Geeraerd model with log-linear secondary models for D and SL fitted to the isothermal inactivation of *Listeria innocua* in acidified TSB and coconut water using a one-step approach.

Parameter	Acidified TSB	Coconut Water
log N_0 (log CFU/mL)	5.34 ± 0.11	5.30 ± 0.23
a (min)	0.58 ± 0.07	0.12 ± 0.31
b (min/°C)	0.20 ± 0.01	0.23 ± 0.05
log D_{55} (log min)	0.46 ± 0.03	0.46 ± 0.04
z (°C)	4.94 ± 0.25	5.00 ± 0.24

These differences in shoulder length are likely due to the composition of the media. The shoulder is often interpreted as the representation of thermal inactivation being a "multi-hit" process, with the microbial cells being able to resist the treatment for some time before inactivation is effective [35]. This resistance is often linked to the physiological state of cells (e.g., membrane porosity), which is dependent on the media composition [36,37]. Therefore, the composition of acidified TSB (which is richer than coconut water) is likely to contain elements that increase the thermal resistance of *L. innocua*.

Figure 2 also illustrates the relationship between the model parameters of the Geeraerd model and the treatment temperature. As expected, the *D*-value showed a log-linear relationship with temperature, as is common in this type of model (Figure 2A). Moreover, the shoulder length also showed a log-linear trend with temperature (Figure 2B). This result is interesting, because the shoulder length is often considered more complex than the *D*-value, as it often depends on additional factors such as the cell history.

The log-linear relationships illustrated in Figure 2 justify the development of log-linear secondary models for both parameters. Table 1 provides the parameter estimates of these secondary models fitted to the complete dataset (for each media) using a one-step model fitting approach. The model parameters agree with the previous interpretation regarding media effects. The estimates of z and D_{ref} are practically the same on both media, confirming that the inactivation rate during the exponential phase in acidified TSB is practically the same as for coconut water. On the other hand, the intercept of the secondary model for the shoulder length (a) is higher in acidified TSB than in coconut water, whereas the slope (b) is lower, indicating that the shoulder length is longer on acidified TSB than in coconut water. Therefore, acidified TSB would be an adequate surrogate media to study the isothermal inactivation of *L. innocua* in coconut water.

3.2. Inactivation of Listeria innocua in TSB and Coconut Water under Dynamic Conditions

A principal challenge when using the Geeraerd model is the extrapolation of the shoulder length estimated under isothermal conditions to dynamic heating conditions. Although the relationship between the shoulder length and the treatment temperature has been successfully described using a log-linear model (Figure 2; Equation (3)), dynamic predictions require the definition of the initial value for the ideal substance C (C_0). Despite the relationship between SL and C_0 being well-defined by the identity in Equation (7), combining this equation with the secondary models for SL and D (Equations (2) and (3)) results in the nonlinear relationship between C_0 and the treatment temperature illustrated in Figure 3. This raises the question of what value of C_0 to use for the model predictions, as the Geeraerd model requires the definition of a unique value for this parameter.

In this study, we use the maximum (C_0 = 0.6 in acidified TSB; C_0 = 0.46 in coconut water) and minimum estimated values (C_0 = 0; equivalent to the Bigelow model) as a way to generate a reasonable envelope for the microbial response based on the experimental data. These values are combined with the parameters of the secondary model for the temperature (Table 1) to predict the microbial response under dynamic environmental conditions.

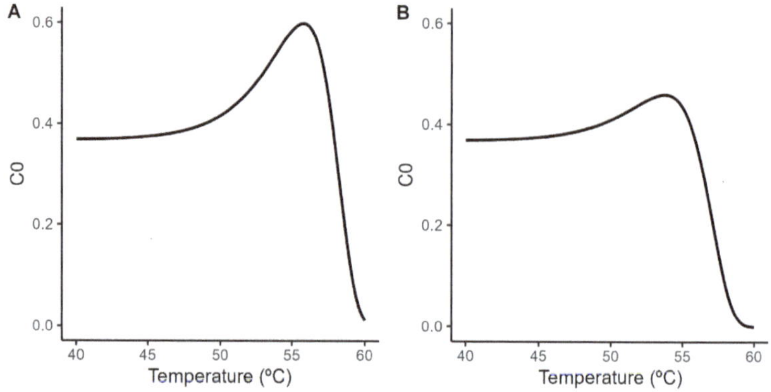

Figure 3. Relationship between initial value of the theoretical substance *C* and the treatment temperature according to the secondary model for the shoulder length (Equation (3)) and the relationship between both variables (Equation (7)) in acidified TSB (**A**) and coconut water (**B**).

As illustrated in Figures 4 and 5, the models fitted to isothermal conditions fail to predict the microbial response under dynamic heating. This deviation between predictions based on isothermal experiments and results under dynamic conditions is often reported in the scientific literature and can be attributed to various mechanisms [20,38–41].

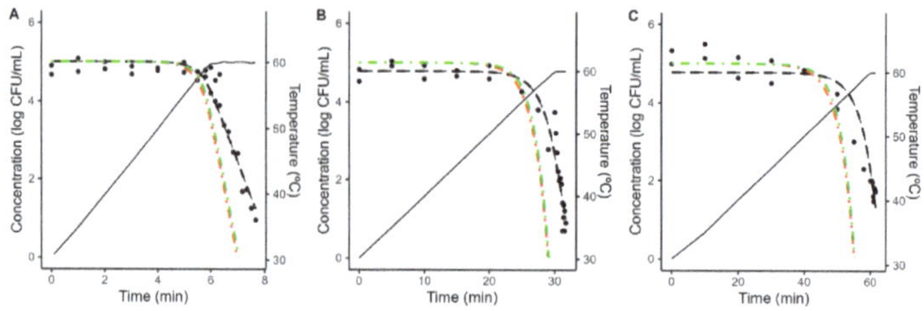

Figure 4. Comparison between the predictions of the Geeraerd model fitted to isothermal data considering the maximum possible value of C_0 (green, dash-dot line) no shoulder (red, dash-dot line) against the experimental data (•) obtained during dynamic heating (black, solid line) in acidified TSB with a heating rate of 5 (**A**), 1 (**B**) or 0.5 (**C**) °C/min. The plots also show the fit of the Bigelow model fitted directly to the dynamic data (black, dashed line).

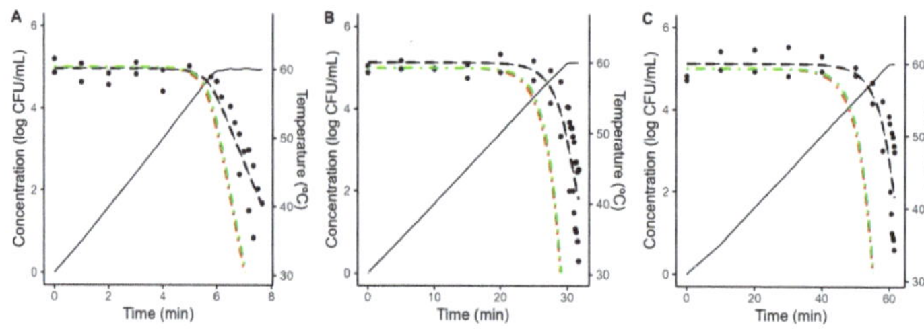

Figure 5. Comparison between the predictions of the Geeraerd model fitted to isothermal data considering the maximum possible value of C_0 (green, dash-dot line) no shoulder (red, dash-dot line) against the experimental data (•) obtained during dynamic heating (black, solid line) in coconut water with a heating rate of 5 (**A**), 1 (**B**) or 0.5 (**C**) °C/min. The plots also show the fit of the Bigelow model fitted directly to the dynamic data (black, dashed line).

One possibility is arguing for an initial bacterial resistance, using similar arguments as for the shoulder in isothermal inactivation curves. However, as illustrated in Figures 4 and 5, even when the shoulder is introduced using the worst-case scenario for C_0, the effect on the model predictions is minimal. Hence, an initial bacterial resistance is unlikely to be the cause for the deviation between predictions based on isothermal data and dynamic observations.

An alternative justification that has recently gained interest within the scientific community is the phenomenon known as "stress acclimation". This hypothesis considered that bacterial cells adapt dynamically to the initial (sub-lethal) stages of the treatment, activating a stress response that increases their thermal resistance [18,42,43]. Stress acclimation is often evidenced by lower heating rates resulting in a large deviation between the model predictions and the observations, due to the bacterial population having more time to adapt to the treatment.

To obtain further insight into the microbial response under dynamic conditions, the Bigelow model was fitted directly to the dynamic data, fixing the *z*-value to the one obtained under dynamic conditions to simplify the comparison between different conditions. As illustrated in Figures 4 and 5, the model fitted to dynamic conditions is able to describe the overall trend in the microbial population. Table 2 reports the parameter values estimated for each condition. It shows that a reduction on the heating rate from 5 to 0.5 °C/min results in approximately a three-fold increase in the *D*-value (5.66 to 17.99 min in acidified TSB; 6.23 to 18.77 min in coconut water).

Table 2. Estimates of the *D*-value at 55 °C (estimate ± standard error) from data obtained under dynamic heating conditions with different heating rates. To ease the interpretation, the *z*-value was fixed to the one obtained under isothermal conditions (4.99 °C on acidified TSB, 5.00 °C in coconut water).

Medium	Heating Rate (°C/Min)	D_{55} (Min)
Acidified TSB	0.5	17.99 ± 1.17
	1	10.44 ± 0.07
	5	5.66 ± 0.23
Coconut water	0.5	18.77 ± 2.15
	1	11.64 ± 1.10
	5	6.23 ± 2.04

This result is compatible with the assumption that stress acclimation would be responsible for the differences in the response of *L. innocua* under isothermal and dynamic conditions, as lower heating rates would allow higher stress acclimation due to cells being exposed to sublethal temperatures for a longer time.

The parameters reported in Table 2 provide further arguments to consider acidified TSB as a suitable surrogate for the thermal inactivation of *L. innocua* in coconut water. Even in dynamic treatments where stress acclimation seems relevant, the *D*-values obtained in acidified TSB are practically the same as those observed in coconut water. This, combined with the results observed under isothermal conditions, confirms that this media can be used as a reliable method for the design of mild pasteurization treatments targeting *L. innocua* in this product. They would need to be combined with refrigeration to inhibit the growth of high heat-resistant, spore forming bacteria.

The comparison between the *D*-values obtained under isothermal and dynamic conditions showed comparable results to those obtained in similar studies. Considering that isothermal experiments resulted in an estimate of the *D*-value of 2.88 min at 55 °C (Table 1), the lowest heating rate would induce a ~6.5 fold increase in the *D*-value. This value is comparable to those reported by Clemente-Carazo et al. [44], who reported a ~7 fold increase in the *D*-value of three strains of *L. monocytogenes* under dynamic conditions on buffered TSB and milk. This result is of high relevance due to the use of a different bacterial strain and media in this study (in fact, a media of different pH), as it could indicate that there could be an overall maximum stress acclimation attainable by a bacterial population that would be independent of the bacterial strain and media. Nonetheless, this hypothesis would need to be studied further in the future.

While non-thermal methods have proven effective in reducing bacterial pathogens in coconut water [13,15–17], it is worth noting that these investigations were conducted on a small laboratory scale (volumes of 5–10 mL) and employed varying process parameters. Therefore, a direct comparison of their efficacy with the current study, which simulates industrial dynamic treatment conditions, is not feasible.

Although the model is based on experimental data for *L. innocua*, there is scientific evidence of this microbial species being a suitable surrogate for the thermal inactivation of *L. monocytogenes* [22]. Hence, the model will be a useful tool to establish safe, mild heat treatments for coconut water. It could also be the basis for future studies that determine optimum heating profiles using a multifactorial approach that combines food safety with nutritional and environmental aspects.

4. Conclusions

Mild heat treatments can be an alternative to preserve high quality coconut water, but a careful design of the heating conditions, in order to avoid microbial stress adaptation, should be carefully selected. Precise knowledge of heat inactivation parameters modelled appropriately is a key aspect for designing safe thermal treatments. This research underscores the need for specific modelling approaches to account for stress acclimation under dynamic temperature conditions when targeting the inactivation of *Listeria* spp., facilitating the design of safe pasteurization processes for coconut water.

Author Contributions: Conceptualization, G.A.G.-T., A.G., P.S.F. and A.P.; methodology, G.A.G.-T., A.G., P.S.F. and A.P.; software, A.G.; validation, A.G. and A.P.; formal analysis, A.G., A.I. and R.W.-Z.; investigation, G.A.G.-T., A.G., R.W.-Z. and A.P.; resources, G.A.G.-T., R.W.-Z. and P.S.F.; data curation, A.G.; writing—original draft preparation, G.A.G.-T., A.G., P.S.F. and A.P.; writing—review and editing, R.W.-Z. and A.I.; visualization, G.A.G.-T., A.G., P.S.F. and A.P.; supervision, G.A.G.-T., A.G., P.S.F. and A.P.; project administration, G.A.G.-T. and P.S.F.; funding acquisition, A.G. and P.S.F. All authors have read and agreed to the published version of the manuscript.

Funding: This research was funded by the Spanish Ministry of Science and Research, MCIN/AEI/10.13039/501100011033 grant number PID2020-116318RBC32. Alberto Garre is funded by a Ramon y Cajal Fellowship (RYC- 2021-034612-I). Gerardo González Tejedor acknowledges the financial support of the National Research System (SNI) of SENACYT, Panama.

Data Availability Statement: The data used to support the findings of this study can be made available by the corresponding authors upon request.

Conflicts of Interest: The authors declare no conflict of interest.

References

1. Sriram, K.P.; Mangrolia, U.; Osborne, W.J. Isolation and characterization of culturable indigenous endophytic bacteria in the tender coconut. *Food Biotechnol.* **2020**, *34*, 228–242. [CrossRef]
2. Tan, T.C.; Cheng, L.H.; Bhat, R.; Rusul, G.; Easa, A.M. Composition, physicochemical properties and thermal inactivation kinetics of polyphenol oxidase and peroxidase from coconut (*Cocos nucifera*) water obtained from immature, mature and overly-mature coconut. *Food Chem.* **2014**, *142*, 121–128. [CrossRef] [PubMed]
3. Tuyekar, S.N.; Tawade, B.S.; Singh, K.S.; Wagh, V.S.; Vidhate, P.K.; Yevale, R.P.; Gaikwad, S.; Kale, M. An overview on coconut water: As a multipurpose nutrition. *Int. J. Pharm. Sci. Rev. Res.* **2021**, *68*, 63–70. [CrossRef]
4. Yong, J.W.H.; Ge, L.; Ng, Y.F.; Tan, S.N. The chemical composition and biological properties of coconut (*Cocos Nucifera* L.) water. *Molecules* **2009**, *14*, 5144–5164. [CrossRef]
5. Mordor Intelligence. *Coconut Water Market Size and Share Analysis—Growth Trends and Forecasts (2023–2028)*; Mordor Intelligence: Hyderabad, India, 2023.
6. Manna, K.; Khan, A.; Das, D.K.; Bandhu Kesh, S.; Das, U.; Ghosh, S.; Sharma Dey, R.; Das Saha, K.; Chakraborty, A.; Chattopadhyay, S.; et al. Protective effect of coconut water concentrate and its active component shikimic acid against hydroperoxide mediated oxidative stress through suppression of NF-κB and activation of Nrf2 pathway. *J. Ethnopharmacol.* **2014**, *155*, 132–146. [CrossRef] [PubMed]

7. Bhagya, D.; Prema, L.; Rajamohan, T. Therapeutic effects of tender coconut water on oxidative stress in fructose fed insulin resistant hypertensive rats. *Asian Pac. J. Trop. Med.* **2012**, *5*, 270–276. [CrossRef] [PubMed]
8. Rao, S.S.; Najam, R. Coconut water of different maturity stages ameliorates inflamatory processes in model of inflammation. *J. Intercult. Ethnopharmacol.* **2016**, *5*, 244–249. [CrossRef] [PubMed]
9. Zulaikhah, S.T. Health benefits of tender coconut water (Tcw). *Int. J. Pharm. Sci. Res.* **2019**, *10*, 474–480.
10. DebMandal, M.; Mandal, S. Coconut (*Cocos nucifera* L.: *Arecaceae*): In health promotion and disease prevention. *Asian Pac. J. Trop. Med.* **2011**, *4*, 241–247. [CrossRef]
11. Anurag, P.; Rajamohan, T. Cardioprotective effect of tender coconut water in experimental myocardial infarction. *Plant Foods Hum. Nutr.* **2003**, *58*, 1–12. [CrossRef]
12. Campbell-Falck, D.; Thomas, T.; Falck, T.M.; Tutuo, N.; Clem, K. The intravenous use of coconut water. *Am. J. Emerg. Med.* **2000**, *18*, 108–111. [CrossRef] [PubMed]
13. Naik, M.; Sunil, C.K.; Rawson, A.; Venkatachalapathy, N. Tender coconut water: A review on recent advances in processing and preservation. *Food Rev. Int.* **2022**, *38*, 1215–1236. [CrossRef]
14. Walter, E.H.M.; Kabuki, D.Y.; Esper, L.M.R.; Sant'Ana, A.S.; Kuaye, A.Y. Modelling the growth of *Listeria monocytogenes* in fresh green coconut (*Cocos nucifera* L.) water. *Food Microbiol.* **2009**, *26*, 653–657. [CrossRef] [PubMed]
15. Basak, S.; Jha, T.; Chakraborty, S. Pasteurization of tender coconut water by pulsed light treatment: Microbial safety, enzymatic inactivation, and impact on physicochemical properties. *Innov. Food Sci. Emerg. Technol.* **2023**, *84*, 103302. [CrossRef]
16. Maguluri, R.K.; Nettam, P.; Chaudhari, S.R.; Yannam, S.K. Evaluation of UV-C LEDs efficacy for microbial inactivation in tender coconut water. *J. Food Process. Preserv.* **2021**, *45*, e15727. [CrossRef]
17. Mahnot, N.K.; Mahanta, C.L.; Keener, K.M.; Misra, N.N. Strategy to achieve a 5-log *Salmonella* inactivation in tender coconut water using high voltage atmospheric cold plasma (HVACP). *Food Chem.* **2019**, *284*, 303–311. [CrossRef]
18. Georgalis, L.; Yeak, K.Y.C.; Tsimpou, C.; Fernandez, P.S.; Wells-Bennik, M.; Garre, A. Disentangling the contributions of initial heterogeneities and dynamic stress adaptation to nonlinearities in bacterial survival curves. *Food Res. Int.* **2023**, *173*, 113385. [CrossRef]
19. Dolan, K.D.; Valdramidis, V.P.; Mishra, D.K. Parameter estimation for dynamic microbial inactivation: Which model, which precision? *Food Control* **2013**, *29*, 401–408. [CrossRef]
20. Hassani, M.; Cebrián, G.; Mañas, P.; Condón, S.; Pagán, R. Induced thermotolerance under nonisothermal treatments of a heat sensitive and a resistant strain of *Staphylococcus aureus* in media of different pH. *Lett. Appl. Microbiol.* **2006**, *43*, 619–624. [CrossRef]
21. Garre, A.; Huertas, J.P.; González-Tejedor, G.A.; Fernández, P.S.; Egea, J.A.; Palop, A.; Esnoz, A. Mathematical quantification of the induced stress resistance of microbial populations during non-isothermal stresses. *Int. J. Food Microbiol.* **2018**, *266*, 133–141. [CrossRef]
22. Fairchild, T.M.; Foegeding, P.M. A proposed nonpathogenic biological indicator for thermal inactivation of *Listeria monocytogenes*. *Appl. Environ. Microbiol.* **1993**, *59*, 1247–1250. [CrossRef] [PubMed]
23. Silva-Angulo, A.B.; Zanini, S.F.; Rodrigo, D.; Rosenthal, A.; Martinez, A. Growth kinetics of *Listeria innocua* and *Listeria monocytogenes* under exposure to carvacrol and the occurrence of sublethal damage. *Food Control* **2014**, *37*, 336–342. [CrossRef]
24. Saucedo-Reyes, D.; Marco-Celdrán, A.; Pina-Pérez, M.C.; Rodrigo, D.; Martínez-López, A. Modeling survival of high hydrostatic pressure treated stationary- and exponential-phase *Listeria innocua* cells. *Innov. Food Sci. Emerg. Technol.* **2009**, *10*, 135–141. [CrossRef]
25. Conesa, R.; Andreu, S.; Fernández, P.S.; Esnoz, A.; Palop, A. Nonisothermal heat resistance determinations with the thermoresistometer Mastia. *J. Appl. Microbiol.* **2009**, *107*, 506–513. [CrossRef] [PubMed]
26. Geeraerd, A.H.; Herremans, C.H.; Van Impe, J.F. Structural model requirements to describe microbial inactivation during a mild heat treatment. *Int. J. Food Microbiol.* **2000**, *59*, 185–209. [CrossRef]
27. Peñalver-Soto, J.L.; Garre, A.; Esnoz, A.; Fernández, P.S.; Egea, J.A. Guidelines for the design of (optimal) isothermal inactivation experiments. *Food Res. Int.* **2019**, *126*, 108714. [CrossRef]
28. Garre, A.; Fernández, P.S.; Lindqvist, R.; Egea, J.A. Bioinactivation: Software for modelling dynamic microbial inactivation. *Food Res. Int.* **2017**, *93*, 66–74. [CrossRef]
29. Garre, A.; Clemente-Carazo, M.; Fernández, P.S.; Lindqvist, R.; Egea, J.A. Bioinactivation FE: A free web application for modelling isothermal and dynamic microbial inactivation. *Food Res. Int.* **2018**, *112*, 353–360. [CrossRef]
30. Fernández, A.; Ocio, M.J.; Fernández, P.S.; Rodrigo, M.; Martinez, A. Application of nonlinear regression analysis to the estimation of kinetic parameters for two enterotoxigenic strains of *Bacillus cereus* spores. *Food Microbiol.* **1999**, *16*, 607–613. [CrossRef]
31. R Core Team. *A Language and Environment for Statistical Computing*; R Foundation for Statistical Computing: Vienna, Austria, 2022.
32. Moré, J.J. The Levenberg-Marquardt algorithm: Implementation and theory. In *Numerical Analysis*; Springer: Berlin/Heidelberg, Germany, 1978; pp. 105–116. [CrossRef]
33. Elzhov, T.V.; Mullen, K.M.; Spiess, A.-N.; Bolker, B. minpack.lm: R interface to the Levenberg-Marquardt Nonlinear Least-Squares Algorithm Found in MINPACK, Plus Support for Bounds. 2016. Available online: https://cran.r-project.org/web/packages/minpack.lm/minpack.lm.pdf (accessed on 4 August 2023).
34. Baranyi, J.; Roberts, T.A.; McClure, P. A non-autonomous differential equation to model bacterial growth. *Food Microbiol.* **1993**, *10*, 43–59. [CrossRef]

35. Ruiz, V.; Alonso, R.; Salvador, M.; Condón, S.; Condón-Abanto, S. Impact of shoulders on the calculus of heat sterilization treatments with different bacterial spores. *Food Microbiol.* **2021**, *94*, 103663. [CrossRef] [PubMed]
36. Álvarez-Ordóñez, A.; Fernández, A.; López, M.; Bernardo, A. Relationship between membrane fatty acid composition and heat resistance of acid and cold stressed *Salmonella* Senftenberg CECT 4384. *Food Microbiol.* **2009**, *26*, 347–353. [CrossRef] [PubMed]
37. Álvarez-Ordóñez, A.; Fernández, A.; López, M.; Arenas, R.; Bernardo, A. Modifications in membrane fatty acid composition of *Salmonella* Typhimurium in response to growth conditions and their effect on heat resistance. *Int. J. Food Microbiol.* **2008**, *123*, 212–219. [CrossRef] [PubMed]
38. Garre, A.; Egea, J.A.; Iguaz, A.; Palop, A.; Fernandez, P.S. Relevance of the induced stress resistance when identifying the critical microorganism for microbial risk assessment. *Front. Microbiol.* **2018**, *9*, 1663. [CrossRef] [PubMed]
39. Stasiewicz, M.J.; Marks, B.P.; Orta-Ramirez, A.; Smith, D.M. Modeling the effect of prior sublethal thermal history on the thermal inactivation rate of *Salmonella* in ground Turkey. *J. Food Prot.* **2008**, *71*, 279–285. [CrossRef]
40. Huertas, J.P.; Ros-Chumillas, M.; Garre, A.; Fernández, P.S.; Aznar, A.; Iguaz, A.; Esnoz, A.; Palop, A. Impact of heating rates on *Alicyclobacillus acidoterrestris* heat resistance under non-isothermal treatments and use of mathematical modelling to optimize orange juice processing. *Foods* **2021**, *10*, 1496. [CrossRef]
41. Janssen, M.; Verhulst, A.; Valdramidis, V.; Devlieghere, F.; Van Impe, J.F.; Geeraerd, A.H. Inactivation model equations and their associated parameter values obtained under static acid stress conditions cannot be used directly for predicting inactivation under dynamic conditions. *Int. J. Food Microbiol.* **2008**, *128*, 136–145. [CrossRef]
42. Garre, A.; González-Tejedor, G.A.; Aznar, A.; Fernández, P.S.; Egea, J.A. Mathematical modelling of the stress resistance induced in *Listeria monocytogenes* during dynamic, mild heat treatments. *Food Microbiol.* **2019**, *84*, 103238. [CrossRef]
43. Valdramidis, V.P.; Geeraerd, A.H.; Van Impe, J.F. Stress-adaptive responses by heat under the microscope of predictive microbiology. *J. Appl. Microbiol.* **2007**, *103*, 1922–1930. [CrossRef]
44. Clemente-Carazo, M.; Cebrián, G.; Garre, A.; Palop, A. Variability in the heat resistance of *Listeria monocytogenes* under dynamic conditions can be more relevant than that evidenced by isothermal treatments. *Food Res. Int.* **2020**, *137*, 109538. [CrossRef]

foods

MDPI

Article

Use of Doehlert Matrix as a Tool for High-Throughput Screening of Organic Acids and Essential Oils on Miniaturized Pork Loins, Followed by Lab-Scale Validation That Confirmed Tested Compounds Do Not Show Synergistic Effects against *Salmonella* Typhimurium

Cristina Resendiz-Moctezuma [1], Arianna P. L. Fonville [1], Bailey N. Harsh [2], Matthew J. Stasiewicz [1] and Michael J. Miller [1,*]

[1] Food Science and Human Nutrition Department, University of Illinois at Urbana-Champaign, 1302 W. Pennsylvania Ave., Urbana, IL 61801, USA; cr638@cornell.edu (C.R.-M.); mstasie@illinois.edu (M.J.S.)

[2] Animal Sciences Department, University of Illinois at Urbana-Champaign, 1503 Maryland Dr., Urbana, IL 61801, USA; bharsh2@illinois.edu

* Correspondence: mille216@illinois.edu; Tel.: +1-217-244-1973

Abstract: The many possible treatments and continuously changing consumer trends present a challenge when selecting antimicrobial interventions during pork processing. Thirty-five potential antimicrobials were screened at commercial working concentrations by individually adding them to miniaturized (69 cm^3) disks of pork loin ends, followed by inoculation with *Salmonella* Typhimurium ATCC 19585. Two organic acids and nine essential oils significantly inhibited *Salmonella* counts on pork ($p < 0.05$). However, six compounds that represent different levels of significance ($p < 0.05$–$p < 0.0001$) were selected as independent variables to build a Response Surface Methodology model based on a Doehlert matrix (Doehlert Matrix—RSM): lactic acid 1.25%, formic acid 0.25%, cumin 0.25%, clove 0.25%, peppermint 0.5%, and spearmint 0.5%. The goal of the Doehlert Matrix—RSM was to study single and paired effects of these antimicrobials on the change in *Salmonella* over 24 h. The Doehlert Matrix—RSM model predicted that lactic acid, formic acid, cumin, peppermint, and spearmint significantly reduced *Salmonella* when added alone, while no significant interactions between these antimicrobials were found. A laboratory-scale validation was carried out on pork loin end slices, which confirmed the results predicted by the model. While this screening did not identify novel synergistic combinations, our approach to screening a variety of chemical compounds by implementing a miniaturized pork loin disk model allowed us to identify the most promising antimicrobial candidates to then formally design experiments to study potential interactions with other antimicrobials.

Keywords: *Salmonella*; pork; antimicrobials; essential oils; Response Surface Methodology; Design of Experiment; Doehlert matrix

Citation: Resendiz-Moctezuma, C.; Fonville, A.P.L.; Harsh, B.N.; Stasiewicz, M.J.; Miller, M.J. Use of Doehlert Matrix as a Tool for High-Throughput Screening of Organic Acids and Essential Oils on Miniaturized Pork Loins, Followed by Lab-Scale Validation That Confirmed Tested Compounds Do Not Show Synergistic Effects against *Salmonella* Typhimurium. *Foods* **2023**, *12*, 4034. https://doi.org/10.3390/foods12214034

Academic Editor: Boce Zhang

Received: 29 September 2023
Revised: 24 October 2023
Accepted: 30 October 2023
Published: 5 November 2023

1. Introduction

Salmonella remains a relevant foodborne pathogen for the pork industry [1]. During 2020, the European Union reported that pork and pork products were the second most commonly encountered food vehicles in confirmed salmonellosis outbreaks [2]. Therefore, pork carcass washes are currently one of the most used practices in the industry to prevent *Salmonella* Typhimurium contamination on pork meat. Currently, hot water and organic acids are the most commonly used agents to spray carcasses and cuts of pork meat. However, according to Phase II of the "Raw Pork Products Exploratory Sampling Program" by the FSIS, 8.1% of the 894 samples and 25.3% of the 1088 samples tested positive for *Salmonella* spp. on intact cuts and comminuted product, respectively [3]. Additionally, a 2016 study compared the efficacy of different washes and conditions (acid concentration, acid temperature, and water temperature) to reduce *Salmonella* contamination on the surface of pork and found that all of the treatments were equally effective [4]. Therefore, additional multi-hurdle approaches (use of multiple antimicrobial practices to achieve pathogen inhibition [5]) are needed to ensure the safety of pork meat.

Organic acids have been used as an effective antimicrobial against *Salmonella* in pork. Current literature suggests that lactic acid (LA), formic acid (FA), and peracetic acid (PA) can significantly reduce *Salmonella* counts on pork [1,6–8]. Food preservatives like potassium sorbate (PS) and sodium propionate (SOP) have also proved to reduce counts of Gram-negative bacteria [9,10]. All of these compounds have GRAS (Generally Recognized as Safe) status, making them suitable for use as antimicrobials in the meat industry. However, consumers are interested in the use of clean-label antimicrobial compounds, like essential oils (EOs). EOs are hydrophobic liquids mainly composed of aliphatic aldehydes, alcohols, and esters [11], and have previously showed promising antimicrobial effects against *Salmonella* in fruit, vegetable, and meat systems [12]. However, due to their dose- and matrix-dependent response, studies differ in the efficacy of EOs. For example, a recent 2020 study compared the effect of 0.2% and 0.4% oregano oil against *Salmonella* Typhimurium in broth [13]. Results showed 0.2% oregano oil did not significantly reduce bacterial counts, while 0.4% reduced *Salmonella* below the limit of detection. In the same study, oregano oil was used to inhibit *Salmonella* on chicken meat, where >0.8% oregano oil was necessary to significantly reduce *Salmonella* counts after 30 s at 4 °C. Furthermore, another study reported the effect of oregano oil against *Salmonella* on minced lamb and found that 0.6% oregano oil was enough to significantly reduce *Salmonella* counts [14]. Thus, these differences in effective levels in broth, chicken, and sheep meat exemplify the dose- and matrix-dependent response aspect of EOs.

To the best of our knowledge, no study has comprehensively screened the effect of multiple antimicrobials (like the ones mentioned), and their combinations, directly on pork to achieve a multi-hurdle approach. Response Surface Methodology (RSM) is a powerful and time- and cost-effective tool that can be used to comprehensively screen a high number of independent variables, such as antimicrobials, and their effect on a response, such as *Salmonella* cell counts, while performing a reduced number of experiments. Contrary to full factorial designs, RSM utilizes Design of Experiments (DOE) software to determine the least amount of runs needed to build a second-order regression model [15,16]. The objective of this study was to use RSM to study the effect of combinations of effective antimicrobial compounds to inhibit *Salmonella* Typhimurium cell counts on pork loin ends destined for comminuted products. The proposed screening method represents a real-life case scenario since (i) the tested substances were applied at regulatory acceptable levels, (ii) their effect was assessed against a food-borne relevant *Salmonella* Typhimurium strain, and (iii) all studies were performed directly on pork loin ends.

2. Materials and Methods

Antimicrobial selection was carried out by identifying antimicrobials that are currently being used by the pork industry (FA, LA, and PA), chemical compounds that have shown antimicrobial activity in foods (PS and SOP), and chemical compounds that can potentially elicit inhibitory effects based on their chemical properties (EOs due to their phenolic compound content). Thirty-five compounds were selected to be screened for antimicrobial activity against *Salmonella* on pork.

2.1. Bacterial Strains and Inoculum Preparations

Salmonella enterica serovar Typhimurium (ATCC 19585) is a clinical isolate. The inoculum of *Salmonella* Typhimurium was prepared by inoculating Luria–Bertani (LB) broth (Fisher Scientific, Hampton, NH, USA) from glycerol stocks stored at −80 °C. The bacterium was grown overnight in 3 mL of LB broth for 18 h at 37 °C with shaking. The overnight culture was then serially diluted in phosphate-buffered saline (PBS) solution (137 mM

NaCl, 2.7 mM KCl, and 10 mM phosphate buffer E404; VWR International, Radnor, PA, USA) to reach the desired inoculum.

2.2. Antimicrobial Preparation of Organic Acids and Salts

The organic acids used in this study were lactic acid (VWR International, Radnor, PA, USA), peracetic acid (Sigma-Aldrich, St. Louis, MO, USA), and formic acid (Sigma-Aldrich, MA, USA). The salts used as antimicrobials were potassium sorbate (Sigma-Aldrich, MA, USA) and sodium propionate (Sigma-Aldrich, MA, USA). All solutions were prepared by diluting the compounds into distilled water to a final volume of 10 mL and were prepared fresh on the day of the experiment. See Table 1 for the final concentration of each compound on the pork disks.

Table 1. Antimicrobial compounds tested on pork across different evaluations.

Antimicrobial	Code	Concentration (% *v*/*w*)		
		Initial Screening	**Doehlert Matrix—RSM**	**Lab-Scale Validation**
Formic acid	FA	1.5 and 3.0	0–1.25	3.0
Lactic Acid	LA	2.5 and 5.0	0–0.25	5.0
Peracetic acid	PA	0.02 and 0.04	NA	NA
Potassium sorbate	PS	5.0 and 10	NA	NA
Sodium propionate	SOP	2.5 and 5.0	NA	NA
Essential oils *	EOs	1.0	0–0.5	1.0

* Tested individually; see Table 2 for list of tested oils and abbreviations.

Table 2. Essential oil library screened for their individual antimicrobial activity against *Salmonella* Typhimurium ATCC 19585 on miniaturized pork disks.

#	Essential Oil	#	Essential Oil	#	Essential Oil
1	Almond	11	Dillweed	21	Orange
2	Ambrette	12	Eucalyptus	22	**Peppermint (PPP)** *
3	Anise star	13	Fenugreek	23	Peru
4	Black pepper	14	Fusel	24	Pine needle
5	Cassia	15	Garlic	25	Rosemary
6	Cinnamon	16	Ginger	26	Sandalwood
7	**Clove (CLV)** *	17	Jasmin	27	**Spearmint (SPT)** *
8	Cocoa extract	18	Juniper berry	28	Tea tree
9	Cognac	19	Lavender	29	Thyme
10	**Cumin (CMN)** *	20	Lemon	30	Ylang-ylang

* Oils that were selected to be used as independent variables in the Doehlert matrix—RSM. Food-grade oils, 99% purity.

2.3. Antimicrobial Preparation of EOs

An EO library composed of 30 oils (Sigma-Aldrich, Burlington, MA, USA) (Table 2) was tested for antimicrobial effects on the pork disk. The oils were emulsified to increase the affinity of the antimicrobial solution to the food matrix [17–19]. The emulsion was prepared as described before [17]. Briefly, the continuous phase of the emulsion was prepared with 3% Tween-80 (VWR International, Radnor, PA, USA) as a surfactant agent, 5% inulin from chicory (Sigma-Aldrich, MA, USA) as an emulsion stabilizer, and PBS. The EOs were added at a final concentration of 0.25–1% (based on the mass of the pork disk, Table 1), by preparing a stock solution of the antimicrobials, with a constant target volume of 400 μL. The solution was then covered with Parafilm (Bemis Company Inc., Terre Haute, IN, USA) to prevent volatilization during sonication. The EO solutions were sonicated (Branson 450 Digital Sonifier with probe, Marshall Scientific, Hampton, NH, USA) at 30% wave amplitude for 10 min in an ice bath. Emulsions were prepared fresh on the day of the experiment, and subsequently applied to the pork disk.

2.4. Sample Preparation—Miniature Pork Disks

Pork loin ends in vacuum-sealed bags were obtained from Rantoul Foods in Rantoul, IL. After the initial 24 h holding period after slaughter, the loins were harvested by processor staff. The loin ends were separated from the loin manually and placed in a vacuum-sealed bag. Then, pork loin ends were collected by laboratory staff and immediately transferred to the laboratory (within 30 min) in an insulated bag to maintain the temperature of the meat below 10 °C. Upon receiving, the pork loin ends were refrigerated at 4 °C before further processing. Pork loin ends were sliced into 12 mm thick slices using a commercial deli meat slicer (Big Bite 8-1/2″ meat slicer, LEM products, West Chester, OH, USA). The slices were then aseptically cut into cubes of 69 cm^3 (0.55 ± 0.10 g approximately) by using a stainless steel blade with a grid of 7.6 by 7.6 mm (Zhejiang Fullstar Houseware Co., Taizhou, Zhejiang, China). Pork disks were then placed inside a 15 mL conical tube (VWR International, Radnor, PA, USA) for further experimentation.

2.5. Initial Antimicrobial Screening

Each antimicrobial compound was tested individually on a pork disk to screen for single antimicrobial effects. Antimicrobial concentrations were calculated based on the weight of the pork disk. Once the pork disks had been placed inside conical tubes as described above, disks were inoculated by directly adding 150 µL of diluted *Salmonella* Typhimurium ATCC 19585 overnight culture on the surface of the meat to reach a final inoculum on the pork disk of approximately 4-log (CFU/g). The disks were incubated at 4 °C for 30 min to allow for cell attachment. Then, 400 µL of antimicrobial solutions was added onto the surface of the pork disk. Pork disks were incubated again at 4 °C for 30 min. Two controls were prepared: (i) a pork disk that was uninoculated to identify any *Salmonella* spp. contamination on the meat before treatment, and (ii) an inoculated disk not treated with antimicrobials to determine the inoculum concentration. Control disks were also incubated at 4 °C for 24 h.

After 24 h, the pork disks were diluted 1:10 and 1:100 in PBS before spiral plating (Neutec group, Farmingdale, NY, USA) onto GranuCult Xylose Lysine Deoxycholate (XLD) agar (Sigma-Aldrich, Burlington, MA, USA) for selective *Salmonella* recovery. Plates were incubated at 37 °C for 24 h. Results are reported as the change in *Salmonella* cell counts after 24 h compared to the inoculum (cell counts at 0 h, Δ Cell counts).

Results were analyzed using GraphPad Prism 9 (GraphPad Software Inc., San Diego, CA, USA). A one-way ANOVA followed by Dunnett's multiple comparisons test was performed to determine statistically significant differences ($p < 0.05$) between EOs and the untreated group. This experiment was performed in biological duplicate, that is, an independent set of pork disks for each replicate.

2.6. Design of Experiments to Build RSM Model

After initial screening, six of the eleven antimicrobial compounds that were most effective at reducing *Salmonella* Typhimurium counts were taken for further analysis. The selected antimicrobials were two organic acids that significantly inhibited *Salmonella* counts during the initial screening, and four of the nine EOs that showed inhibitory effects. Four EOs were chosen to represent the width of statistical significance found during the initial screening ($p < 0.05$–$p < 0.0001$).

A Doehlert matrix design was chosen as the DOE to build an RSM model (Doehlert matrix—RSM) due to its buildability characteristics [16,20–24]. A Doehlert uniform shell design for six factors was developed using MODDE 13 software (Sartorius AG, Gottingen, Lower Saxony, Germany). The software interface requires users to input independent variables and their range, as well as the model's response. The model presented here included six independent variables (LA 0–1.25%, FA 0–0.25%, CLV 0–0.25%, CMN 0–0.25%, PPP 0–0.5%, and SPT 0–0.25%) and the selected response was expressed as the change in *Salmonella* cell counts after 24 h compared to the inoculum (Δ Cell counts) in log CFU/g. MODDE 13 automatically generates an experiment matrix that includes the necessary experiments to build a second-order linear model to explain the effect of single and paired interactions between the independent variables on the response. The experiment matrix (Table 3) included 42 experiments, 1 central point in triplicate (to calculate pure error), and 1 untreated control (inoculated yet non-treated disk) in duplicate.

Table 3. Designed Doehlert matrix for six antimicrobials (LA, FA, CLV, CMN, PPP, and SPT). Experimental data represent the mean of three biological replicates. Predicted data were calculated based on a linear regression model fitted to the experimental data.

Exp. No.	Antimicrobial Doehlert Matrix (%)						Experimental Data (log CFU/g)			Predicted Data (log CFU/g)	
	LA	FA	CMN	PPP	CLV	SPT	*Salmonella* Inoculum	*Salmonella* Cell Counts after 24 h	Δ *Salmonella* Cell Counts	Δ *Salmonella* Cell Counts	Residuals (Actual—Predicted)
1	1.25	0.13	0.13	0.25	0.13	0.25	4.05	3.1	−0.95	−0.88	−0.07
2	0.94	0.23	0.13	0.25	0.13	0.25	4.05	3.08	−0.97	−0.84	−0.13
3	0.94	0.16	0.23	0.25	0.13	0.25	4.05	3.07	−0.98	−1	0.02
4	0.94	0.16	0.15	0.45	0.13	0.25	4.05	3.15	−0.9	−0.96	0.06
5	0.94	0.16	0.15	0.29	0.22	0.25	4.05	3.12	−0.93	−1.05	0.12
6	0.94	0.16	0.15	0.29	0.14	0.44	4.05	3.21	−0.84	−0.94	0.1
7	0	0.13	0.13	0.25	0.13	0.25	4.05	3.38	−0.68	−0.42	−0.26
8	0.31	0.02	0.13	0.25	0.13	0.25	4.05	3.74	−0.31	−0.41	0.1
9	0.31	0.09	0.02	0.25	0.13	0.25	4.05	3.82	−0.23	−0.39	0.16
10	0.31	0.09	0.1	0.05	0.13	0.25	4.05	3.76	−0.29	−0.31	0.02
11	0.31	0.09	0.1	0.21	0.03	0.25	4.05	3.73	−0.32	−0.48	0.16
12	0.31	0.09	0.1	0.21	0.11	0.06	4.05	3.9	−0.15	−0.21	0.06
13	0.94	0.02	0.13	0.25	0.13	0.25	4.05	3.37	−0.68	−0.69	0.01
14	0.94	0.09	0.02	0.25	0.13	0.25	4.05	3.59	−0.46	−0.52	0.06
15	0.94	0.09	0.1	0.05	0.13	0.25	4.05	3.63	−0.43	−0.51	0.08
16	0.94	0.09	0.1	0.21	0.03	0.25	4.05	3.28	−0.77	−0.7	−0.07
17	0.94	0.09	0.1	0.21	0.11	0.06	4.05	3.56	−0.5	−0.48	−0.02
18	0.63	0.2	0.02	0.25	0.13	0.25	4.05	3.47	−0.58	−0.57	−0.01
19	0.63	0.2	0.1	0.05	0.13	0.25	4.05	3.72	−0.34	−0.48	0.14
20	0.63	0.2	0.1	0.21	0.03	0.25	4.05	3.57	−0.49	−0.52	0.03
21	0.63	0.2	0.1	0.21	0.11	0.06	4.05	3.62	−0.44	−0.51	0.07
22	0.63	0.13	0.2	0.05	0.13	0.25	4.05	3.23	−0.83	−0.76	−0.07

Table 3. *Cont.*

Exp. No.	Antimicrobial Doehlert Matrix (%)						Experimental Data (log CFU/g)			Predicted Data (log CFU/g)	
	LA	FA	CMN	PPP	CLV	SPT	*Salmonella* Inoculum	*Salmonella* Cell Counts after 24 h	Δ *Salmonella* Cell Counts	Δ *Salmonella* Cell Counts	Residuals (Actual—Predicted)
23	0.63	0.13	0.2	0.21	0.03	0.25	4.05	3.59	−0.46	−0.61	0.15
24	0.63	0.13	0.2	0.21	0.11	0.06	4.05	3.24	−0.82	−0.72	−0.1
25	0.63	0.13	0.13	0.41	0.03	0.25	4.05	3.57	−0.48	−0.58	0.1
26	0.63	0.13	0.13	0.41	0.11	0.06	4.05	3.58	−0.47	−0.52	0.05
27	0.63	0.13	0.13	0.25	0.21	0.06	4.05	3.37	−0.69	−0.61	−0.08
28	0.31	0.23	0.13	0.25	0.13	0.25	4.05	3.37	−0.69	−0.65	−0.04
29	0.31	0.16	0.23	0.25	0.13	0.25	4.05	3.47	−0.59	−0.66	0.07
30	0.31	0.16	0.15	0.45	0.13	0.25	4.05	3.29	−0.77	−0.68	−0.09
31	0.31	0.16	0.15	0.29	0.22	0.25	4.05	3.49	−0.57	−0.8	0.23
32	0.31	0.16	0.15	0.29	0.14	0.44	4.05	3.34	−0.71	−0.75	0.04
33	0.63	0.05	0.23	0.25	0.13	0.25	4.05	3.43	−0.62	−0.75	0.13
34	0.63	0.05	0.15	0.45	0.13	0.25	4.05	3.18	−0.88	−0.7	−0.18
35	0.63	0.05	0.15	0.29	0.22	0.25	4.05	3.47	−0.58	−0.67	0.09
36	0.63	0.05	0.15	0.29	0.14	0.44	4.05	3.14	−0.92	−0.82	−0.1
37	0.63	0.13	0.05	0.45	0.13	0.25	4.05	3.29	−0.77	−0.79	0.02
38	0.63	0.13	0.05	0.29	0.22	0.25	4.05	3.42	−0.63	−0.57	−0.06
39	0.63	0.13	0.05	0.29	0.14	0.44	4.05	3.24	−0.81	−0.85	0.04
40	0.63	0.13	0.13	0.09	0.22	0.25	4.05	3.5	−0.55	−0.52	−0.03
41	0.63	0.13	0.13	0.09	0.14	0.44	4.05	3.31	−0.75	−0.61	−0.14
42	0.63	0.13	0.13	0.25	0.04	0.44	4.05	3.24	−0.81	−0.78	−0.03
43	0.63	0.13	0.13	0.25	0.13	0.25	4.05	3.31	−0.74	−0.65	−0.09
44	0.63	0.13	0.13	0.25	0.13	0.25	4.05	3.24	−0.82	−0.65	−0.17
45	0.63	0.13	0.13	0.25	0.13	0.25	4.05	3.08	−0.98	−0.65	−0.33
46	0	0	0	0	0	0	4.05	4.23	0.18	0.23	−0.05
47	0	0	0	0	0	0	4.05	4.28	0.23	0.23	0

As shown in Table 3, the experiment matrix requires the addition of all six antimicrobials to all 42 treatments. Considering that each antimicrobial was selected due to its individual inhibitory effect, antimicrobial concentrations had to be adjusted to prevent bacterial counts resulting from each treatment to be below the limit of detection (LOD). When the RSM models are fitted to a second-order linear model, actual bacterial counts need to be included. Counts below the LOD cause the model to reduce its prediction accuracy, since the data would not accurately reflect the effect of different antimicrobial doses on the response [24]. Therefore, the antimicrobial concentrations used for the Doehlert matrix—RSM were lower than those used during the initial screening (Table 1).

Experiments were carried out following the experiment matrix in triplicate. However, each replicate was individually fitted to a second-order model to confirm that each replicate had the statistical quality parameters necessary to build an RSM model. The following parameters were considered to determine whether the obtained experimental data had enough quality to build the model: residual standard deviation (RSD), model regression p-value, lack of fit test, and linearity. Finally, the average of three replicates was used to generate the regression models that met quality parameters to reduce the RSD.

MODDE 13 software was used to analyze and fit the Doehlert matrix—RSM. The full quadratic model was reduced by deleting the quadratic effects of antimicrobials from the model since none of these showed to be significant ($p > 0.1$) to the model.

2.7. Laboratory-Scale Validation of Antimicrobial Combination on Pork Slices

Pork loin ends were collected and sliced as mentioned in Section 2.3. The lab-scale validation protocol is based on the methodology previously described [7,25] with modifications. The laboratory-scale validation was carried out on pork slices with a surface area of approximately 50 cm^2 (48.0 $\pm$ 3.0 g approximately). Slices were placed inside a Biosafety Cabinet (BSC) (1300 Series A2, Thermo Fisher Scientific, Waltham, MA, USA) lined with sanitized aluminum foil. Then, 200 µL of stationary phase *Salmonella* Typhimurium ATCC 19585 inoculum, diluted to 10^7 CFU/mL (See Section 2.1), was added onto the slice to reach a desired inoculum concentration of approximately 4-log (CFU/g). Then, a sterile L-shaped rod (VWR International, Radnor, PA, USA) was used to distribute the inoculum over the surface of the slice. Slices were allowed to air-dry for 10 min before further treatment. Preliminary experiments showed that a 7-log inoculum resulted in a desired final concentration of 4-log CFU/g after spraying slices with 25 mL of DI water (non-treated control).

Sanitized Z-shaped metal hangers (Walmart Inc., Bentonville, AR, USA) were used to hang pork slices from a sanitized metal rod inside a BSC. An autoclave-safe bin was placed under the hanging slices to collect drippings from the meat. Antimicrobial treatments were added by using the same concentrations used in the original screening, also referred to as "working industry concentrations" (Table 1).

A handheld gardening sprayer (Scotts Miracle-Gro Company, Marysville, OH, USA) was used to deliver 25 mL of antimicrobial solutions to the inoculated surface of the pork slice at a flow rate of approximately 4.1 mL/s. Pork slices were immediately transferred to an air-tight container (Sterilite Co., Townsend, MA, USA) where the slices hung from zip-ties during incubation at 4 °C for 24 h. Control slices used were non-inoculated (to identify any *Salmonella* spp. contamination on the meat before treatment) and untreated (inoculated slices used to determine the inoculum concentration).

For enumeration, slices were placed inside a Whirl-Pak bag (Whirl-Pak Filtration Group, Fort Atkinson, WI, USA) before adding enough PBS to reach an exact 1:5 dilution. Slices were then homogenized using a stomacher (Seward Stomacher 80 Biomaster Laboratory Blender, Seward Inc., Bohemia, NY, USA) at "normal" speed for 1 min. Slices were serially diluted to reach 1:10 and 1:100 dilutions with PBS and enumerated as described in 2.5.

3. Results

3.1. High-Throughput Screening of Potential Antimicrobial Compounds Was Achieved Using a Miniaturized Pork Loin End Model

A miniaturized pork model was developed by placing 69 cm^3 cubes of pork loin ends into test tubes. This miniaturized food model allowed us to effectively screen many potential antimicrobial compounds directly on pork destined for comminuted products. Results of the initial screening of organic acids and common food preservatives at different levels (Table 1) can be found in Figure 1. LA was analyzed at 5% and 2.5% ($p < 0.0001$ and $p = 0.001$, respectively); results showed both concentrations to be statistically significant from the untreated control. Similarly, FA was analyzed at 3% and 1.5% ($p < 0.0001$ and $p = 0.0002$, respectively), where both concentrations were also found to be significantly different from the control group. Other screened compounds (PA, PS, and SOP) were not significantly different ($p > 0.05$) from the untreated control group, which might be due to the concentrations used in this study. These results confirm that LA and FA can be used as single antimicrobial treatments to reduce the *Salmonella* Typhimurium cell counts in pork.

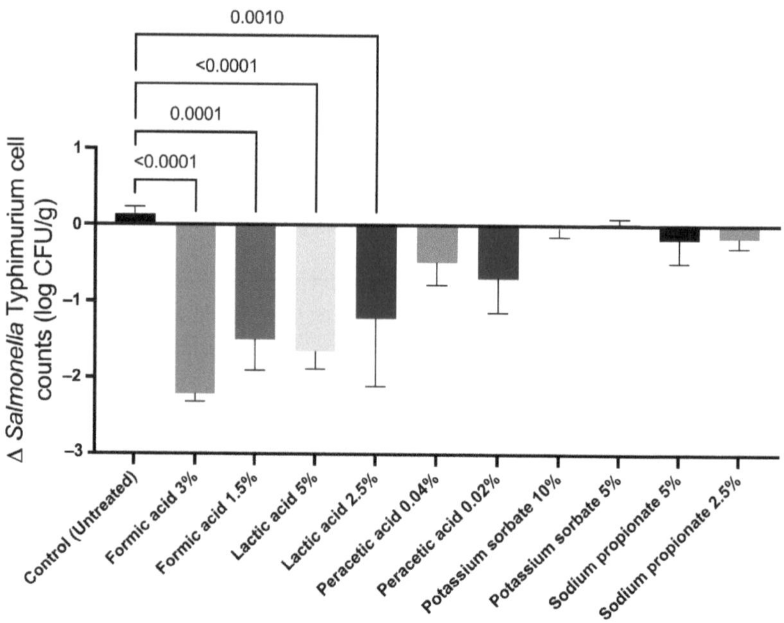

Figure 1. Antimicrobial screening of chemically polar compounds that were surface-applied to miniaturized pork disks previously inoculated with *Salmonella* Typhimurium ATCC 19585. Results are presented as the difference in cell counts after 24 h incubation at 4 °C; experiments were performed in biological triplicate. Antimicrobial treatments tested were formic acid (FA), lactic acid (LA), peracetic acid (PA), potassium sorbate (PS), and sodium propionate (SOP). Antimicrobial treatment concentrations are represented as % (v/w). Statistically significant treatments that had $p < 0.05$ from untreated sample have specific p-values annotated, while all other compounds had $p > 0.05$. Statistically significant compounds were identified through a one-way ANOVA followed by Dunnett's multiple comparisons test.

A library of 30 EOs (Table 2) was also screened for their potential use as antimicrobial treatments against *Salmonella* Typhimurium on miniaturized pork loin ends (Figure 2). The screening showed nine treatments of the 30 EOs used had significantly lower *Salmonella* counts than the untreated control group. CMN showed the highest antimicrobial activity against *Salmonella* Typhimurium ($p < 0.0001$), compared to the control group. CLV, dillweed, and PPP also showed significantly reduced *Salmonella* Typhimurium counts ($p = 0.002, 0.006$, and 0.006, respectively), followed by Cognac and SPT ($p = 0.006$ and 0.003, respectively), when compared to the untreated control group. Finally, cassia, eucalyptus, and thyme were also significant ($p = 0.013, 0.014$, and 0.042, respectively) when compared to the untreated control group for the reduction in *Salmonella* cell counts on a miniaturized pork disk.

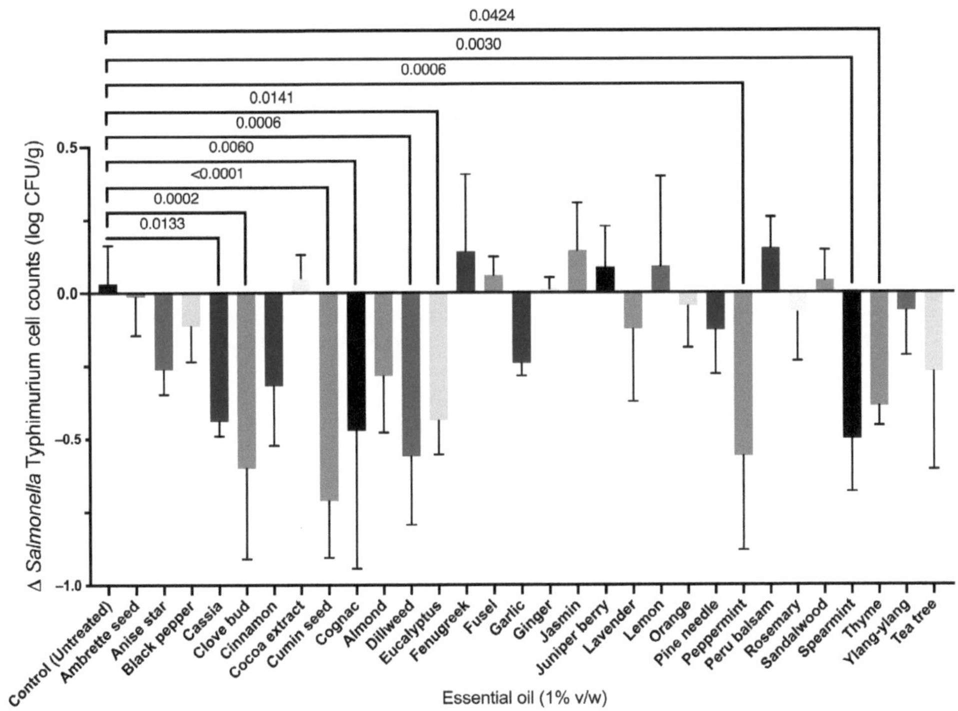

Figure 2. Antimicrobial screening of EO library. EOs were surface-applied to miniaturized pork disks previously inoculated with *Salmonella* Typhimurium ATCC 19585. Results are presented as the difference in cell counts after 24 h incubation at 4 °C. Experiments were carried out in biological duplicate. Statistically significant EOs were identified through a one-way ANOVA followed by Dunnett's multiple comparisons test. Statistically significant treatments that had $p < 0.05$ from untreated sample have specific p-values annotated, while all other compounds had $p > 0.05$.

Based on these results, six antimicrobial compounds were selected to be analyzed for possible interactions when applied in combination to pork. The selected antimicrobials represent the width of significance that was found during the initial screening. Nine EOs showed statistical significance that ranged from $p < 0.05$ to $p < 0.0001$. Therefore, six EOs were chosen to build a Doehlert matrix—RSM model, in addition to the two organic acids that showed significant inhibitory effects. The selected antimicrobial compounds were LA, FA, CMN, CLV, PPP, and SPT.

3.2. A Doehlert Matrix—RSM Model Was Developed to Screen 2-Way Interactions between Antimicrobials and No Synergistic Interactions Were Identified

The Doehlert matrix was selected as the DOE for our analysis due to its buildability capacity. The Doehlert matrix relies on the development of an experiment matrix (Table 3) that contains the minimum number of experiments necessary to build a complete RSM linear regression model. However, The Doehlert matrix allows for the continuous addition of independent variables (antimicrobial compounds, in this case) even after the RSM model has been fitted [16,23]. This feature allows researchers to build into the already acquired knowledge from the first set of experiments by performing a few extra treatments without having to completely build a new experiment matrix. To the best of our knowledge, no other study has used Doehlert matrix—RSM to identify interactions solely between antimicrobial compounds to control bacterial growth on a food matrix.

The Doehlert matrix was built by using six independent variables (LA, FA, CMN, CLV, SPT, and PPP) at different concentrations (see Table 1), and one dependent variable, which was the change in *Salmonella* cell counts 24 h at 4 °C after treatment (represented as Δ *Salmonella* Typhimurium cell counts). As described in Section 2.6, the doses of antimicrobials were adjusted to build the Doehlert matrix—RSM model using actual bacterial counts (instead of <2.01 log CFU/g, LOD of spiral plating), which increases the accuracy of the model's predictions [24]. For all antimicrobials, the lowest concentration used was 0%.

The Doehlert matrix was used to gather experimental data that were then fitted into an RSM model. The obtained RSM linear regression model showed to be statistically significant ($p < 0.0001$), no lack of fit was detected ($p = 0.248$), and the RSD was 0.154 log CFU/g (Table 4). The obtained R^2 value showed that 83% of the variance in the change in *Salmonella* cell counts can be explained by the dependent variables in the model (antimicrobial compounds). Therefore, the fitted Doehlert matrix—RSM model was appropriate for the analysis of antimicrobial interactions on the miniature pork loin end disks.

Table 4. Analysis of variance (ANOVA) for the fitted Doehlert matrix—RSM model. Doehlert matrix—RSM model consisted of six independent variables (LA 1.25%, FA 0.25%, CLV 0.25%, CMN 0.25%, PPP 0.5%, and SPT 0.5%) on the difference in *Salmonella* Typhimurium cell counts after 24 h at 4 °C.

Source	DF	Sum of Squares	Mean Square	F	*p*
Regression	21	2.91	0.14	5.80	<0.001
Residual	25	0.60	0.02		
Lack of fit	22	0.57	0.03	2.48	0.25
Pure error	3	0.03	0.01		
Total corrected	46	3.50	0.08		
R^2	0.83	S.D.	0.28		
Adjusted R^2	0.69	RSD	0.15		

The experimentally obtained data for the Doehlert matrix were compared to the data predicted by the Doehlert matrix—RSM model (Table 3). The obtained residuals' average was 6.7×10^{-4} log CFU/g, which suggests a strong correlation between the experimental and the predicted data. The Doehlert matrix—RSM linear regression model identified LA, CMN, PPP, and SPT as statistically significant ($p = 0.0005$, 0.003, 0.008, and 0.004, respectively) to the model when applied alone to the miniature pork disks (Table 5). Statistically significant terms in the model mean that varying the concentration of those antimicrobials has a significant effect on the change in *Salmonella* Typhimurium cell counts on pork. These results are to be expected, since the antimicrobials in the model (LA, FA, CMN, CLV, SPT, and PPP) showed significant effects against *Salmonella* in the initial screening results ($p = 0.0010$, 0.0001, <0.0001, 0.0002, 0.0030, and 0.0006, respectively) where they were tested individually (Figures 1 and 2). Additionally, the model found CLV not to be significant to the model ($p = 0.119$). Furthermore, the 15 combinations that result from combining six antimicrobials in pairs showed no significant differences in the model ($p > 0.1$), thus suggesting that the Doehlert matrix—RSM model did not identify any two-way combination of antimicrobials to be significantly synergistic or antagonistic against *Salmonella* cell counts on pork meat.

Table 5. Doehlert matrix—RSM linear regression for all terms in the model and their corresponding coefficients and *p*-values (column 1–3). Fitted Doehlert matrix—RSM was used to predict *Salmonella* Typhimurium cell counts by using the model's highest antimicrobial concentrations (column 4). Laboratory-scale validation results and statistical analysis for all terms in the final Doehlert matrix—RSM model (column 5).

Term	Coefficient (β) *	*p* §	Predicted Δ *Salmonella* Cell Counts at Highest Doehlert Matrix—RSM Conc. **	Actual Δ *Salmonella* Cell Counts at Validation Conc. **¶
Constant	0.23	1.75×10^{-20}	0.23	0.09 [a]
LA	−0.32	0	−0.16	−1.23 [bc]
FA	−0.44	0.07	0.12	−2.19 [c]
CMN	−2.73	0.003	−0.45	−0.62 [ab]
PPP	−0.12	0.008	0.17	−0.49 [ab]
CLV	3.79	0.12	1.18	−0.56 [ab]
SPT	−2.53	0.004	−1.03	−0.68 [ab]
LA*FA	0.7	0.761	−0.27	−0.72 [ab]
LA*CMN	−1.86	0.47	−0.85	−0.68 [ab]
LA*PPP	−0.19	0.888	−0.22	−0.83 [ab]
LA*CLV	0.11	0.968	0.78	−0.50 [ab]
LA*SPT	0.48	0.719	−1.43	−0.61 [ab]
FA*CMN	1.37	0.91	−0.56	−1.01 [b]
FA*PPP	−1.29	0.841	0.06	−1.01 [b]
FA*CLV	−15.28	0.251	1.07	−1.06 [b]
FA*SPT	4.75	0.477	−1.14	−1.21 [bc]
CMN*PPP	8.14	0.197	−0.51	−0.27 [ab]
CMN*CLV	−14.61	0.261	0.5	−0.37 [ab]
CMN*SPT	8.01	0.226	−1.72	−0.68 [ab]
PPP*CLV	−6.47	0.304	1.12	−0.55 [ab]
PPP*SPT	−1.94	0.545	−1.09	−0.61 [ab]
CLV*SPT	2.98	0.631	−0.08	−0.41 [ab]

* Coefficient values are unscaled and can be used to predict the change in *Salmonella* Typhimurium cell counts by multiplying them by any desired antimicrobial concentration within the limits of the model (see Tables 1 and 2 for specific concentrations, and abbreviations). § Significant $p < 0.1$ represented in bold. ** Changes in *Salmonella* Typhimurium cell counts are represented as log CFU/g. ¶ Terms not connected by the same letter are statistically significant ($\alpha = 0.05$) based on ANOVA analysis followed by a Tukey–Kramer multiple comparison test.

3.3. Lab-Scale Validation Confirmed Doehlert Matrix—RSM Model Results: Combining Organic Acids with EOs Reduces the Antimicrobial Activity of Organic Acids

A lab-scale validation was carried out to confirm the Doehlert matrix—RSM model predictions. The validation experiments included six antimicrobials tested alone, fifteen two-way combinations of these, and an untreated control group. Specific antimicrobial concentrations can be found in Table 1. The lab-scale validation was performed on pork loin slices that were left hanging for 24 h at 4 °C after the antimicrobial treatments were sprayed onto the surface of the meat. Results of the lab-scale validation (Table 5, column 5) showed that the untreated control group increased *Salmonella* Typhimurium cell counts by 0.09 log CFU/g. Results also showed that LA, FA, FA*CMN, FA*PPP, FA*CLV, and FA*SPT were significantly different from the untreated control group ($p = 0.007$, <0.001, 0.040, 0.042, 0.027, and 0.008, respectively).

However, since the objective of the validation experiments was to confirm the results obtained by the Doehlert matrix—RSM model, the model predictions were compared to the results obtained by the validation experiments. Regardless of LA being significant ($p = 0.007$) when added alone to the pork slices, none of the two-way combinations of antimicrobials that include LA (LA*FA, LA*CMN, LA*PPP, LA*CLV, and LA*SPT) were significantly different from the control group ($p > 0.05$). Furthermore, LA*FA, LA*CMN, LA*PPP, LA*CLV, and LA*SPT are not significantly different from LA alone ($p = 0.914$, 0.847, 0.989, 0.706). Therefore, these results suggest that (i) adding LA alone is significantly better than no treatment, (ii) combining LA with any of the other tested antimicrobials is not significantly different from adding LA alone, and (iii) two-way combinations that include LA do not significantly change *Salmonella* cell counts compared to non-treated pork slices.

On the other hand, the combination of FA*SPT is not statistically different from adding FA alone ($p = 0.099$). Combinations of FA with CMN, PPP, and CLV were statistically different from adding FA alone ($p = 0.022$, 0.021, and 0.002, respectively) and from the untreated control ($p = 0.040$, 0.042, and 0.027, respectively). Additionally, the combination of FA with LA is not statistically different from the untreated control group ($p = 0.303$), thus suggesting that FA alone or FA combined with SPT are the most effective treatments for reducing *Salmonella* cell counts on pork slices, followed by FA*CMN, FA*PPP, and FA*CLV.

Finally, the lab-scale validation showed that CMN, PPP, CLV, and SPT alone were not significantly different from the untreated control group ($p = 0.569$, 0.801, 0.388, and 0.519, respectively). Similarly, two-way combinations between EOs (CMN*PPP, CMN*CLV, CMN*SPT, PPP*CLV, PPP*SPT, and CLV*SPT) did not show significant differences from the untreated control group ($p > 0.05$). Lab-scale results confirmed the previously obtained results from the Doehlert matrix—RSM model. Briefly, FA and LA are effective antimicrobials that can be used alone to reduce *Salmonella* Typhimurium cell counts on pork. Furthermore, no synergistic interactions were identified between the tested organic acids and EOs to control *Salmonella* cell counts.

4. Discussion

Industry efforts to reduce *Salmonella* spp. cell counts on pork are mostly based on spraying organic acids directly onto the surface of the meat. However, due to lack of evidence of these kinds of treatments being significantly better than others (hot water sprays, dipping, higher acid concentrations, etc.) [4], concerns about negative impacts on product's quality [9], and new consumer trends [26], an innovative approach to efficiently screen the antimicrobial potential of a large number of compounds was implemented and validated in this study. Our approach to screen a variety of chemically different compounds by implementing a miniaturized pork loin disk model allowed us to reduce the number of potential antimicrobial compounds from 35 to 11. A Doehlert matrix was used to develop an RSM model with the capacity to identify significant interactions between antimicrobials. The results showed no statistically significant interactions between tested antimicrobials added in pairs to pork meat to reduce *Salmonella* Typhimurium cell counts. Finally, a lab-scale validation that mimics the handling of pork cuts in a processing plant was implemented to validate the results from Doehlert matrix—RSM. Validation results confirmed the Doehlert matrix—RSM model's predictions and provided insight into two-way combinations of the tested antimicrobial compounds. Figure 3 lays out the overall progression of experiments carried out in the present study, as well as specific results from each experimental stage.

4.1. High-Throughput Screening of EOs and Other Antimicrobials Found Most Do Not Show Antimicrobial Activity against Salmonella Typhimurium in Miniaturized Pork at Maximum Working Industry Concentrations

Considering the overwhelming number of potential antimicrobial compounds available, this study implemented a time-efficient method to evaluate antimicrobial activity against *Salmonella* Typhimurium in pork. The results obtained by the proposed screening method represent a real-life case scenario since (i) the tested substances were added at regulatory acceptable concentrations, (ii) their effect was assessed against a food-borne relevant *Salmonella* Typhimurium strain, and (iii) the studies were performed directly on the food matrix of interest, pork meat. The use of EOs as antimicrobials against *Salmonella* Typhimurium has been reported before in broth media. A 2015 study screened 21 EOs using the agar disk diffusion method against 10 strains of *Salmonella* sp. [27]. Results showed that CLV, cinnamon, oregano, thyme, bergamot, orange, cajeput, and sage oil were significantly effective against *Salmonella* spp. However, our results did not identify cinnamon, thyme, or orange to be effective using miniaturized pork loin disks. Another study in 2009 reported the effect of dipping pork loins into LA and acetic acid to reduce *Salmonella* sp. cell counts, as well as the effect of these acids in combination with supercritical carbon dioxide [28]. Results from the treatments with 3% LA showed a significant reduction in bacterial cell counts of 0.86 log CFU/cm^2. However, LA treatments at 2.5% and 5% showed higher *Salmonella* Typhimurium cell reductions (1.51 and 1.66 log CFU/g, respectively) in the present study. These differences might be due to the considerably higher inoculum (7.02 log CFU/cm^2) and to the inoculated samples being incubated overnight before antimicrobial treatments [28], which increases bacterial cell attachment, making antimicrobial treatments less effective. In 2022, the effect of adding LA and a lytic phage to a marinade (pH = 6.8) to reduce *Salmonella* sp. cell counts on pork loins was studied [29]. Results showed that 2.5% LA did not significantly inhibit *Salmonella* sp. compared to the inoculum (3.90 log CFU/cm^2) after 1 h in the marinade. However, in the present study, 2.5% LA significantly reduced *Salmonella* counts by 1.2 log CFU/g (p = 0.001) compared to the control group. A possible reason for these two results to differ from each other might be that the higher pH and salt concentration of the marinade acts as a buffer, which impedes LA from dissociating in the matrix, making it less effective against bacteria [9]. These results emphasize the importance of screening antimicrobial treatments directly on the matrix of interest and designing experiments with the desired mode of application in mind.

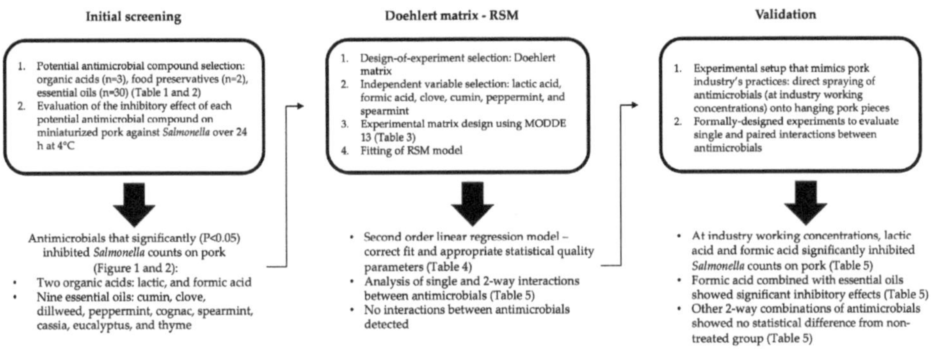

Figure 3. Experimental overview and specific results of the three major sections in the present study. Briefly, the results from the initial screening were used to design the Doehlert matrix—RSM. Similarly, the results from the Doehlert matrix—RSM were then used to set up the validation experiments.

4.2. Formally Designed Experiments Can Be Used to Efficiently Screen Antimicrobial Treatments against *Salmonella* Typhimurium on Miniaturized Pork Disks

In the current study, we implemented an RSM model based on a Doehlert matrix (Doehlert matrix—RSM) to analyze the effect of six antimicrobials added alone or in pairs to control *Salmonella* Typhimurium cell counts on pork. RSM is an empirical modelling approach that is commonly used to optimize processes [16]. Central composite design (CCD) is the DOE most used for RSM models [30]; however, CCD is not the most efficient DOE. The Doehlert matrix uses the least number of experiments for the same number of variables, compared to CCD and Box–Behnken designs. Furthermore, the Doehlert matrix allows users to sequentially add new variables into the model by recycling the original DOE [23]. A 2001 study used a Doehlert matrix—RSM model to inhibit *Escherichia coli* through the combined effect of water activity (a_w), pH, and nisin in peptone water [20]. Authors reported that all three of the selected factors significantly impacted *E. coli* cell counts. Regardless of the model showing a significant lack of fit, the linear regression model was used to optimize *E. coli* inhibition with good correlation (R^2 = 0.947). Doehlert matrix—RSM has been extensively used in engineering, chemistry, and food analysis fields [22,31,32]. However, to the best of our knowledge, no other study has been published recently that implements a Doehlert matrix for food safety applications.

Due to the nature of the Doehlert matrix, the independent variables in the model are applied together to every experiment in the matrix. Such an approach showed disadvantages in the current study, since the antimicrobial concentrations that showed inhibitory effects during initial screening were lowered significantly in the experiment matrix (Table 1). As expected, decreasing the concentration of the antimicrobials negatively affected their antimicrobial activity. For example, the Doehlert matrix—RSM predicted 0.25% FA to increase *Salmonella* cell counts (Table 5) regardless of screening results identifying 3% FA as the most efficient treatment (Figure 1). Furthermore, it is important to note that the Doehlert matrix—RSM developed in this study is a feasible tool to identify antimicrobial combinations that have a synergistic effect when combined against *Salmonella* Typhimurium. However, since the model requires all antimicrobials to be added at the same time onto the matrix, antimicrobial concentrations must be lowered significantly. Therefore, a secondary validation test must be implemented to find concentrations of synergistic antimicrobials that potentiate their effect on the food matrix, which is a potential disadvantage for the implementation of Doehlert matrix—RSM as a screening tool. However, a different DOE can be implemented to overcome the disadvantage of the Doehlert matrix, while still being more efficient than the traditional one-off experiment approach.

4.3. *Pork Slices Were Used in a Lab-Scale Validation That Suggests EOs Decrease the Antimicrobial Activity of Tested Organic Acids against Salmonella Typhimurium*

Results from Doehlert matrix—RSM were validated using a lab-scale validation. Current pork industry practices consist of spraying antimicrobials on the meat. Therefore, validation studies of antimicrobial efficacy should aim to mimic industry practices. A recent study [33] showed the effect of different meat surfaces on the antimicrobial activity of 3% LA and 0.04% PA. Pre-rigor skin-on pork carcasses were surface-inoculated with *Salmonella*, to a concentration of 5–6 log CFU/g. Antimicrobial treatments were applied by using an industrial sprayer. Results showed no significant differences in *Salmonella* counts between samples treated with LA or PA and the ambient temperature water control when applied to the skin-on surface of the carcass. However, LA and PA were significantly effective when applied to the inside of the body cavity of pork carcasses, which is mainly lean tissue. The results from the present study also showed that LA added alone to pork loins (mostly lean tissue) significantly inhibited *Salmonella* cell counts. However, it has been reported before that bacteria are more susceptible to organic acids when cells are attached to fatty tissue due to the reduced water activity. Less moisture reduces the chance of the acid being diluted or buffered by other components present on the surface of the matrix [9]. Therefore, antimicrobial interventions should be designed and selected considering the composition of the product that the treatment will be applied to. Furthermore, the current study showed no synergistic interactions between treatments, which is similar to what has been reported before. A study in 2019 reported the effect of six EOs, six organic acids, and their salts, alone and in combination, against *Salmonella* sp. in broth [34]. Results showed FA and CLV to be individually effective. However, two-way combinations between all tested compounds showed no synergistic effects; only additive effects were identified. Similarly, another study in 2005 combined five EOs and five organic acids against *Salmonella* Typhimurium in broth [35]. Results showed no synergistic effects between antimicrobials. However, the antimicrobial combinations were able to reduce bacteria cell counts. There is scarce information on the effect of combining antimicrobials directly on pork. However, this study provides insight into a possible antagonism from combining certain treatments in a lab-scale setting to inhibit *Salmonella* Typhimurium cell counts on pork.

The current study presents an effective approach for antimicrobial screening on pork. Despite the limitations of the study, this is a viable option to study interactions between antimicrobials. The results from the present study also showcase that simply combining two effective antimicrobial compounds does not always lead to higher inhibition, which can potentially save money to industry when trying to implement novel antimicrobial interventions. Furthermore, the approach presented here represents a laboratory-scale validation that can easily be scaled up to industry, since the experimental design is based on current industry practices. Finally, the present study shows that antimicrobial screening should be designed and selected considering the composition of the product surface to which it is applied.

5. Conclusions

The current study implemented a high-throughput screening tool to screen new antimicrobial compounds as they become relevant for the pork industry directly on pork loin ends. Furthermore, once potential antimicrobials have been identified, a formally designed model can predict interactions between antimicrobials, or lack thereof. The model showed that LA, FA, CMN, PPP, and SPT can be used to efficiently inhibit *Salmonella* counts on pork. However, no two-way combinations between these antimicrobials showed significant antimicrobial inhibition. The predictions from the model were validated by spraying antimicrobials on pork loin end slices. The results obtained in this study showcase the importance of testing antimicrobials directly on the food matrix of interest, while keeping in mind the possible application of such interventions. This study exemplifies that simply combining antimicrobials that are individually effective does not necessarily lead to desired food safety outcomes. Product- and surface-specific antimicrobial treatments that prioritize practical implementation and ensure the desired quality parameters of the product are needed in the pork industry to reduce possible contamination of raw meat with foodborne bacteria. Further studies should focus on the implementation of RSM models for antimicrobial screening on food matrices that use other DOE to compare their efficacy to that of the Doehlert matrix.

Author Contributions: Conceptualization, M.J.M., M.J.S., B.N.H. and C.R.-M.; Formal analysis, C.R.-M.; Funding acquisition, M.J.M.; Investigation, C.R.-M.; Methodology, A.P.L.F. and C.R.-M.; Project administration, M.J.M.; Resources, M.J.S.; Supervision, M.J.M. and M.J.S.; Validation, C.R.-M.; Writing—original draft, C.R.-M.; Writing—review and editing, M.J.M., M.J.S., B.N.H. and C.R.-M. All authors have read and agreed to the published version of the manuscript.

Funding: This work was supported by the Meat and Poultry Research and Education Association and by the National Council for Science and Technology (CONAHCYT), Mexico (Doctoral student fellowship program).

Data Availability Statement: The data used to support the findings of this study can be made available by the corresponding author upon request.

Acknowledgments: This work was supported by the Meat and Poultry Research and Education Association and by the National Council for Science and Technology (CONAHCYT), Mexico (Doctoral student fellowship program). Any opinions, findings, conclusions, or recommendations expressed in this publication (or audiovisual) are those of the author(s) and do not necessarily reflect the view of the funding agencies. We thank Paola Corea, Michael O'Connell, and Aurora Miranda for their support during the experimental stages of this project.

Conflicts of Interest: The authors declare no conflict of interest. The funders had no role in the design of the study, analyses, data analysis, manuscript writing, or in the decision to publish the results.

References

1. Baer, A.A.; Miller, M.J.; Dilger, A.C. Pathogens of Interest to the Pork Industry: A Review of Research on Interventions to Assure Food Safety. *Compr. Rev. Food Sci. Food Saf.* **2013**, *12*, 183–217. [CrossRef]
2. Teklemariam, A.; Al-Hindi, R.; Albiheyri, R.; Alharbi, M.; Alghamdi, M.; Filimban, A.; Mutiri, A.; Al-Alyani, A.; Alseghayer, M.; Almaneea, A.; et al. Human Salmonellosis: A Continuous Global Threat in the Farm-to-Fork Food Safety Continuum. *Foods* **2023**, *12*, 1756. [CrossRef]
3. Raw Pork Products Exploratory Sampling Program | Food Safety and Inspection Service. Available online: http://www.fsis.usda.gov/science-data/sampling-program/raw-pork-products-exploratory-sampling-program (accessed on 15 June 2023).
4. Totton, S.C.; Glanville, J.M.; Dzikamunhenga, R.S.; Dickson, J.S.; O'Connor, A.M. Systematic Review of the Magnitude of Change in Prevalence and Quantity of *Salmonella* after Administration of Pathogen Reduction Treatments on Pork Carcasses. *Anim. Health Res. Rev.* **2016**, *17*, 39–59. [CrossRef]

5. Mukhopadhyay, S.; Gorris, L.G.M. Hurdle Technology. In *Encyclopedia of Food Microbiology*, 2nd ed.; Batt, C.A., Tortorello, M.L., Eds.; Academic Press: Oxford, UK, 2014; pp. 221–227, ISBN 978-0-12-384733-1.
6. Eastwood, L.C.; Taylor, T.M.; Savell, J.W.; Gehring, K.B.; Arnold, A.N. Efficacy of Antimicrobial Interventions in Reducing *Salmonella enterica*, Shiga Toxin-Producing *Escherichia coli*, *Campylobacter*, and *Escherichia coli* Biotype I Surrogates on Non-Chilled and Chilled, Skin-on and Skinless Pork. *Meat Sci.* **2021**, *172*, 108309. [CrossRef]
7. Gonzalez, S.V.; Nair, M.N.; Belk, K.E.; Geornaras, I. Efficacy of Antimicrobial Spray Treatments in Reducing *Salmonella enterica* Populations on Chilled Pork. *J. Food Prot.* **2023**, *86*, 100068. [CrossRef]
8. Van Ba, H.; Seo, H.-W.; Seong, P.-N.; Kang, S.-M.; Cho, S.-H.; Kim, Y.-S.; Park, B.-Y.; Moon, S.-S.; Kang, S.-J.; Choi, Y.-M.; et al. The Fates of Microbial Populations on Pig Carcasses during Slaughtering Process, on Retail Cuts after Slaughter, and Intervention Efficiency of Lactic Acid Spraying. *Int. J. Food Microbiol.* **2019**, *294*, 10–17. [CrossRef]
9. Mani-López, E.; García, H.S.; López-Malo, A. Organic Acids as Antimicrobials to Control *Salmonella* in Meat and Poultry Products. *Food Res. Int.* **2012**, *45*, 713–721. [CrossRef]
10. Wang, C.; Yang, J.; Zhu, X.; Lu, Y.; Xue, Y.; Lu, Z. Effects of *Salmonella* Bacteriophage, Nisin and Potassium Sorbate and Their Combination on Safety and Shelf Life of Fresh Chilled Pork. *Food Control* **2017**, *73*, 869–877. [CrossRef]
11. Schneider, G.; Steinbach, A.; Putics, Á.; Solti-Hodován, Á.; Palkovics, T. Potential of Essential Oils in the Control of *Listeria monocytogenes*. *Microorganisms* **2023**, *11*, 1364. [CrossRef] [PubMed]
12. Calo, J.R.; Crandall, P.G.; O'Bryan, C.A.; Ricke, S.C. Essential Oils as Antimicrobials in Food Systems—A Review. *Food Control* **2015**, *54*, 111–119. [CrossRef]
13. Mohan, A.; Purohit, A.S. Anti-*Salmonella* Activity of Pyruvic and Succinic Acid in Combination with Oregano Essential Oil. *Food Control* **2020**, *110*, 106960. [CrossRef]
14. Govaris, A.; Solomakos, N.; Pexara, A.; Chatzopoulou, P.S. The Antimicrobial Effect of Oregano Essential Oil, Nisin and Their Combination against *Salmonella* Enteritidis in Minced Sheep Meat during Refrigerated Storage. *Int. J. Food Microbiol.* **2010**, *137*, 175–180. [CrossRef]
15. Dussault, D.; Vu, K.D.; Lacroix, M. Development of a Model Describing the Inhibitory Effect of Selected Preservatives on the Growth of *Listeria monocytogenes* in a Meat Model System. *Food Microbiol.* **2016**, *53*, 115–121. [CrossRef]
16. Witek-Krowiak, A.; Chojnacka, K.; Podstawczyk, D.; Dawiec, A.; Pokomeda, K. Application of Response Surface Methodology and Artificial Neural Network Methods in Modelling and Optimization of Biosorption Process. *Bioresour. Technol.* **2014**, *160*, 150–160. [CrossRef]
17. Dávila-Rodríguez, M.; López-Malo, A.; Palou, E.; Ramírez-Corona, N.; Jiménez-Munguía, M.T. Antimicrobial Activity of Nanoemulsions of Cinnamon, Rosemary, and Oregano Essential Oils on Fresh Celery. *LWT* **2019**, *112*, 108247. [CrossRef]
18. Paudel, S.K.; Bhargava, K.; Kotturi, H. Antimicrobial Activity of Cinnamon Oil Nanoemulsion against *Listeria monocytogenes* and *Salmonella* spp. on Melons. *LWT* **2019**, *111*, 682–687. [CrossRef]
19. Schmidt, S.J.; Fontana, A.J., Jr. Appendix E: Water Activity Values of Select Food Ingredients and Products. In *Water Activity in Foods*; John Wiley & Sons, Ltd.: Hoboken, NJ, USA, 2007; pp. 407–420, ISBN 978-0-470-37645-4.
20. Cerrutti, P.; Terebiznik, M.R.; De Huergo, M.S.; Jagus, R.; Pilosof, A.M.R. Combined Effect of Water Activity and pH on the Inhibition of *Escherichia coli* by Nisin. *J. Food Prot.* **2001**, *64*, 1510–1514. [CrossRef] [PubMed]
21. Bezerra, M.A.; Santelli, R.E.; Oliveira, E.P.; Villar, L.S.; Escaleira, L.A. Response Surface Methodology (RSM) as a Tool for Optimization in Analytical Chemistry. *Talanta* **2008**, *76*, 965–977. [CrossRef] [PubMed]
22. Cerqueira, U.M.F.M.; Bezerra, M.A.; Ferreira, S.L.C.; de Jesus Araújo, R.; da Silva, B.N.; Novaes, C.G. Doehlert Design in the Optimization of Procedures Aiming Food Analysis—A Review. *Food Chem.* **2021**, *364*, 130429. [CrossRef]
23. Ferreira, S.; Santos, W.; Quintella, C.; Neto, B.; Bosque-Sendra, J. Doehlert Matrix: A Chemometric Tool for Analytical Chemistry—Review. *Talanta* **2004**, *63*, 1061–1067. [CrossRef]
24. Myers, R.H.; Montgomery, D.C.; Anderson-Cook, C.M. *Response Surface Methodology: Process and Product Optimization Using Designed Experiments*; John Wiley & Sons: Hoboken, NJ, USA, 2016; ISBN 978-1-118-91603-2.
25. Castillo, A.; Lucia, L.M.; Goodson, K.J.; Savell, J.W.; Acuff, G.R. Comparison of Water Wash, Trimming, and Combined Hot Water and Lactic Acid Treatments for Reducing Bacteria of Fecal Origin on Beef Carcasses. *J. Food Prot.* **1998**, *61*, 823–828. [CrossRef] [PubMed]
26. Delgado-Pando, G.; Ekonomou, S.I.; Stratakos, A.C.; Pintado, T. Clean Label Alternatives in Meat Products. *Foods* **2021**, *10*, 1615. [CrossRef]
27. Mazzarrino, G.; Paparella, A.; Chaves-López, C.; Faberi, A.; Sergi, M.; Sigismondi, C.; Compagnone, D.; Serio, A. *Salmonella enterica* and *Listeria monocytogenes* Inactivation Dynamics after Treatment with Selected Essential Oils. *Food Control* **2015**, *50*, 794–803. [CrossRef]
28. Choi, Y.M.; Kim, O.Y.; Kim, K.H.; Kim, B.C.; Rhee, M.S. Combined Effect of Organic Acids and Supercritical Carbon Dioxide Treatments against Nonpathogenic *Escherichia coli*, *Listeria monocytogenes*, *Salmonella Typhimurium* and *E. coli* O157:H7 in Fresh Pork. *Lett. Appl. Microbiol.* **2009**, *49*, 510–515. [CrossRef]
29. Li, S.; Konoval, H.M.; Marecek, S.; Lathrop, A.A.; Pokharel, S. *Salmonella* spp. Response to Lytic Bacteriophage and Lactic Acid on Marinated and Tenderized Raw Pork Loins. *Foods* **2022**, *11*, 879. [CrossRef]

30. Azevedo, A.N.; Buarque, P.R.; Cruz, E.M.O.; Blank, A.F.; Alves, P.B.; Nunes, M.L.; Santana, L.C.L.d.A. Response Surface Methodology for Optimisation of Edible Chitosan Coating Formulations Incorporating Essential Oil against Several Foodborne Pathogenic Bacteria. *Food Control* **2014**, *43*, 1–9. [CrossRef]
31. Narenderan, S.T.; Meyyanathan, S.N.; Karri, V.V.S.R. Experimental Design in Pesticide Extraction Methods: A Review. *Food Chem.* **2019**, *289*, 384–395. [CrossRef]
32. Mousavi, L.; Tamiji, Z.; Khoshayand, M.R. Applications and Opportunities of Experimental Design for the Dispersive Liquid–Liquid Microextraction Method—A Review. *Talanta* **2018**, *190*, 335–356. [CrossRef] [PubMed]
33. Bonilla, K.P.; Vega, D.; Maher, J.; Najar-Villareal, F.; Kang, Q.; Trinetta, V.; O'Quinn, T.G.; Phebus, R.K.; Gragg, S.E. Validation of Commercial Antimicrobial Intervention Technologies to Control *Salmonella* on Skin-on Market Hog Carcasses and Chilled Pork Wholesale Cuts. *Food Control* **2023**, *151*, 109829. [CrossRef]
34. Gómez-García, M.; Sol, C.; de Nova, P.J.G.; Puyalto, M.; Mesas, L.; Puente, H.; Mencía-Ares, Ó.; Miranda, R.; Argüello, H.; Rubio, P.; et al. Antimicrobial Activity of a Selection of Organic Acids, Their Salts and Essential Oils against Swine Enteropathogenic Bacteria. *Porc. Health Manag.* **2019**, *5*, 32. [CrossRef]
35. Nazer, A.I.; Kobilinsky, A.; Tholozan, J.-L.; Dubois-Brissonnet, F. Combinations of Food Antimicrobials at Low Levels to Inhibit the Growth of *Salmonella* Sv. Typhimurium: A Synergistic Effect? *Food Microbiol.* **2005**, *22*, 391–398. [CrossRef]

foods

Article

Variability in Physicochemical Parameters and Its Impact on Microbiological Quality and Occurrence of Foodborne Pathogens in Artisanal Italian Organic Salami

Frédérique Pasquali [1], Antonio Valero [2], Arícia Possas [2,*], Alex Lucchi [1], Cecilia Crippa [1], Lucia Gambi [1], Gerardo Manfreda [1] and Alessandra De Cesare [3]

1 Department of Agricultural and Food Sciences, Alma Mater Studiorum—University of Bologna, 40126 Bologna, Italy; frederique.pasquali@unibo.it (F.P.); alex.lucchi3@unibo.it (A.L.); cecilia.crippa2@unibo.it (C.C.); lucia.gambi@gmail.com (L.G.); gerardo.manfreda@unibo.it (G.M.)
2 Department of Food Science and Technology, University of Cordoba, Agrifood Campus of International Excellence ceiA3, Campus Rabanales, 14014 Córdoba, Spain; bt2vadia@uco.es
3 Department of Veterinary Medical Sciences, Alma Mater Studiorum—University of Bologna, 40126 Bologna, Italy; alessandra.decesare@unibo.it
* Correspondence: g12mepoa@uco.es

Abstract: Artisanal salami is produced in small-scale production plants, where the lack of full automation might result in higher variability in food intrinsic properties. The aim of the present study was to evaluate the inter- and intra-batch variability in physicochemical parameters and its impact on microbial quality and occurrence of foodborne pathogens on 480 samples collected from six batches of an artisanal Italian production of organic salami. Relatively high total bacterial counts (TBC) were found on the surface of the table in the stuffing room (4.29 ± 0.40 log cfu/cm^2). High loads of *Enterobacteriaceae* in the meat mixture of batch 2 and TBC in batch 5 were associated with a higher occurrence of bacterial pathogens. During ripening, water activity (a_w) and pH failed to reach values lower than 0.86 and 5.3, respectively. Six *Staphylococcus aureus* and four *Listeria monocytogenes* isolates were collected from the salami meat mixture during ripening and the processing environment. A total of 126 isolates of *Enterobacteriaceae* were characterized at a species level, with *Escherichia coli, Klebsiella pneumoniae, Enterobacter cloacae,* and *Citrobacter freundii* isolated from the final products. Results suggest the relevance of first steps of production in terms of the hygiene of raw materials and handling during stuffing procedures, especially when the physicochemical parameters of the final products do not reach values that represent hurdles for foodborne pathogens.

Keywords: pH; water activity; microbial quality; food safety; *Staphylococcus aureus; Listeria monocytogenes; Klebsiella pneumoniae*

Citation: Pasquali, F.; Valero, A.; Possas, A.; Lucchi, A.; Crippa, C.; Gambi, L.; Manfreda, G.; De Cesare, A. Variability in Physicochemical Parameters and Its Impact on Microbiological Quality and Occurrence of Foodborne Pathogens in Artisanal Italian Organic Salami. *Foods* **2023**, *12*, 4086. https://doi.org/10.3390/foods12224086

Academic Editor: Jan Mei Soon

Received: 6 October 2023
Revised: 29 October 2023
Accepted: 6 November 2023
Published: 10 November 2023

1. Introduction

An increased demand for artisanal foods has been observed in the last decades. These products are generally elaborated locally in small-scale, family-based companies perceived as producing healthier and more ethical food [1,2]. Due to the large variety of local, small-scale productions, microbial data on each specific product are scarce. Moreover, in small-scale food productions, the reduced automation results in variability in physicochemical and microbiological parameters of the final product [3,4].

Italian salami falls within the category of dry fermented sausages with a ripening period longer than 4 weeks and a water activity lower than 0.90 [5]. In Italy, a wide variety of artisanal salami recipes exist with ripening times from 3 to 6 months [6,7]. In traditional salami, the addition of salt and nitrates inhibits Gram-negative spoilage bacteria and enhances coagulase-negative staphylococci (CNS) during fermentation and ripening [7]. CNS use oxygen contributing to a reduction in the redox potential, which in turn inhibits aerobic bacteria in favor of lactic acid bacteria (LAB). Acidification due to LAB growth and the decrease in water activity during natural ripening represent the hurdles reducing the risk of bacterial pathogen growth in fermented sausages [8]. However, the size of these technological hurdles might vary in different productions. In particular, in organic productions, the absence of starter cultures and of nitrates might potentially impact the effective acidification of the food matrix and reduction in Gram-negative and Gram-positive bacterial pathogens.

Shiga toxin-producing *Escherichia coli* (STEC) as well as *Listeria monocytogenes*, *Salmonella* spp., and *Staphylococcus aureus* were isolated from fermented productions [8–10]. *L. monocytogenes* was detected in a French plant producing dry sausages, its equipment, and in 10% of the final products [11]. Higher occurrences of 13–15%, 16%, 42%, and 60% were registered in Italy, Spain, Greece, and Portugal, respectively [2,12–15]. *Salmonella* was detected in Italy in three dry fermenting processing plants with a prevalence of 16.7% in food and 5.8% in the environment [10]. Additionally, *S. aureus* was detected in raw meat, casing, semifinished, and finished fermented sausages produced in both a nitrite and a nitrite-free production in Spain [9]. Although unusual, outbreaks associated with the consumption of contaminated fermented sausages have been described. *E. coli* O157:H7, O103:H25, and O26:H11 have been identified as etiologic agents in outbreaks involving fermented pork and beef meat salami as vehicles in United States in 1994, in Sweden in 2002, in Italy in 2004, in Norway in 2008, and in Denmark in 2018 [16–20]. Similarly, fermented sausages contaminated by *Salmonella* spp. were associated with outbreaks in Germany in 2001, Spain in 2011, and more recently in the United States in 2022, linked to salami sticks [21–23].

The aim of the present study was to evaluate the variability in physicochemical parameters and its impact on the microbiological quality and occurrence of foodborne pathogens in an Italian artisanal factory producing an organic pig-meat salami over a year of sampling. Based on results, potential routes of contamination are suggested.

2. Materials and Methods

2.1. Experimental Design

One artisanal Italian factory was sampled over 10 months (July 2020–May 2021). The factory produces an organic salami made of pig meat with no addition of nitrites/nitrates or starter cultures and with a ripening period of 6 months. Regarding processing, after the stuffing of a mixture of meat, spices, and salt into natural casing, salami were dried at 10 °C and RH 55–60% for 7–9 days and then stored in controlled environment for 5 weeks at 10 °C and RH 70% for maturation. Afterwards, salami were stored in a cellar up to 6 months (Figure 1). Raw materials were sampled, along with intermediate and final products, as well as the processing environment. A total of 420 samples were analyzed, namely, meat mixture (n = 30), salami from the drying room (n = 30), salami in the maturation room (n = 30), and salami after 10 (n = 30), 18 (n = 30), and 28 (n = 30) weeks of ripening in the storage room. Swabs from the surface of a table in the stuffing room (n = 30) as well as the walls (n = 90) and drains (n = 90) were also collected from the stuffing, drying, and maturation rooms. In addition, swabs from the stuffing machine (n = 30) were gathered. For environmental samples, sterile cotton swabs (Copan Italia, Brescia, Italy) were moistened in 10 mL of saline solution (0.9% NaCl) and then used to swab a 100 cm^2 area. These samples were collected before disinfection and cleaning. Five sample units per matrix (food and environment) per batch were tested. Overall, six batches were analyzed: batch 1 (stuffing on the 1 July 2020), batch 2 (23 September 2020), batch 3 (7 October 2020), batch 4 (21 October 2020), batch 5 (4 November 2020), and batch 6 (18 November 2020).

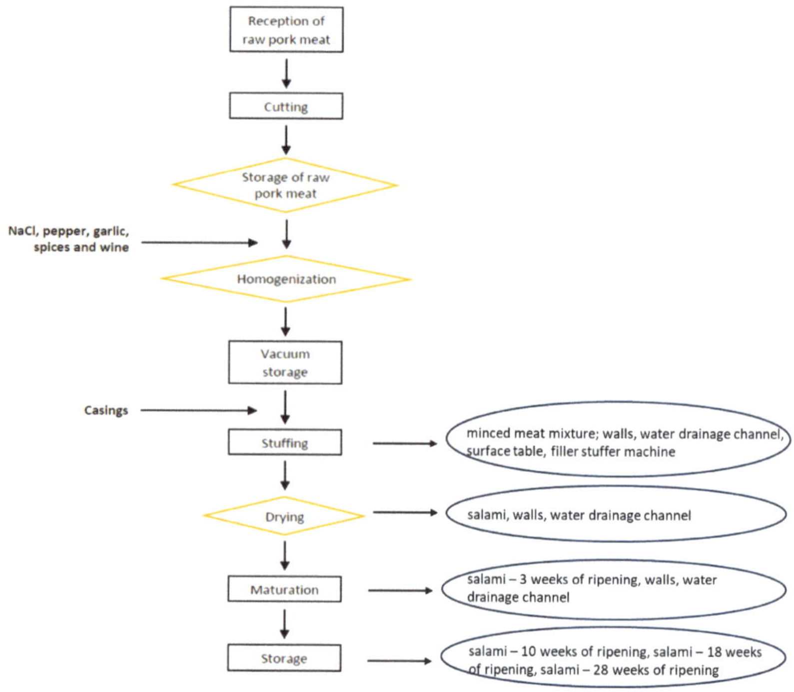

Figure 1. Production flowchart of the Italian organic salami. Sampling spots in the processing environment are indicated in round circles.

2.2. *Microbiological and Physicochemical Analyses*

Total bacterial count (TBC) (ISO 4833-2), water activity (ISO 21807), pH (ISO 2917), and the occurrence of *L. monocytogenes* (ISO 11290-1), coagulase positive Staphylococci (ISO 6888-1), verotoxigenic *E. coli* (VTEC) (ISO 16649), and *Salmonella* (ISO 6579) were investigated in all 480 samples [24–30]. In particular for VTEC, after the isolation of *E. coli* on Tryptone Bile X-GLUC Agar (TBX, Thermo Scientific, Milan, Italy), a PCR for the identification of Shiga toxin-encoding genes was applied as previously described [31]. Additionally, lactic acid bacteria (LAB) (ISO 15214) and *Enterobacteriaceae* (ISO 21528-2) were enumerated in raw materials and semifinished and finished products [32,33]. Moreover, to characterize *Enterobacteriaceae* at a species level, 25 g of food sample was diluted in 225 mL of Buffer Peptone Water (BPW, Thermo Scientific, Milan, Italy) and incubated for 24 h at 37 °C. BPW pre-enriched cultures were then streaked on MacConkey agar (Thermo Scientific) and incubated for 24 h at 37 °C. Five colonies per plate (both lactose fermenting and non-lactose fermenting) were submitted to biochemical test (RapID ONE System and RapID STAPH PLUS System, Thermo Scientific) and PCR for confirmation [31,34–36]. One confirmed isolate per species per sample was retained.

2.3. *Data Analysis and Modelling*

Data analysis and modeling were carried out using R Studio v4.2.2. Statistical comparisons were made between microbial counts from various food batches and environmental samples through ANOVA (with a significance level of $p \leq 0.05$). To identify homogeneous groups, we applied Tukey's HSD test. Additionally, we created boxplots to visually represent the variation in microbial counts within and between batches of salami samples stored at different temperatures throughout their shelf life.

Longitudinal data sets, encompassing TBC and *Enterobacteriaceae* counts, collected from various sources such as environmental surfaces, food contact surfaces, meat mixture, and the finished salami within the drying room were analyzed using generalized linear mixed models (GLMMs). These models were adjusted using the R packages *lme4* [37] and *nlme* to account for inter-batch variability, treating it as a random effect. Main effects considered the influence of the variables "Stage" (mixing, stuffing, drying, and ripening) according to the form:

$$Y_{is(b)} = (\beta_0 + u_b) + Stage_s + \varepsilon_{is(b)}, \quad (1)$$

where $Y_{is(b)}$ represents the count (in log cfu/g) of a specific microbial group (TBC and *Enterobacteriaceae*), determined within the processing stage denoted as s, belonging to the batch labeled as b; β_0 is the model inter-cept which can undergo random shifts denoted as u_b, associ-ated with the specific batch b; $Stage_s$ represents a particular processing stage s; and $\varepsilon_{is(b)}$ rep-resents the error associated with the microbial count i determined within the processing stage s, belonging to batch b. Sam-ples, including environmental and drain swabs, meat mixture, and finished salami collected from each processing stage, were integrated into another main effects model as follows:

$$Y_{is(b)} = (\beta_0 + u_b) + Stage_s(Environment_i) + \varepsilon_{is(b)}, \quad (2)$$

where $Stage_s(Envionment_i)$ represents a sample denoted as i, which was obtained from the processing stage s within a specific batch b. The variability between batches within a fac-tory was quantified by assessing the squared standard deviation of the random effects, while it was assumed that errors also conformed to a normal distribu-tion.

3. Results

3.1. Enumeration of Total Bacterial Count, Lactic Acid Bacteria, and Enterobacteriaceae

Regarding environmental swabs collected at the processing plant, manhole samples presented the highest loads of TBC compared to the other tested environmental sites (Figure 2). More specifically, manhole samples in the drying and ripening rooms (SWD and SWR) showed the highest TBC, with a mean value of 7.08 ± 0.29 log cfu/cm^2 and 7.05 ± 0.61 log cfu/cm^2, respectively. No statistically significant differences ($p > 0.05$) were observed for TBC between batches, also supported by the high intra-batch variability. Interestingly, the surface swab of the table in the stuffing room (STM) showed TBC 4.29 ± 0.40 log cfu/cm^2 higher than those found in the other swabs of the same room, notably the stuffing machine swab (SM, 2.77 ± 0.99 log cfu/cm^2) and the wall swab (SEM, 3.24 ± 1.32 log cfu/cm^2) (Figure 2).

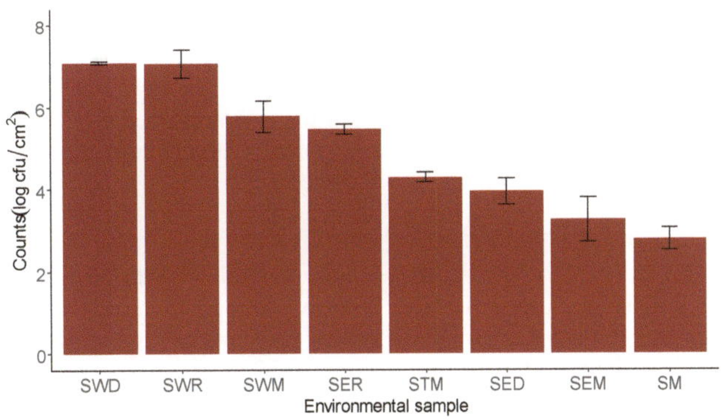

Figure 2. Total bacterial count (TBC) (log cfu/cm^2) on environmental samples collected at the processing plant. SWD: manhole swab—drying room; SWR: manhole swab—ripening room; SWM: manhole swab—stuffing room; SER: wall swab—ripening room; STM: surface swab—stuffing room; SED: wall swab—drying room; SEM: wall swab—stuffing room; SM: minced-meat machine swab—stuffing room (mean ± standard deviation of 6 batches).

Regarding the meat mixture used to produce salami, high intra- and inter-batch variability for the different groups evaluated was observed, especially for *Enterobacteriaceae* (Figure 3). Higher *Enterobacteriaceae* counts were detected in meat mixture samples of batch 2 compared to the other batches, except for batch 1 ($p \leq 0.05$). In addition, significantly higher LAB counts were registered in samples of batch 2. Higher TBC counts in the meat mixture samples of batch 5 were observed compared to the other batches, except for batch 2 ($p \leq 0.05$).

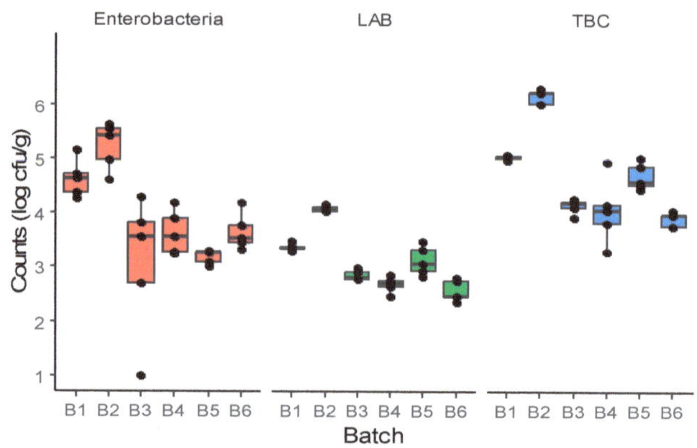

Figure 3. *Enterobacteriaceae*, lactic acid bacteria (LAB), and total bacteria counts (TBC) of meat mixture samples (n = 5) belonging to six different batches.

Regarding salami samples, slight increases in TBC and LAB counts were associated with a decrease in *Enterobacteriaceae* load in all batches (Figure 4). Interestingly, the increase in TBC and LAB reached maximum levels in salami at 3 or 10 weeks of ripening (depending on the batch), after which a decrease was observed up to 28 weeks of ripening. The increase in TBC in batch 1, for example, started at 8.58 log cfu/g in minced meat mixture and reached 9.75 log cfu/g in salami after 10 weeks of ripening, and slightly decreased to 8.77 log cfu/g in the final product (28 weeks of ripening). Similarly, in batch 1, LAB loads increased from 8.53 to 9.34 log cfu/g (10 weeks of ripening) and reached 8.10 log cfu/g in the final product. Regarding *Enterobacteriaceae*, loads showed a decreasing trend all across the ripening process, starting from 4.56 and reaching 1.09 log cfu/g in the final product in batch 1 (Figure 4). No statistically significant differences were observed between batches in TBC and LAB loads in salami after 28 weeks of ripening ($p > 0.05$). In the final product of all batches, the load of *Enterobacteriaceae* was lower or close to the detection limit of 1 log cfu/g, suggesting that the ripening process was effective in reducing the risk for human health related to the potential occurrence of foodborne pathogens included in this bacterial family.

3.2. Physicochemical Parameters (pH and a_w)

As expected, the pH decreased along with the increase in LAB up to 3–10 weeks of ripening, after which it increased along with a reduction in LAB (Figures 4 and 5). Initial pH in the salami in the drying room ranged from 5.28 (batch 1) to 5.58 (batch 5). During ripening, the pH decreased up to 5.3 in all batches except batch 5. However, after 10 weeks of ripening, the pH increased in all batches, reaching values ranging from 5.87 (batch 4) to 6.25 (batch 3) in the final product (Figure 5).

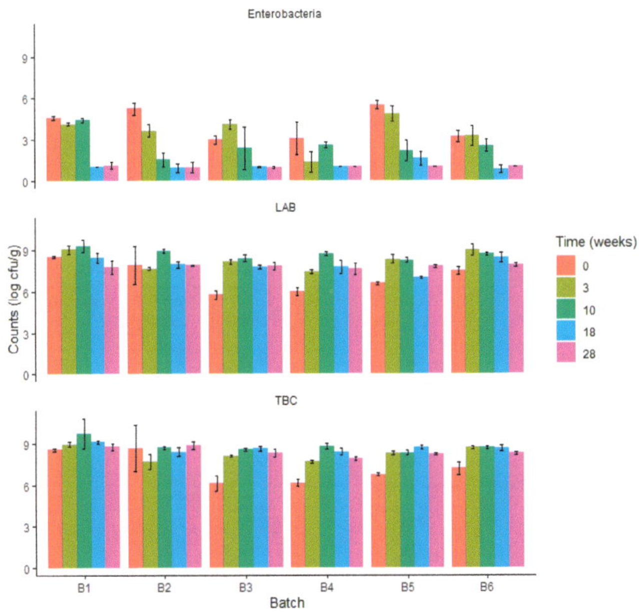

Figure 4. *Enterobacteriaceae*, lactic acid bacteria (LAB), and total bacterial counts (TBC) of salami samples analyzed over the production process: drying room (time = 0) and ripening (3 to 28 weeks) for the six batches evaluated (n = 5 for each batch).

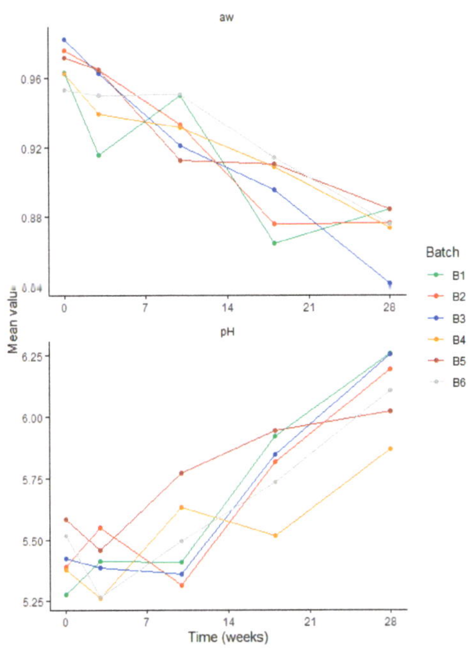

Figure 5. Water activity (a_w) and pH of salami samples analyzed over the production process: drying room (time = 0) and ripening (3 to 28 weeks) for the six batches evaluated (n = 5 for each batch).

Regarding water activity, a constant decrease was observed from values ranging from 0.953 (batch 6) to 0.982 (batch 3) in the salami in the drying room to values ranging from 0.842 (batch 3) to 0.884 (batch 5) in salami after 28 weeks of ripening (Figure 5).

3.3. Generalized Linear Mixed Models

Generalized linear mixed models were created to explore whether variations in the production stage and the environment could account for some of the differences observed between batches in TBC and *Enterobacteriaceae* counts throughout the salami production process. The outcomes of this investigation are presented in Tables 1 and 2.

Table 1. Estimations of model parameters in linear mixed models with random effects, investigating the influence of "Stage" and "Sample" variables on total bacteria counts (TBC) in Italian salami, alongside assessments of between-batch variability.

Model	Parameters	Estimate (SE)	t-Value	Pr > |t|
Main effects: Stage [1]	**Random effects (σ)**			
	Batch in factory	0.504	-	-
	Residual	1.827	-	-
	Fixed effects			
	Intercept	3.153 (0.313)	10.085	<0.001
	Drying	2.941 (0.304)	9.674	<0.001
	Stuffing	0.845 (0.304)	2.780	0.005
	Ripening	4.327 (0.272)	15.916	<0.001
Main effects: Sample [2]	**Random effects (σ)**			
	Batch in factory	0.503	-	-
	Residual	1.390	-	-
	Fixed effects			
	Intercept	7.256 (0.335)	21.615	<0.001
	Sample: MB	−3.783 (0.367)	−10.318	<0.001
	Sample: SBR_28	1.148 (0.367)	3.130	0.002
	Sample: SBR_18	1.117 (0.367)	3.047	0.002
	Sample: SBR_10	0.092 (0.367)	0.251	0.801
	Sample: SBR_3	0.992 (0.367)	2.705	0.007
	Sample: SEM	−4.421 (0.367)	−12.058	<0.001
	Sample: SED	−3.311 (0.367)	−9.031	<0.001
	Sample: SER	−1.796 (0.367)	−4.899	<0.001
	Sample: SM	−4.574 (0.367)	−12.473	<0.001
	Sample: STM	−3.681 (0.367)	−10.039	<0.001
	Sample: SWD	−0.173 (0.367)	−0.472	0.637
	Sample: SWM	−1.517 (0.367)	−4.136	<0.001
	Sample: SWR	−0.202 (0.367)	−0.550	0.583

[1] TBC in mixing stage was set as reference category. [2] TBC in salami samples of drying room was set as reference category. MB: meat mixture; SEM: wall swab (stuffing room); SED: wall swab (drying room); SER: wall swab (ripening room); SM: minced-meat machine swab (stuffing room); STM: surface swab (stuffing room); SWD: water drainage swab (drying room); SWM: water drainage swab (stuffing room); SWR: water drainage swab (ripening room); SBR: ripened salami (3, 10, 18, and 28 weeks).

Results from GLMMs indicated that the TBC in the drying, stuffing, and ripening phases were significantly higher than the TBC found in the meat mixture ($p \leq 0.05$). Specifically, the ripening step produced the largest increase (4.33 ± 0.27 log cfu/g), which can be partially attributed to the increase in LAB populations. During ripening, a positive effect was predicted for TBC counts in 18- and 28-week ripened salami (Table 1). Regarding *Enterobacteriaceae* counts, there was a significant negative effect after mixing, indicating the decrease in microbial counts during the drying, stuffing, and ripening phases. A significant decrease in *Enterobacteriaceae* populations was predicted during salami ripening (Table 2). Significant negative effects were found between most of the analyzed surfaces and TBC and *Enterobacteriaceae* counts in the salami samples in the drying room, thus indicating an increase in microbial populations in the finished salami samples. Nevertheless, TBC counts from the water drainages (ripening and drying rooms) were similar to the TBC levels in the salami samples in the drying room ($p > 0.05$, Table 1). Regarding the inter-batch variability, random effects indicate a higher inter-batch variability for TBC in comparison to *Enterobacteriaceae*.

Table 2. Estimations of model parameters in linear mixed models with random effects, evaluating the impact of "Stage" and "Sample" variables on *Enterobacteriaceae* in Italian salami, and including assessments of between-batch variability.

Model	Parameters	Estimate (SE)	t-Value	Pr > |t|
Main effects: Stage [1]	**Random effects (σ)**			
	Batch in factory	0.193	-	-
	Residual	1.654	-	-
	Fixed effects			
	Intercept	1.952 (0.228)	8.576	<0.001
	Drying	−0.588 (0.275)	−2.132	0.034
	Stuffing	−1.337 (0.275)	−4.849	<0.001
	Ripening	−0.658 (0.246)	−2.670	0.008
Main effects: Sample [2]	**Random effects (σ)**			
	Batch in factory	0.503	-	-
	Residual	1.390	-	-
	Fixed effects			
	Intercept	4.095 (0.171)	24.010	<0.001
	Sample: MB	−0.190 (0.188)	−1.011	0.313
	Sample: SBR_28	−3.096 (0.188)	−16.506	<0.001
	Sample: SBR_18	−3.045 (0.188)	−16.233	<0.001
	Sample: SBR_10	−1.926 (0.188)	−10.267	<0.001
	Sample: SBR_3	−0.547 (0.188)	−2.909	0.004
	Sample: SEM	−4.095 (0.188)	−21.830	<0.001
	Sample: SED	−4.095 (0.188)	−21.830	<0.001
	Sample: SER	−4.095 (0.188)	−21.830	<0.001
	Sample: SM	−4.095 (0.188)	−21.830	<0.001
	Sample: STM	−2.247 (0.188)	−11.978	<0.001
	Sample: SWD	−4.095 (0.188)	−21.830	<0.001
	Sample: SWM	−4.095 (0.188)	−21.830	<0.001
	Sample: SWR	−4.095 (0.188)	−21.830	<0.001

[1] *Enterobacteriaceae* in mixing stage was set as reference category. [2] *Enterobacteriaceae* in salami samples in the drying room was set as reference category. MB: meat mixture; SEM: wall swab (stuffing room); SED: wall swab (drying room); SER: wall swab (ripening room); SM: minced-meat machine swab (stuffing room); STM: surface swab (stuffing room); SWD: water drainage swab (drying room); SWM: water drainage swab (stuffing room); SWR: water drainage swab (ripening room); SBR: ripened salami (3, 10, 18, and 28 weeks).

3.4. Occurrence of Bacterial Pathogens

Regarding bacterial pathogens, none of the 480 samples collected were positive for VTEC or *Salmonella* spp. Four isolates were positive for *L. monocytogenes* and six for *S. aureus*, eight for *S. warneri*, one for *S. capitis*, and one for *S. xilosus*. Regarding *Enterobacteriaceae*, the following species were identified: *Klebsiella oxytoca* and *K. pneumoniae* (33), *E. coli* (30), *Citrobacter freundii* (26), *Enterobacter cloacae* (16), and *Routella planticola* (1) (Table S1).

Bacteria were collected from mixed meat and salami during and at the end of the ripening period, and environmental swabs were gathered during production. In particular, pathogens were mostly found in the meat mixture (22), the surface of the table in the stuffing room (15), and in the salami in the drying room (25) and after 3 weeks of ripening (19) (Table S2). Following ripening, the number of pathogens decreased from 25 in salami in the drying room to 5 and 12 in salami at 18 and 28 weeks of ripening, respectively (Table S2). One *Enterobacter cloacae* strain, six *E. coli*, two *Citrobacter freundii*, and two *K. pneu-*

moniae were found in the finished product, suggesting that although the ripening process was effective in reducing the risk associated with the presence of bacterial pathogens, it was not enough to guarantee the lack of bacterial pathogens in the final food product. Although the water activity in the final product was lower than 0.90, data on the increased pH suggest the need to better control this parameter, for example, by considering the possibility of shortening the ripening period from 28 to 18 weeks.

4. Discussion

In the present study, the effect of pH and water activity on the number of bacterial indicators of the process's hygiene and the occurrence of foodborne pathogens was assessed in an artisanal Italian production of organic salami.

pH is an essential parameter in fermented meat products. The acid hurdle is crucial for the control of the safety of the product [38,39]. pH values below 5.3 are essential to inhibit the growth of Gram-positive bacteria such as *S. aureus*. The pH of the majority of Mediterranean-style fermented sausages is approximately 4.5/5.4, which has several beneficial effects on both the shelf life and the manufacturing process [39,40]. However, the final pH can rise up to 6.0/6.7 in some low-acid fermented sausages (e.g., Soudjouk, Fuet).

In the present study, the pH decreased below 5.3 after 10 weeks of ripening, after which an increase to values higher than 6 in the final products was achieved. These results might be associated with the lack of standardized starter cultures and the different growth rates of autochthonous LAB groups in each batch. In salami prepared with starter cultures of dairy origin, pH 5.3 was reached after 3 weeks of ripening [41]. Neutral values of pH in the final product might reduce the acid hurdles and thus enhance the growth of bacteria. In the present study, although the number of *Enterobacteriaceae* was lower than the detection limit in most final products, bacterial pathogens were detected.

Water activity values lower than 0.90 and 0.86 are essential to control the growth of Gram-positive and Gram-negative bacterial pathogens, respectively [5,42]. In the final product of the present study, all batches showed a water activity value lower than 0.90, but only one (batch 3) fell under the 0.86 limit (batch 3, 0.842), suggesting a potential hurdle for *S. aureus* control. Although not found in the final product, *S. aureus* was found in raw materials, salami in the drying room, and salami after 3 weeks of ripening. This finding suggests the relevance of a high microbiological quality of raw materials, especially when pH and water activity hurdles cannot be fully maintained.

Besides bacterial pathogens being found in the final product, the detection of pathogens in the processing environment is of concern since potential events of cross-contamination might occur involving both workers and the food product before commercialization.

The ability of *L. monocytogenes* to survive on several environmental stress conditions boost the likelihood of detecting this foodborne pathogen in contaminated ready-to-eat products of meat, fish, and dairy-origin products [43,44]. Recently, several *L. monocytogenes* outbreaks were associated with ready-to-eat meat product contamination from Italy and worldwide [45–47], among which a recent US outbreak involved Italian-type salami, mortadella, and prosciutto [48]. Despite *L. monocytogenes* not being detected in the final products, it was found in the processing environment of batch 2 (manhole of the drying room). Therefore, a more intensive control of the contamination routes in the salami processing plant might be necessary. Likewise, *S. aureus* is able to tolerate a wide range of environmental conditions, including pH ranges from 4.5 to 9.0 and NaCl concentrations up to 9%. Many strains have been recently recovered from dry-cured meat processing facilities and related products [49,50]. The finding of six *S. aureus* strains from all batches but the first one, as well as from different sources (including meat mixture, processing surfaces, and in salami of the drying and ripening rooms), suggests that different strains have been introduced in the food processing chain through raw materials, hygiene failures, or food handlers.

Enterobacteriaceae represented the predominant family recovered across the salami facility, resulting in 6 species and 106 strains confirmed overall (84% prevalence). Among

Enterobacteriaceae, *E. coli* corresponded to the dominant species, exhibiting a remarkable predominance in the meat mixture and associated environment up to the drying and ripening of the product. Moreover, it represented the bacterial species isolated in the highest proportion from the final product (n = 6 strains). Additionally, *K. oxytoca* and *K. pneumoniae* as well as *C. freundii* and *E. cloacae* were recovered from five to six batches; these were associated with the environment and food matrices of the meat mixture and the drying room, then persisting in the salami throughout the ripening up to the final products, where few strains were detected. Though strains belonging to enterohaemorrhagic *E. coli* (EHEC) have been attributed to foodborne outbreaks associated with fermented sausage consumption [51], the persistence of commensal *Escherichia* spp. strains has been already pointed out in spontaneously fermented sausages of the Mediterranean area [52]. Correspondingly, *Klebsiella* spp. were detected from minced pork meat during fermentation and raw pork sausages before ripening in Belgium and Spain [53,54], as well as *E. cloacae* and *Citrobacter* spp. during the ripening of traditional fermented sausages [53,55]. Considering that these bacteria have already been involved in nosocomial infections [56–58], further investigations should be addressed to verify their potential role as contributors to food-related infections.

Whether introduced from raw materials or environmental contamination due to poor hygienic conditions, enhanced hurdle technologies, disinfection measures, and training should be properly adopted to control the growth of potentially pathogenic or undesirable bacteria along the whole food chain. In the present study, the meat mixture and the surface of the table in the stuffing room were spotted as principal potential sources of contamination. The meat mixture might have been contaminated at the first steps of the production during the cutting and addition of spices, or before that, during the primary production. Unfortunately, no samples from pig carcasses were collected; therefore, no speculations can be formulated about carcass hygiene. However, since the same person owns both the primary production of the organic pig and the food production of organic salami, cross-contamination cannot be ruled out.

The adoption of tailor-made biosecurity plans including the hygiene of farms and workers has been described as an effective measure to reduce the risk of occurrence of foodborne pathogens in pig farms and the dissemination of those bacteria to humans through direct contact [59]. In addition, special attention should be paid to hygiene procedures for surfaces in direct contact with food [60]. The sampling of surfaces in direct contact with food should be prioritized, and results obtained from different batches should be compared in order to identify deviations and take corrective actions [60]. Besides the implementation of hygiene procedures, food safety training has been described as particularly effective in small-scale facilities. In particular, food safety training programs, which incorporate both knowledge and behavior-based training, were described as the most effective in commercial food services [61].

5. Conclusions

High inter-batch variability was detected in the physicochemical and microbiological parameters of organic salami produced in an artisanal factory, confirming process standardization to be a challenge in small-scale, not fully automized, production facilities. Higher TBC in the meat mixture and on the surface of the table in the stuffing room were associated with a higher occurrence of bacterial pathogens, suggesting TBC to be a good predictor of the microbial quality of the final product. This predictor might be of particular relevance, especially when the protocol of production cannot guarantee the acid and desiccation hurdles essential for biohazard control. In these conditions, enhanced hygiene measures and training could be effective control measures against the growth of potentially pathogenic or undesirable bacteria along the whole food chain.

Supplementary Materials: The following supporting information can be downloaded at: https://www.mdpi.com/article/10.3390/foods12224086/s1, Table S1: Results of the species identification of bacterial food-borne pathogens and pathogens of clinical importance from the 126 isolates in the six tested batches (B1–B6); Table S2: Results of the species identification of bacterial food-borne pathogens and pathogens of clinical importance from the 126 isolates in relation to the type of sample.

Author Contributions: Conceptualization: F.P.; Methodology: F.P., C.C., L.G. and A.L.; Software: A.V. and A.P.; Validation: F.P., C.C., L.G. and A.L.; Formal Analysis: F.P., A.V., A.P., G.M. and A.D.C.; Investigation: C.C., L.G. and A.L.; Resources: F.P., G.M. and A.D.C.; Data curation: F.P., C.C., L.G. and A.L.; Writing—Original Draft: F.P. and A.V.; Writing—Review and Editing: A.P., G.M. and A.D.C.; Visualization: F.P., A.V. and A.P.; Supervision: A.D.C.; Project administration: F.P. and A.D.C.; Funding acquisition; F.P., G.M. and A.D.C. All authors have read and agreed to the published version of the manuscript.

Funding: This research was funded by EU H2020 PRIMA ARTISANEFOOD project "Innovative Bio-interventions and Risk Modeling Approaches for Ensuring Microbial Safety and Quality of Mediterranean Artisanal Fermented Foods" and project PID2019-108420RB-C31-ASEQURA financed by the Spanish Government (Ministry of Science and Innovation, research).

Data Availability Statement: The original contributions presented in the study are included in the article. Further inquiries can be directed to the corresponding authors.

Conflicts of Interest: The authors declare no conflict of interest. The funders had no role in the design of the study; in the collection, analyses, or interpretation of data; in the writing of the manuscript; or in the decision to publish the results.

References

1. Almli, V.L.; Verbeke, W.; Vanhonacker, F.; Næs, T.; Hersleth, M. General image and attribute perceptions of traditional food in six European countries. *Food Qual. Prefer.* **2011**, *22*, 129–138. [CrossRef]
2. Cirne, C.T.; Tunick, M.H.; Trout, R.E. The chemical and attitudinal differences between commercial and artisanal products. *NPJ Sci. Food* **2019**, *3*, 19. [CrossRef] [PubMed]
3. Roccato, A.; Uyttendaele, M.; Barrucci, F.; Cibin, V.; Favretti, M.; Cereser, A.; Cin, M.D.; Pezzuto, A.; Piovesana, A.; Longo, A.; et al. Artisanal Italian salami and soppresse: Identification of control strategies to manage microbiological hazards. *Food Microbiol.* **2017**, *61*, 5–13. [CrossRef] [PubMed]
4. Jaramillo-Bedoya, E.; Trujillo-Alzate, Y.A.; Ocampo-Ibáñez, I.D. Surveillance of Fresh Artisanal Cheeses Revealed High Levels of *Listeria monocytogenes* Contamination in the Department of Quindío, Colombia. *Pathogens* **2021**, *10*, 1341. [CrossRef]
5. Lücke, F.K. Fermented sausages. In *The Microbiology of Fermented Foods*, 2nd ed.; Wood, B.J.B., Ed.; Blackie Academic and Profession: Glasgow, UK, 1998; Chapter 14; Volume 2, pp. 441–483.
6. Comi, G.; Urso, R.; Iacumin, L.; Rantsiou, K.; Cattaneo, P.; Cantoni, C.; Cocolin, L. Characterisation of naturally fermented sausages produced in the North East of Italy. *Meat Sci.* **2005**, *69*, 381–392. [CrossRef]
7. Halagarda, M.; Wójciak, K.M. Health and safety aspects of traditional European meat products. A review. *Meat Sci.* **2022**, *184*, 108623. [CrossRef]
8. Meloni, D. Presence of *Listeria monocytogenes* in Mediterranean-Style Dry Fermented Sausages. *Foods* **2015**, *4*, 34–50. [CrossRef]
9. Gonzales-Barron, U.; Cadavez, V.; Pereira, A.P.; Gomes, A.; Araújo, J.P.; Saavedra, M.J.; Estevinho, L.; Butler, F.; Pires, P.; Dias, T. Relating physicochemical and microbiological safety indicators during processing of linguiça, a Portuguese traditional dry-fermented sausage. *Food Res. Int.* **2015**, *78*, 50–61. [CrossRef]
10. Piras, F.; Spanu, C.; Mocci, A.M.; Demontis, M.; Santis, E.P.L.; Scarano, C. Occurrence and traceability of *Salmonella* spp. in five Sardinian fermented sausage facilities. *Ital. J. Food Saf.* **2019**, *8*, 8011. [CrossRef]
11. Thévenot, D.; Delignette-Muller, M.L.; Christieans, S.; Vernozy-Rozand, C. Fate of *Listeria monocytogenes* in experimentally contaminated French sausages. *Int. J. Food Microbiol.* **2005**, *101*, 189–200. [CrossRef]
12. De Cesare, A.; Mioni, R.; Manfreda, G. Prevalence of *Listeria monocytogenes* in fresh and fermented Italian sausages and ribotyping of contaminating strains. *Int. J. Food Microbiol.* **2007**, *120*, 124–130. [CrossRef] [PubMed]
13. Martin, B.; Garriga, M.; Aymerich, T. Prevalence of *Salmonella* spp. and *Listeria monocytogenes* at small-scale Spanish factories producing traditional fermented sausages. *J. Food Prot.* **2011**, *74*, 812–815. [CrossRef] [PubMed]
14. Gounadaki, A.S.; Skandamis, P.N.; Drosinos, E.H.; Nychas, G.J. Microbial ecology of food contact surfaces and products of small-scale facilities producing traditional sausages. *Food Microbiol.* **2008**, *25*, 313–323. [CrossRef] [PubMed]
15. Ferreira, V.; Barbosa, J.; Silva, J.; Felício, M.T.; Mena, C.; Hogg, T.; Gibbs, P.; Teixeira, P. Characterisation of alheiras, traditional sausages produced in the North of Portugal, with respect to their microbiological safety. *Food Control* **2007**, *18*, 436–440. [CrossRef]

16. Alexander, E.R.; Boase, J.; Davis, M.; Kirchner, L.; Osaki, C.; Tanino, T.; Samadpour, M.; Tarr, P.; Goldoft, M.; Lankford, S.; et al. *Escherichia-coli* O157/H7 linked to commercially distributed dry-cured salami–Washington and California, 1994. *J. Am. Med. Assoc.* **1995**, *273*, 985–986.
17. Sartz, L.; De Jong, B.; Hjertqvist, M.; Plym-Forshell, L.; Alsterlund, R.; Löfdahl, S.; Osterman, B.; Ståhl, A.; Eriksson, E.; Hansson, H.B.; et al. An outbreak of *Escherichia coli* O157:H7 infection in southern Sweden associated with consumption of fermented sausage; aspects of sausage production that increase the risk of contamination. *Epidemiol. Infect.* **2008**, *136*, 370–380. [CrossRef]
18. Conedera, G.; Mattiazzi, E.; Russo, F.; Chiesa, E.; Scorzato, I.; Grandesso, S.; Bessegato, A.; Fioravanti, A.; Caprioli, A. A family outbreak of *Escherichia coli* O157 haemorrhagic colitis caused by pork meat salami. *Epidemiol. Infect.* **2007**, *135*, 311–314. [CrossRef]
19. Schimmer, B.; Nygard, K.; Eriksen, H.M.; Lassen, J.; Lindstedt, B.A.; Brandal, L.T.; Kapperud, G.; Aavitsland, P. Outbreak of haemolytic uraemic syndrome in Norway caused by stx2-positive *Escherichia coli* O103:H25 traced to cured mutton sausages. *BMC Infect. Dis.* **2008**, *8*, 41. [CrossRef]
20. Ethelberg, S.; Smith, B.; Torpdahl, M.; Lisby, M.; Boel, J.; Jensen, T.; Nielsen, E.M.; Mølbak, K. Outbreak of non-O157 Shiga toxin-producing *Escherichia coli* infection from consumption of beef sausage. *Clin. Infect. Dis.* **2009**, *48*, e78–e81. [CrossRef]
21. Bremer, V.; Leitmeyer, K.; Jensen, E.; Metzel, U.; Meczulat, H.; Weise, E.; Werber, D.; Tschaepe, H.; Kreienbrock, L.; Glaser, S.; et al. Outbreak of *Salmonella* Goldcoast infections linked to consumption of fermented sausage, Germany 2001. *Epidemiol. Infect.* **2004**, *132*, 881–887. [CrossRef]
22. Arnedo-Pena, A.; Sabater-Vidal, S.; Herrera-León, S.; Bellido-Blasco, J.B.; Silvestre-Silvestre, E.; Meseguer-Ferrer, N.; Yague-Muñoz, A.; Gil-Fortuño, M.; Romeu-García, A.; Moreno-Muñoz, R. An outbreak of monophasic and biphasic *Salmonella* Typhimurium, and *Salmonella* Derby associated with the consumption of dried pork sausage in Castellon (Spain). *Enfermedades Infecc. Microbiol. Clin.* **2016**, *34*, 544–550. [CrossRef] [PubMed]
23. Salmonella Outbreak Linked to Salami Sticks. Available online: https://www.cdc.gov/salmonella/i45-10-21/index.html (accessed on 26 September 2023).
24. *4833-2:2013*; Microbiology of the Food Chain. Horizontal Method for the Enumeration of Microorganisms—Part 2: Colony Count at 30 °C by the Surface Plating Technique. ISO: Geneva, Switzerland, 2013.
25. *21807:2004*; Microbiology of Food and Animal Feeding Stuffs. Determination of Water Activity. ISO: Geneva, Switzerland, 2004.
26. *2917:1999*; International Standards Meat and Meat Products. Measurement of the pH (Reference Method). ISO: Geneva, Switzerland, 1999.
27. *11290–11291:2017*; Microbiology of the Food Chain. Horizontal Method for the Detection and Enumeration of Listeria monocytogenes and of Listeria spp.—Part 1: Detection Method. ISO: Geneva, Switzerland, 2017.
28. *ISO 6888-1/A1:2004*; Microbiology of Food and Animal Feeding Stuffs. Horizontal Method for the Enumeration of Coagulase-Positive Staphylococci (Staphylococcus aureus and other species)—Part 1: Technique using Baird-Parker agar medium. ISO: Geneva, Switzerland, 2004.
29. *16649:2001*; Microbiology of the Food Chain—Horizontal Method for the Enumeration of Beta-Glucuronidase-Positive Escherichia coli—Part 2: Colony-Count Technique at 44 Degrees C Using 5-Bromo-4-chloro-3-indolyl beta-D-glucuronide. ISO: Geneva, Switzerland, 2001.
30. *6579–1:2017*; Microbiology of the Food Chain. Horizontal Method for the Detection, Enumeration and Serotyping of Salmonella—Part 1: Detection of Salmonella spp. ISO: Geneva, Switzerland, 2017.
31. Perelle, S.; Dilasser, F.; Grout, J.; Fach, P. Detection by 5′-nuclease PCR of Shiga-toxin producing *Escherichia coli* O26, O55, O91, O103, O111, O113, O145 and O157:H7, associated with the world's most frequent clinical cases. *Mol. Cell Probes* **2004**, *18*, 185–192. [CrossRef] [PubMed]
32. *15214:1998*; Microbiology of Food and Animal Feeding Stuffs. Horizontal Method for the Numeration of Mesophilic Lactic Acid Bacteria—Colony-Count Technique at 30 Degrees C. ISO: Geneva, Switzerland, 1998.
33. *21528–2:2017*; Microbiology of the Food Chain. Horizontal Method for the Detection and Enumeration of Enterobacteriaceae—Part 2: Colony-Count Technique. ISO: Geneva, Switzerland, 2017.
34. Wesley, I.V.; Harmon, K.M.; Dickson, J.S.; Schwartz, A.R. Application of a multiplex polymerase chain reaction assay for the simultaneous confirmation of *Listeria monocytogenes* and other *Listeria* species in turkey sample surveillance. *J. Food Prot.* **2002**, *65*, 780–785. [CrossRef] [PubMed]
35. Chander, Y.; Ramakrishnan, M.A.; Jindal, N.; Hanson, K.; Goyal, S.M. Differentiation of *Klebsiella pneumoniae* and *K. oxytoca* by multiplex polymerase chain reaction. *Int. J. Appl. Res. Vet. M.* **2011**, *9*, 138.
36. Saraiva, M.M.; De Leon, C.M.; Santos, S.C.; Stipp, D.T.; Souza, M.M.; Santos Filho, L.; Gebreyes, W.A.; Oliveira, C.J. Accuracy of PCR targeting different markers for *Staphylococcus aureus* identification: A comparative study using matrix-assisted laser desorption/ionization time-of-flight mass spectrometry as the gold standard. *J. Vet. Diagn. Investig.* **2018**, *30*, 252–255. [CrossRef] [PubMed]
37. Bates, D.; Mächler, M.; Bolker, B.; Walker, S. Fitting Linear Mixed-Effects Models Using Lme4. *J. Stat. Soft.* **2015**, *67*, 1–48. [CrossRef]
38. Ordóñez, J.A.; de la Hoz, L. Mediterranean products. In *Handbook of Fermented Meat and Poultry*; Toldrà, F., Ed.; Blackwell Publishing: Ames, IA, USA, 2007; pp. 333–347.

39. Jofré, A.; Aymerich, T.; Garriga, M. Improvement of the food safety of low acid fermented sausages by enterocins A and B and high pressure. *Food Control* **2009**, *20*, 179–184. [CrossRef]
40. Greco, M.; Mazzette, R.; De Santis, E.P.; Corona, A.; Cosseddu, A.M. Evolution and identification of lactic acid bacteria isolated during the ripening of Sardinian sausages. *Meat Sci.* **2005**, *69*, 733–739. [CrossRef]
41. Cenci-Goga, B.T.; Ranucci, D.; Miraglia, D.; Cioffi, A. Use of starter cultures of dairy origin in the production of *Salame nostrano*, an Italian dry-cured sausage. *Meat Sci.* **2008**, *78*, 381–390. [CrossRef]
42. Ewald, S.; Notermans, S. Effect of water activity on growth and enterotoxin D production of *Staphylococcus aureus*. *Int. J. Food Microbiol.* **1988**, *6*, 25–30. [CrossRef]
43. Osek, J.; Lachtara, B.; Wieczorek, K. *Listeria monocytogenes*—How This Pathogen Survives in Food Production Environments. *Front. Microbiol.* **2022**, *13*, 866462. [CrossRef] [PubMed]
44. Pasquali, F.; Palma, F.; Guillier, L.; Lucchi, A.; De Cesare, A.; Manfreda, G. *Listeria monocytogenes* Sequence Types 121 and 14 Repeatedly Isolated within One Year of Sampling in a Rabbit Meat Processing Plant: Persistence and Ecophysiology. *Front. Microbiol.* **2018**, *9*, 596. [CrossRef] [PubMed]
45. Althaus, D.; Jermini, M.; Giannini, P.; Martinetti, G.; Reinholz, D.; Nüesch-Inderbinen, M.; Lehner, A.; Stephan, R. Local Outbreak of *Listeria monocytogenes* Serotype 4b Sequence Type 6 Due to Contaminated Meat Pâté. *Foodborne Pathog. Dis.* **2017**, *14*, 219–222. [CrossRef] [PubMed]
46. Duranti, A.; Sabbatucci, M.; Blasi, G.; Acciari, V.A.; Ancora, M.; Bella, A.; Busani, L.; Centorame, P.; Cammà, C.; Conti, F.; et al. A severe outbreak of listeriosis in central Italy with a rare pulsotype associated with processed pork products. *J. Med. Microbiol.* **2018**, *67*, 1351–1360. [CrossRef] [PubMed]
47. Gelbíčová, T.; Zobaníková, M.; Tomáštíková, Z.; Van Walle, I.; Ruppitsch, W.; Karpíšková, R. An outbreak of listeriosis linked to turkey meat products in the Czech Republic, 2012–2016. *Epidemiol. Infect.* **2018**, *146*, 1407–1412. [CrossRef] [PubMed]
48. Outbreak of Listeria Infections Linked to Deli Meats. Available online: https://www.cdc.gov/listeria/outbreaks/delimeat-10-20/index.html (accessed on 26 September 2023).
49. Pérez-Boto, D.; D'Arrigo, M.; García-Lafuente, A.; Bravo, D.; Pérez-Baltar, A.; Gaya, P.; Medina, M.; Arqués, J.L. *Staphylococcus aureus* in the Processing Environment of Cured Meat Products. *Foods* **2023**, *12*, 2161. [CrossRef]
50. Gelbíčová, T.; Brodíková, K.; Karpíšková, R. Livestock-associated methicillin-resistant *Staphylococcus aureus* in Czech retailed ready-to-eat meat products. *Int. J. Food Microbiol.* **2022**, *374*, 109727. [CrossRef]
51. Holck, A.L.; Axelsson, L.; Rode, T.M.; Høy, M.; Måge, I.; Alvseike, O.; L'abée-Lund, T.M.; Omer, M.K.; Granum, P.E.; Heir, E. Reduction of verotoxigenic *Escherichia coli* in production of fermented sausages. *Meat Sci.* **2011**, *89*, 286–295. [CrossRef]
52. Barbieri, F.; Tabanelli, G.; Montanari, C.; Dall'Osso, N.; Šimat, V.; Smole Možina, S.; Baños, A.; Özogul, F.; Bassi, D.; Fontana, C.; et al. Mediterranean Spontaneously Fermented Sausages: Spotlight on Microbiological and Quality Features to Exploit Their Bacterial Biodiversity. *Foods* **2021**, *10*, 2691. [CrossRef]
53. Charmpi, C.; Van Reckem, E.; Sameli, N.; Van der Veken, D.; De Vuyst, L.; Leroy, F. The Use of Less Conventional Meats or Meat with High pH Can Lead to the Growth of Undesirable Microorganisms during Natural Meat Fermentation. *Foods* **2020**, *9*, 1386. [CrossRef]
54. Roig-Sagues, A.X.; Hernandez-Herrero, M.; Lopez-Sabater, E.I.; Rodriguez-Jerez, J.J.; Mora-Ventura, M.T. Histidine Decarboxylase Activity of Bacteria Isolated from Raw and Ripened Salchichón, a Spanish Cured Sausage. *J. Food Prot.* **1996**, *59*, 516–520. [CrossRef] [PubMed]
55. Bogdanović, S.; Stanković, S.; Berić, T.; Tomasevic, I.; Heinz, V.; Terjung, N.; Dimkić, I. Bacteriobiota and Chemical Changes during the Ripening of Traditional Fermented "Pirot 'Ironed' Sausage". *Foods* **2023**, *12*, 664. [CrossRef] [PubMed]
56. Davin-Regli, A.; Pagès, J.M. *Enterobacter aerogenes* and *Enterobacter cloacae*; versatile bacterial pathogens confronting antibiotic treatment. *Front. Microbiol.* **2015**, *6*, 392. [CrossRef] [PubMed]
57. Chang, D.; Sharma, L.; Dela Cruz, C.S.; Zhang, D. Clinical Epidemiology, Risk Factors, and Control Strategies of *Klebsiella pneumoniae* Infection. *Front. Microbiol.* **2021**, *12*, 750662. [CrossRef] [PubMed]
58. Rezaei, M.; Akya, A.; Elahi, A.; Ghadiri, K.; Jafari, S. The clonal relationship among the *Citrobacter freundii* isolated from the main hospital in Kermanshah, west of Iran. *Iran. J. Microbiol.* **2016**, *8*, 175–180.
59. Scollo, A.; Perrucci, A.; Stella, M.C.; Ferrari, P.; Robino, P.; Nebbia, P. Biosecurity and hygiene procedures in pig farms: Effects of a tailor-made approach as monitored by environmental samples. *Animals* **2023**, *13*, 1262. [CrossRef]
60. Agüeria, D.A.; Libonatti, C.; Civit, D. Cleaning and disinfection programmes in food establishments: A literature review on verification procedures. *J. Appl. Microbiol.* **2021**, *131*, 23–35. [CrossRef]
61. McFarland, P.; Checinska Sielaff, A.; Rasco, B.; Smith, S. Efficacy of food safety training in commercial food service. *J. Food Sci.* **2019**, *84*, 1239–1246. [CrossRef]

MDPI AG
Grosspeteranlage 5
4052 Basel
Switzerland
Tel.: +41 61 683 77 34

Foods Editorial Office
E-mail: foods@mdpi.com
www.mdpi.com/journal/foods